AF508667

Advanced Research on Animal Venoms in China

Advanced Research on Animal Venoms in China

Editors

Ren Lai
Qiumin Lü

MDPI • Basel • Beijing • Wuhan • Barcelona • Belgrade • Manchester • Tokyo • Cluj • Tianjin

Editors
Ren Lai
Chinese Academy of Sciences
China

Qiumin Lü
Chinese Academy of Sciences
China

Editorial Office
MDPI
St. Alban-Anlage 66
4052 Basel, Switzerland

This is a reprint of articles from the Special Issue published online in the open access journal *Toxins* (ISSN 2072-6651) (available at: https://www.mdpi.com/journal/toxins/special_issues/Venoms_in_China).

For citation purposes, cite each article independently as indicated on the article page online and as indicated below:

LastName, A.A.; LastName, B.B.; LastName, C.C. Article Title. *Journal Name* **Year**, *Volume Number*, Page Range.

ISBN 978-3-0365-7434-9 (Hbk)
ISBN 978-3-0365-7435-6 (PDF)

Contents

Preface to "Advanced Research on Animal Venoms in China"

Chinese researchers have made many exciting discoveries on animal toxins in recent years. The aim of this reprint is to provide the latest research on the discovery of animal toxins, the mechanism underlining their actions, mining of drug leads from toxins, and diagnosis and treatment of bites or stings by venoms animals in China. We thank all the authors who contributed to the reprint and the researchers who have made significant contributions to the study of animal toxins in China.

Ren Lai and Qiumin Lü

Editors

Editorial

Advanced Research on Animal Venoms in China

Ren Lai * and Qiumin Lu *

Key Laboratory of Animal Models and Human Disease Mechanisms of the Chinese Academy of Sciences/Key Laboratory of Bioactive Peptides of Yunnan Province, Kunming Institute of Zoology, Kunming 650223, China
* Correspondence: rlai@mail.kiz.ac.cn (R.L.); lvqm@mail.kiz.ac.cn (Q.L.)

Citation: Lai, R.; Lu, Q. Advanced Research on Animal Venoms in China. *Toxins* **2023**, *15*, 272. https://doi.org/10.3390/toxins15040272

Received: 29 March 2023
Accepted: 4 April 2023
Published: 6 April 2023

For millennia, scientists, researchers, and the general public have been intrigued by animal venoms due to their potent effects and paradoxical ability to both harm and heal. Animal venoms are complex cocktails of bioactive molecules produced and delivered by various animal species, including snakes, spiders, scorpions, toads, cone snails, centipedes, and even some mammals such as tree shrews and platypi. These molecules are incredibly diverse and specialized, with unique properties that allow venomous animals to capture prey, protect themselves, and compete for resources. Furthermore, many of these toxins have evolved to interact with specific molecular targets (i.e., ion channels) in the nervous, cardiovascular, and musculoskeletal systems of prey or predators. Due to their high potency and specificity to molecular targets, animal venoms are valuable tools for researching ion channel functions and related diseases, which can inform the development of new drugs and therapeutic strategies.

Recently, Chinese researchers have conducted significant work in animal venom research to comprehend the diversity and complexity of these molecules and their potential applications in medicine, biotechnology, and as research tools. Advancements in 'omics' technologies such as proteomics, genomics, and transcriptomics have empowered Chinese researchers to identify and analyze venom-coded genes and proteins, revealing a kaleidoscope of novel peptide toxins with diverse functions and activities.

In this Special Issue of *Toxins*, primary research papers have been assembled that provide the reader with a comprehensive and up-to-date perspective on some of the most recent and dynamic contributions of animal toxins research by Chinese researchers. The purpose of this issue is to provide the latest work by Chinese researchers on the discovery of animal toxins, the mechanism underlining their actions, the mining of drug leads from peptide toxins, and the diagnosis and treatment of bites or stings by venomous animals. Below is a brief synopsis of the 11 papers that make up this Special Issue.

Skin secretions from amphibians contain toxin-like proteins and peptides that play an important role in their physiological and pathological functions. Qingqing Ye and colleagues [1] have reported that βγ-CAT, a Chinese red-belly toad-derived pore-forming toxin-like protein complex, could induce various toxic effects via its membrane perforation process, including membrane binding, oligomerization, and endocytosis. This research reveals a hitherto unrecognized toxicological role of a vertebrate-derived pore-forming toxin-like protein in the nervous system, which causes hippocampus neuronal cells to undergo pyroptosis, which in turn results in cognitive retardation. Additionally, a study conducted by Chuanling Yin et al. [2] has focused on the secretions of tree frogs in order to gain insight into a unique defense mechanism in amphibians. The team has reported the presence of PAX in the secretions of tree frogs (*Hyla japonica*). There is a biological significance to PAX as it inhibits both BKCa and KCNK18 channels, which fire excitatory currents in sensory neurons, causing predators or competitors to feel tingly and buzzy. The presence of bifunctional PAX in frog skin secretion may indicate a unique defensive mechanism that is involved in amphibian adaptation. Moreover, a novel peptide, PM-7, from the frog Polypedates megacephalus has been identified by Siqi Fu and colleagues [3]. The authors demonstrate that the peptide has the capacity to promote wound healing in mice. PM-7 may be a potential candidate in the development of cutting-edge drugs for treating wounds.

The neurotoxic and myotoxic phospholipase A2 toxins found in the venom of Russell's viper (*Daboia siamensis*) can potentially harm motor nerve terminals irreversibly. Antivenoms may only be partially effective against some of these venom components because of the time interval between envenoming and the delivery of the antivenom. Mimi Lay et al. [4] have assessed the effectiveness of Chinese *D. siamensis* antivenom alone and in conjunction with a PLA2 inhibitor, Varespladib, in restoring the in vitro neuromuscular blockade in the chick biventer cervicis nerve-muscle preparation. The authors demonstrate that the antivenom has a shorter window of effectiveness than Varespladib, and that some of the effects of Chinese *D. siamensis* venom may be inhibited by small-molecule inhibitors.

Fan Zhao and colleagues [5] have investigated the key factors responsible for the adverse effects caused by an analgesic-antitumor β-scorpion toxin (AGAP) from scorpion venom on human voltage-gated sodium channels 1.4 and 1.5. Their research explains the mechanism by which the tremendous modification of subtype selectivity might result from the mutation of a single amino acid. This work advances the development of safer and more effective therapies specifically targeting VGSC subtypes and preventing muscle and myocardium toxicity.

The peptides derived from scorpion venom have garnered significant interest as potential anticancer agents due to their unique properties, such as their high specificity and potency toward cancer cells. The anticancer mechanism of Smp24, a peptide derived from the venom of *Scorpio Maurus palmatus*, against HepG2 cells has been described by Tienthanh Nguyen et al. [6] as being related to cell membrane breakdown and mitochondrial malfunction, suppressing cell viability by inducing cell death, cycle arrest, and autophagy. In addition, Ruiyin Guo and colleagues [7] show that the anticancer activity of Smp24, an antimicrobial peptide derived from Egyptian scorpion *Scorpio maurus palmatus*, on A549 cells is associated with the development of apoptosis, autophagy, and cell cycle arrest via dysfunctional mitochondria and ROS buildup. Moreover, Ruiyin Guo et al. [8] have also reported that the production of membrane defects and cytoskeleton disruption by Smp24, a peptide from Egyptian scorpion *Scorpio maurus palmatus*, leads to a strong anticancer effect in a xenograft mouse model of A549. These reports provided new insights into the anticancer mechanism of Smp24, which may aid in the future development of therapy for lung cancer cells.

The transcriptomes of the venom glands of *Mesobuthus martensii* from various populations and genders have been analyzed by Zhiyong Di and colleagues [9]. They have also examined the expression preferences of various toxin gene families as well as the features of members of toxin gene clusters. Their study will encourage further in-depth research and utilization of scorpions and their toxic resources, which will be helpful for establishing standardization in Quanxie identification and medicinal applications in traditional Chinese medicine.

For the first time, Xuekui Nie and colleagues [10] have integrated proteomic and transcriptome methods to examine the variations in venom composition between captive and wild individuals of Chinese cobras. Under breeding conditions, the venom composition differs between captive and wild individuals, demonstrating the plasticity of venom composition. As a result of this study, we will better understand the mechanism of snakebite intoxication and ensure the safe preparation and administration of traditional antivenom and next-generation drugs against snakebites.

As a biocontrol agent, *Pyemotes zhonghuajia* plays a vital role against pests of the Isoptera, Homoptera, Hymenoptera, Lepidoptera, and Coleoptera families. *P. zhonghuajia* injects toxins into the host (eggs, larvae, pupae, and adults), thereby obtaining nourishment for reproduction. By providing a genome assembly of *P. zhonghuajia*, Yanfei Song et al. [11] have given insights into a detailed description of its toxin-related gene families. Their work will contribute to improving the parasitism efficiency of these mites and provide a basis for the development of new biological pesticides.

Acknowledgments: The editors would like to express their gratitude to all the authors who contributed to this Special Issue, as well as the expert peer reviewers for their diligent and thorough manuscript evaluations.

Conflicts of Interest: The authors declare no conflict of interest.

References

1. Ye, Q.; Wang, Q.; Lee, W.; Xiang, Y.; Yuan, J.; Zhang, Y.; Guo, X. A Pore Forming Toxin-like Protein Derived from Chinese Red Belly Toad Bombina maxima Triggers the Pyroptosis of Hippomal Neural Cells and Impairs the Cognitive Ability of Mice. *Toxins* **2023**, *15*, 191. [CrossRef] [PubMed]
2. Yin, C.; Zeng, F.; Huang, P.; Shi, Z.; Yang, Q.; Pei, Z.; Wang, X.; Chai, L.; Zhang, S.; Yang, S.; et al. The Bi-Functional Paxilline Enriched in Skin Secretion of Tree Frogs (*Hyla japonica*) Targets the KCNK18 and BKCa Channels. *Toxins* **2023**, *15*, 70. [CrossRef] [PubMed]
3. Fu, S.; Du, C.; Zhang, Q.; Liu, J.; Zhang, X.; Deng, M. A Novel Peptide from *Polypedates megacephalus* Promotes Wound Healing in Mice. *Toxins* **2022**, *14*, 753. [CrossRef] [PubMed]
4. Lay, M.; Liang, Q.; Isbister, G.K.; Hodgson, W.C. In Vitro Efficacy of Antivenom and Varespladib in Neutralising Chinese Russell's Viper (*Daboia siamensis*) Venom Toxicity. *Toxins* **2023**, *15*, 62. [CrossRef] [PubMed]
5. Zhao, F.; Fang, L.; Wang, Q.; Ye, Q.; He, Y.; Xu, W.; Song, Y. Exploring the Pivotal Components Influencing the Side Effects Induced by an Analgesic-Antitumor Peptide from Scorpion Venom on Human Voltage-Gated Sodium Channels 1.4 and 1.5 through Computational Simulation. *Toxins* **2023**, *15*, 33. [CrossRef] [PubMed]
6. Nguyen, T.; Guo, R.; Chai, J.; Wu, J.; Liu, J.; Chen, X.; Abdel-Rahman, M.A.; Xia, H.; Xu, X. Smp24, a Scorpion-Venom Peptide, Exhibits Potent Antitumor Effects against Hepatoma HepG2 Cells via Multi-Mechanisms In Vivo and In Vitro. *Toxins* **2022**, *14*, 717. [CrossRef] [PubMed]
7. Guo, R.; Chen, X.; Nguyen, T.; Chai, J.; Gao, Y.; Wu, J.; Li, J.; Abdel-Rahman, M.A.; Chen, X.; Xu, X. The Strong Anti-Tumor Effect of Smp24 in Lung Adenocarcinoma A549 Cells Depends on Its Induction of Mitochondrial Dysfunctions and ROS Accumulation. *Toxins* **2022**, *14*, 590. [CrossRef] [PubMed]
8. Guo, R.; Liu, J.; Chai, J.; Gao, Y.; Abdel-Rahman, M.A.; Xu, X. Scorpion Peptide Smp24 Exhibits a Potent Antitumor Effect on Human Lung Cancer Cells by Damaging the Membrane and Cytoskeleton In Vivo and In Vitro. *Toxins* **2022**, *14*, 438. [CrossRef] [PubMed]
9. Di, Z.; Qiao, S.; Liu, X.; Xiao, S.; Lei, C.; Li, Y.; Li, S.; Zhang, F. Transcriptome Sequencing and Comparison of Venom Glands Revealed Intraspecific Differentiation and Expression Characteristics of Toxin and Defensin Genes in *Mesobuthus martensii* Populations. *Toxins* **2022**, *14*, 630. [CrossRef] [PubMed]
10. Nie, X.; Chen, Q.; Wang, C.; Huang, W.; Lai, R.; Lu, Q.; He, Q.; Yu, X. Venom Variation of Neonate and Adult Chinese Cobras in Captivity Concerning Their Foraging Strategies. *Toxins* **2022**, *14*, 598. [CrossRef] [PubMed]
11. Song, Y.-F.; Yu, L.-C.; Yang, M.-F.; Ye, S.; Yan, B.; Li, L.-T.; Wu, C.; Liu, J.-F. A Long-Read Genome Assembly of a Native Mite in China *Pyemotes zhonghuajia* Yu, Zhang & He (Prostigmata: Pyemotidae) Reveals Gene Expansion in Toxin-Related Gene Families. *Toxins* **2022**, *14*, 571. [PubMed]

Article

Transcriptome Sequencing and Comparison of Venom Glands Revealed Intraspecific Differentiation and Expression Characteristics of Toxin and Defensin Genes in *Mesobuthus martensii* Populations

Zhiyong Di [1,2,*], Sha Qiao [1,2], Xiaoshuang Liu [1,2], Shuqing Xiao [1], Cheng Lei [1], Yonghao Li [1], Shaobin Li [3] and Feng Zhang [1,2,*]

[1] Key Laboratory of Zoological Systematics and Application of Hebei Province, College of Life Sciences, Hebei University, Baoding 071002, China
[2] Institute of Life Science and Green Development, Hebei University, Baoding 071002, China
[3] College of Life Sciences, Yangtze University, Jingzhou 434025, China
* Correspondence: zydi@ustc.edu.cn (Z.D.); scorpionking@whu.edu.cn (F.Z.)

Abstract: *Mesobuthus martensii*, a famous and important Traditional Chinese Medicine has a long medical history and unique functions. It is the first scorpion species whose whole genome was sequenced worldwide. In addition, it is the most widespread and infamous poisonous animal in northern China with complex habitats. It possesses several kinds of toxins that can regulate different ion channels and serve as crucial natural drug resources. Extensive and in-depth studies have been performed on the structures and functions of toxins of *M. martensii*. In this research, we compared the morphology of *M. martensii* populations from different localities and calculated the COI genetic distance to determine intraspecific variations. Transcriptome sequencing by RNA-sequencing of the venom glands of *M. martensii* from ten localities and *M. eupeus* from one locality was analyzed. The results revealed intraspecific variation in the expression of sodium channel toxin genes, potassium channel toxin genes, calcium channel toxin genes, chloride channel toxin genes, and defensin genes that could be related to the habitats in which these populations are distributed, except the genetic relationships. However, it is not the same in different toxin families. *M. martensii* and *M. eupeus* exhibit sexual dimorphism under the expression of toxin genes, which also vary in different toxin families. The following order was recorded in the difference of expression of sodium channel toxin genes: interspecific difference; differences among different populations of the same species; differences between sexes in the same population, whereas the order in the difference of expression of potassium channel toxin genes was interspecific difference; differences between both sexes of same populations; differences among the same sex in different populations of the same species. In addition, there existed fewer expressed genes of calcium channel toxins, chloride channel toxins, and defensins (no more than four members in each family), and their expression differences were not distinct. Interestingly, the expression of two calcium channel toxin genes showed a preference for males and certain populations. We found a difference in the expression of sodium channel toxin genes, potassium channel toxin genes, and chloride channel toxin genes between *M. martensii* and *M. eupeus*. In most cases, the expression of one member of the toxin gene clusters distributed in series on the genome were close in different populations and genders, and the members of most clusters expressed in same population and gender tended to be the different. Twenty-one toxin genes were found with the MS/MS identification evidence of *M. martensii* venom. Since scorpions were not subjected to electrical stimulation or other special treatments before conducting the transcriptome extraction experiment, the results suggested the presence of intraspecific variation and sexual dimorphism of toxin components which revealed the expression characteristics of toxin and defensin genes in *M. martensii*. We believe this study will promote further in-depth research and use of scorpions and their toxin resources, which in turn will be helpful in standardizing the identification and medical applications of Quanxie in traditional Chinese medicine.

Keywords: intraspecific differentiation; *Mesobuthus martensii*; sexual dimorphism; toxin; transcriptome

Citation: Di, Z.; Qiao, S.; Liu, X.; Xiao, S.; Lei, C.; Li, Y.; Li, S.; Zhang, F. Transcriptome Sequencing and Comparison of Venom Glands Revealed Intraspecific Differentiation and Expression Characteristics of Toxin and Defensin Genes in *Mesobuthus martensii* Populations. *Toxins* **2022**, *14*, 630. https://doi.org/10.3390/toxins14090630

Received: 19 June 2022
Accepted: 1 August 2022
Published: 11 September 2022

Publisher's Note: MDPI stays neutral with regard to jurisdictional claims in published maps and institutional affiliations.

Key Contribution: *Mesobuthus martensii* is the most widely distributed and medicinal scorpion species in China. However, its intraspecific differentiation, especially of its toxins, remains a mystery. In this study, we compared the transcriptomes of venom glands of different populations and different genders of *M. martensii* and studied the expression preferences of different toxin gene families in addition to the expression characteristics of toxin gene cluster members.

1. Introduction

Mesobuthus martensii is the most famous member of the family Buthidae (Arachnida: Scorpiones) in Asia. It is known for the widest distribution, the most complex habitats, and the largest biomass among scorpion species from China, as well as its extensive applications in traditional Chinese medicine. The species is distributed to latitude south of 43° N and the north sides of the Yangtze River, with its distribution bordered by the Helan Mountain and the Tengger Desert and Mo Us Desert in the west and north and limited by the sea in the east [1]. In the Chinese medicinal materials market, *M. martensii* is the primary component as one of the most famous animal-derived Chinese herbs, named "Quanxie", which is the same as the stipulation of "Chinese Pharmacopoeia".

Traditional Chinese medicine has played an important role during the outbreak of COVID-19. Quanxie was reported to have antiviral properties. "Xiang-su-hua-zhu Particle" created by Diangui Li, a great master of traditional Chinese medicine, was approved by the Hebei Drug Administration. Similarly, "Niu-huang-qing-nao-kai-qiao-wan" launched by Sihuan Aokang Pharmaceutical Co., Ltd. added Quanxie as a prescription for COVID-19. During the outbreak of SARS in 2003, the Nanshan Zhong medical team added Quanxie to both Kangyan II recipe and Kangyan III recipe (traditional Chinese medicine), with the following results: the days of hospitalization for 71 patients with SARS were 9 to 78 days (mean 27.1 ± 14.4 days). Seventy patients were healed (the healing rate was 98.6%), and one patient died; the mortality rate was 1.4% [2]. Some prescriptions of traditional Chinese medicine for treating plague, such as "Hui-sheng-zhi-bao-dan" in the "Xian-Nian-Ji" published in the Qing Dynasty, "Ji-jiu-wan" and "Bao-ying-duo-ming-dan" in the "National Formulary of Traditional Chinese Medicine", contained Quanxie [3]. Since 1954, Keming Guo, a famous doctor of traditional Chinese medicine, used Quanxie and heat clearing and detoxifying herbs to save 34 patients with severe epidemic encephalitis type B patients. Quanxie has often been used to treat several viral diseases, with curative effects on children's pneumonia and mumps, and inhibitory effects on influenza A virus H1N1, hand-foot-mouth disease of EV71 virus, and other viruses [4–8]. During the clinical application of Traditional Chinese Medicine in the treatment of viral infections, it is impossible to accurately judge the medicinal value of Quanxie as double-blind trials were not conducted; however, there is no doubt that it has long been valued by traditional Chinese medicine in such diseases.

Previous research suggested that the morphology and venom of *M. martensii* show intraspecific variations. Qi et al. (2004) reported differences in the body colors of the populations of *M. martensii* obtained from different localities [9]. For example, specimens collected from the Longhua County in Hebei Province were dark brown tergites, whereas those obtained from Baligou in Henan Province had dark red-brown tergites [9]. In other localities, *M. martensii* could have yellowish or pale yellow tergites [9]. Zhang & Zhu (2009) analyzed the intraspecific morphological differences of *M. martensii* from different localities in China; however, these differences were under subspecies level [10]. The results based on a few specimens with no gender information provided by Zhang & Zhu (2009) [10]. In addition, the LD50 values of *M. martensii* venom from different localities in mice were different: compared with the venom of populations obtained from Henan, Liaoning and Shandong Provinces, the LD50 value of venom of the population procured from Shanxi Province in mice was the smallest [11]. Shi et al. (2013) constructed a Bayesian evolutionary tree for species of *Mesobuthus* in China using COI and three nuclear genes, including 28 populations of *M. martensii*. The study revealed that *M. martensii* formed four branches in China: east China subclade, central north China subclade, isolated east China subclade, and widespread subclade [12]. In addition, Wang et al. (2019) analyzed the relationship between metabolic rate and the environment in 21 localities of *M. martensii* in China and revealed significant differences in the resting metabolic rate between sexes from the same locality and among different populations [13]. Similarly, Gao et al. (2021) reported differences in the expression patterns of toxin genes in different populations from four localities (Hebei, Henan, Shandong, and Shanxi Provinces) and genders (from Gansu Province) of *M. martensii* [14,15]. There exist certain cases of intraspecific variations of other species belonging to the genus *Mesobuthus*. For example, *M. eupeus*, which is a closely related species to *M. martensii*, has been divided into 23 subspecies [16].

A study of intraspecific variations can reveal the diversity of scorpion toxins. For instance, Abdel-Rahman et al. (2009) used polyacrylamide gel electrophoresis (SDS-PAGE) to analyze the venom of *Scorpio maurus palmatus* collected from four geographically isolated localities in Egypt [17]. The results showed differences in the expression of toxins obtained from different localities. Zhao et al. (2010) comprehensively analyzed venom transcriptomes (cDNA library) of the scorpion *Lychas mucronatus* from Hainan and Yunnan Provinces [18], revealing that the venom peptides and proteins of the same scorpion species from different geographical regions are highly diverse. Furthermore, scorpions evolved to adapt to new environments by altering the primary structure and abundance of their venom peptides and proteins. Carcamo-Noriega et al. (2018) investigated the venom collected from two distinct populations of the scorpion *Centruroides sculpturatus* that inhabit different regions of Arizona. The study reported intraspecific variations between venoms mostly in the composition and proportion of the two toxins (CsEv1 and CsEv2) [19].

The study on intraspecific variations in the transcriptome of *M. martensii* has a genomics foundation and a proteomics foundation. Cao et al. (2013) published the whole genome draft of *M. martensii* [20], which is the first sequenced whole genome of the order Scorpiones worldwide. Furthermore, the authors predicted the existence of at least 32,016 protein-coding genes, including 116 neurotoxin genes: 61 NaTx (toxins for sodium channels), 46 KTx (toxins for potassium channels), 5 ClTx (toxins for chloride channels), and 4 CaTx (toxins for ryanodine receptors) genes. Among these, 51 sodium channel toxin genes and potassium channel toxin genes were found to be clustered on 17 scaffold genes. In addition, six defensin genes were discovered by Cao et al. (2013) [20]. Subsequently, 153 fractions were isolated from the *M. martensii* venom by 2-DE, SDS-PAGE, and RP-HPLC, whereas 227 non-redundant protein sequences were unambiguously identified, consisting of 134 previously known and 93 unknown proteins [21]. Among 134 previously known proteins, 115 proteins were first confirmed from the *M. martensii* crude venom, and 19 toxins were confirmed once again, including 43 typical toxins, 7 atypical toxins, 12 venom enzymes, and 72 cell-associated proteins [21]. Li et al. (2016) comprehensively reviewed the diversity of toxin types, structures, and functions of *M. martensii* followed by a study on the genome and proteome of this species, and potential importance of scorpion toxins in medicine, including chronic pain relief, antitumor properties, anti-infection characteristics, and treatment of autoimmune diseases [22–25].

2. Results

2.1. Morphological Intraspecific Differentiation of Mesobuthus martensii

2.1.1. Color

As the color of specimens changed gradually after soaking in alcohol, we observed the live scorpions and fresh specimens obtained from different localities (Figure S1). Populations from arid areas were lighter than those from sub-arid and sub-wet areas. These differences were particularly significant in the carapace, tergites, and segments of metasoma. Most individuals from Luoyang and Baoding had black or black-brown carapace and tergites, whereas individuals from Lanzhou and Helan were yellow-brown in the same parts, which was identical in both sexes, but especially remarkable in females (Figures 1, 2 and S1).

In addition, specimens from Luoyang and Baoding had yellow metasoma, yellowish brown carina, and yellowish appendages, whereas specimens from Lanzhou and Helan were lighter (Figure 2). This is consistent with the finding reported by Qi et al. (2004) [9]. Under the background of different annual precipitation and annual temperature in the eastern and western regions, different habitats were formed in these locations, thus promoting different physiological traits linked with evolutionary fitness among populations.

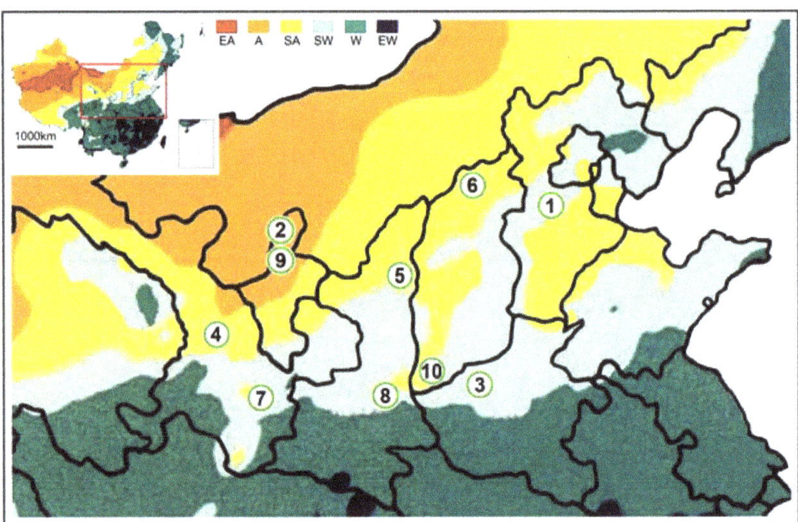

Figure 1. The distribution of populations of *Mesobuthus martensii* and *M. eupeus* from sub-wet areas, sub-arid areas, and arid areas in this study. *M. martensii*: 1, BD, Baoding; 2, HL, Helan Mountain (Helan); LY, Luoyang (3, LYF, females from Luoyang; same as 3′, LYM, males from Luoyang); LZ, Lanzhou (4, LZF, females from Lanzhou); 5, SD, Suide; SZ, Shuozhou (6, SZF, females from Shuozhou; same as 6′-SZM, males from Shuozhou); 7, TS, Tianshui; WN, Weinan (8, WNF, females from Weinan; same as 8′-WNM, males from Weinan); WZ, Wuzhong (9, WZF, females from Wuzhong); 10, YC, Yuncheng. *M. eupeus*: ME, from Yinchuan (11, MEF, females of from Yinchuan; same as 11′, MEM, males from Yinchuan, overlapping with the locality of 2 (HL, Helan)). EA, Extreme arid; A, Arid; SA, Sub-arid; SW, Sub-wet; W, Wet; EW, Extreme wet.

2.1.2. Body Size

Adult individuals from arid and sub-arid areas were smaller than those from sub-wet areas (Figure 3a,b). For example, the average body length of females from Baoding and Weinan was 63.3 mm (10 adults) and 60.7 mm (10 adults), respectively, whereas the average body length of females from Helan and Wuzhong was 54.0 mm (10 adults) and 56.2 mm (10 adults), respectively. The average body length of specimens from Helan, Lanzhou, Suide, and Wuzhong was similar but smaller than populations from wetter or warmer areas. As an exception, females from Tianshui, although living in the sub-wet area, had the smallest average size (average was 52.2 mm in 10 adults) which could be related to lower temperatures and lower humidity as well as a more barren habitat in Tianshui as compared to the other four populations from sub-wet areas (Baoding, Luoyang, Weinan and Yuncheng) (Figure 3a,b).

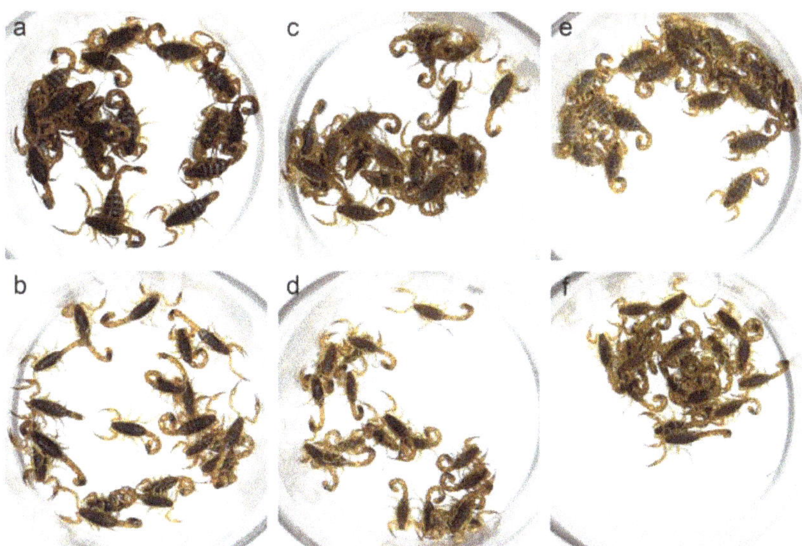

Figure 2. Color differences in *Mesobuthus martensii* populations from Luoyang (sub-wet area), Suide (sub-arid area), and Helan (arid area). (**a,b**) Females and males from Luoyang; (**c,d**) females and males from Suide; and (**e,f**) females and males from Helan.

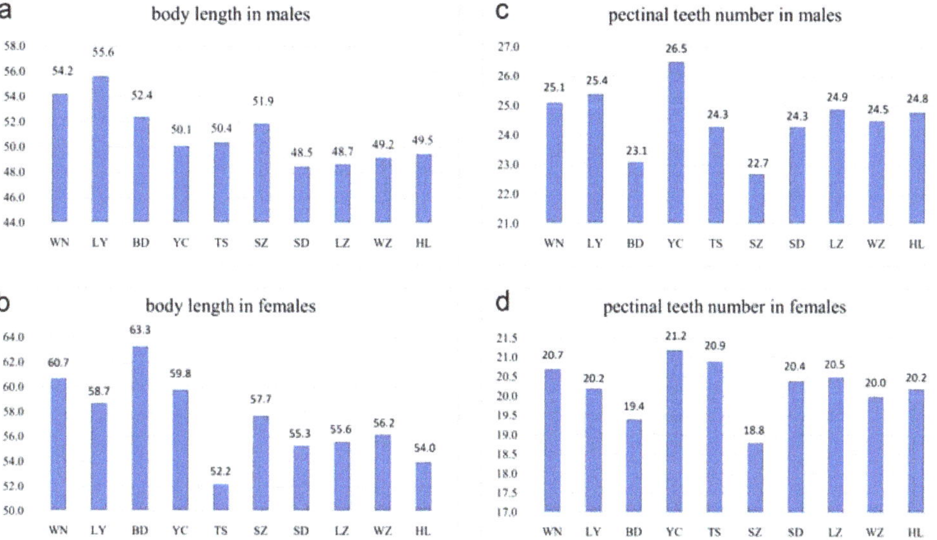

Figure 3. Differences in body length and pectinal teeth number among *Mesobuthus martensii* populations from sub-wet (WN, LY, BD, YC, TS), sub-arid (SZ, SD, LZ), and arid (WZ, HL) areas in China (Tables S1 and S2, means ± SD, $p < 0.001$ using a Kruskal–Wallis H test). (**a**) Body length in males; (**b**) body length in females; (**c**) pectinal teeth number in males; and (**d**) pectinal teeth number in females. BD, Baoding; HL, Helan; LY, Luoyang; LZ, Lanzhou; SD, Suide; SZ, Shuozhou; TS, Tianshui; WN, Weinan; WZ, Wuzhong; YC, Yuncheng.

2.1.3. Pectinal Teeth

Pectinal teeth constitute an important taxonomic feature of the order Scorpiones, especially in the genus *Mesobuthus*. Sun & Sun (2011) reported the difference in the number of pectinal teeth between the two subspecies, *M. caucasicus intermedius* (20–25 in females and 26–30 in males) and *M. caucasicus przewalskii* (15–19 in females and 19–23 in males) [26]. Qi et al. (2004) recorded the pectinal teeth count in *M. martensii*: 21–26 in males and 17–22 in females [9], which could be the common range in populations from Hebei Province, Henan Province, Inner Mongolia Autonomous Region, Liaoning Province, and Shanxi Province.

In this study, the number of pectinal teeth (PTN) was highly diverse among different populations (Figures 3c,d and 4a–d). In males from Helan, PTN was 21–29 (average was 24.8 in 36 pectines), whereas in females, PTN was 18–22 (average was 20.2 in 42 pectines). In males from Wuzhong, PTN was 21–27 (average was 24.5 in 22 pectines) and 18–22 (average was 20.0 in 54 pectines) in females. In males from Yuncheng, PTN was 23–29 (average was 26.5 in 34 pectines) and 19–23 (average was 21.2 in 57 pectines) in females. In males from Weinan, PTN was 22–28 (average was 25.1 in 66 pectines) and 19–22 (average was 20.7 in 46 pectines) in females. Statistically, it indicates that the number of pectinal teeth in arid areas was slightly less than those in sub-wet areas, whereas Baoding is an exception, implying the intraspecific variation in pectines in *M. martensii* (Figures 3c,d and 4a–d).

2.1.4. Large Granules of Fingers

The number of oblique rows of movable finger teeth is another important identification characteristic of *Mesobuthus*. Qi et al. (2004) described movable and fixed fingers in both males and females of *M. martensii* composed of 12 oblique rows of granules [9]. Sun & Sun (2011) reported this characteristic in *M. caucasicus intermedius* (dentate margins of movable and fixed fingers with 12 and 11 oblique rows of granules, respectively) and *M. caucasicus przewalskii* (dentate margins of movable and fixed fingers with 11 and 10 oblique rows of granules, respectively) [26]. We found that slightly different teeth row shapes of *M. martensii* affected the calculation of number of teeth row by researchers. For example, individuals from Weinan and Helan (Figure 4e–l) showed differences in the shape of their oblique teeth rows that affected accurate recording of the number of teeth rows. Instead, we recorded the number of large granules in the lateral sides of movable fingers and fixed fingers (GMN & GFN, Figure 5).

In this study, the number of GMN and GFN in males was 12–14 (average was 12.78 in 36 specimens) and 10–12 (average was 10.75 in 36 specimens) in individuals from Helan, whereas this number was 12–14 (average was 12.95 in 42 specimens) and 10–12 (average was 11.02 in 41 specimens) in females, respectively. In the individuals from Wuzhong, the number of GMN and GFN in males was 12–14 (average was 12.68 in 22 specimens) and 10–11 (average was 10.64 in 22 specimens), but 11–14 (average was 12.76 in 54 specimens) and 10–12 (average was 10.70 in 54 specimens) in females. In individuals from Yuncheng, the number of GMN and GFN in males was 12–15 (average was 13.59 in 34 specimens) and 10–13 (average was 11.59 in 34 specimens), 12–14 (average was 13.44 in 55 specimens) and 10–12 (average was 11.53 in 57 specimens) in females. In individuals from Luoyang, the number of GMN and GFN in males was 11–15 (average was 13.73 in 44 specimens) and 11–12 (average was 11.43 in 44 specimens), 12–14 (average was 13.60 in 48 specimens) and 10–12 (average was 11.29 in 48 specimens) in females. This indicated that the number of large granules in the lateral sides of movable and fixed fingers in arid areas was slightly less than that in sub-wet areas and was similar to pectinal teeth Tianshui is an exception. The difference was significant, and it implies intraspecific variation in *M. martensii*.

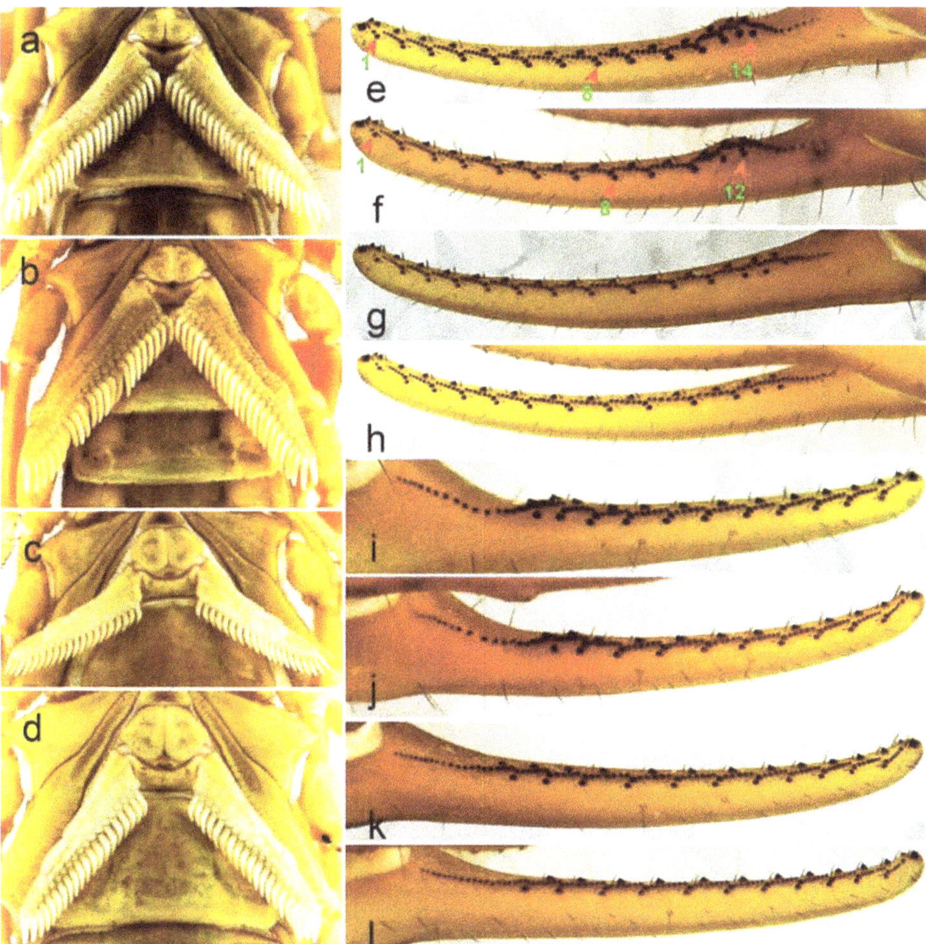

Figure 4. The pectines, moveable fingers, and fixed fingers of *Mesobuthus martensii* from Weinan and Helan. (**a**) Pectines of a male from Weinan; (**b**) pectines of a male from Helan; (**c**) pectines of a female from Weinan; (**d**) pectines of a female from Helan; (**e**) moveable finger of a male from Weinan showing 14 large granules, the numbers "1, 8, 14" are the serial numbers of large granules; (**f**) moveable finger of a male from Helan showing 12 large granules, the numbers "1, 8, 12" are the serial numbers of large granules; (**g**) moveable finger of a female from Weinan showing large granules; (**h**) moveable finger of a female from Helan showing large granules; (**i**) fixed finger of a male from Weinan showing large granules; (**j**) fixed finger of a male from Helan showing large granules; (**k**) fixed finger of a female from Weinan showing large granules; (**l**) fixed finger of a female from Helan showing large granules.

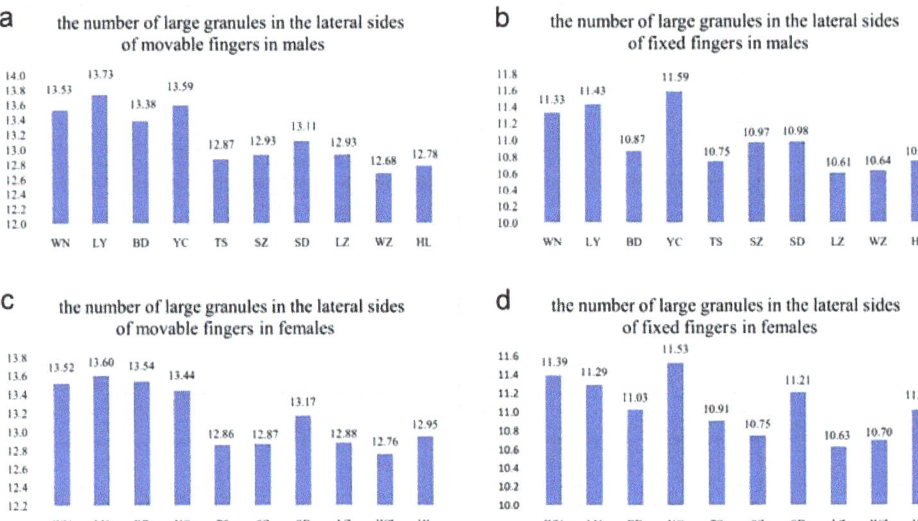

Figure 5. Differences in the number of large granules in the lateral sides of movable and fixed fingers among different *Mesobuthus martensii* populations from sub-wet (WN, LY, BD, YC, TS), sub-arid (SZ, SD, LZ), and arid (WZ, HL) areas in China (Tables S1 and S2: means ± SD, $p < 0.001$ using a Kruskal-Wallis H test). (**a**) The number of large granules in the lateral sides of movable fingers in males; (**b**) number of large granules in the lateral sides of fixed fingers in males; (**c**) the number of large granules in the lateral sides of movable fingers in females; (**d**) number of large granules in the lateral sides of fixed fingers in females. BD, Baoding; HL, Helan; LY, Luoyang; LZ, Lanzhou; SD, Suide; SZ, Shuozhou; TS, Tianshui; WN, Weinan; WZ, Wuzhong; YC, Yuncheng.

2.2. Genetic Distance Revealed That All Populations from the Reported Distribution Range Belong to Mesobuthus martensii

Mirshamsi et al. (2010) reported that the intraspecific genetic distance of *M. eupeus*, the geographical closest relative of *M. martensii*, ranged from 4% to 7% [27]. We analyzed the genetic distance of COI sequences of *M. martensii* from 44 localities and the result was 0.2% to 6.2% (Supplementary Information S1). These 44 localities covered the representative localities of the distribution area of *M. martensii*, supporting that the 10 populations of wild *M. martensii* in this research belonged to the same species.

2.3. Comparison of Venom Gland Transcriptomes Revealed Intraspecific Variations in the Expression of Toxin Genes in Mesobuthus martensii

To reveal the difference in venoms collected from different populations of *M. martensii* from different localities, we sequenced the transcriptomes of mixed venom glands of both sexes from five wild populations (Baoding, Yuncheng, Tianshui, Suide, and Helan). Supplementary Information S2 includes the FPKM of toxin genes of different populations of *M. martensii* and *M. eupeus*.

2.3.1. Expression of Sodium Channel Toxin Genes in Different Populations

Sodium channel scorpion toxins exist widely in the order Scorpiones. It belongs to the long-chain scorpion neurotoxin, which contains 58 to 76 amino acid residues [22,23]. It is a molecular weapon that plays a major role in scorpion venom. When a scorpion injects its venom into its prey, sodium channel toxins make its prey produce spasms or undergo paralysis. The difference in the expression of 30 sodium channel toxin genes (including four putative new members: MMa12627 (BmKNaTx62), MMa29116 (BmKNaTx63), MMa34629

(BmKNaTx64) and MMa38588 (BmKNaTx65)) from different populations showed intraspecific variations (Figure 6a).

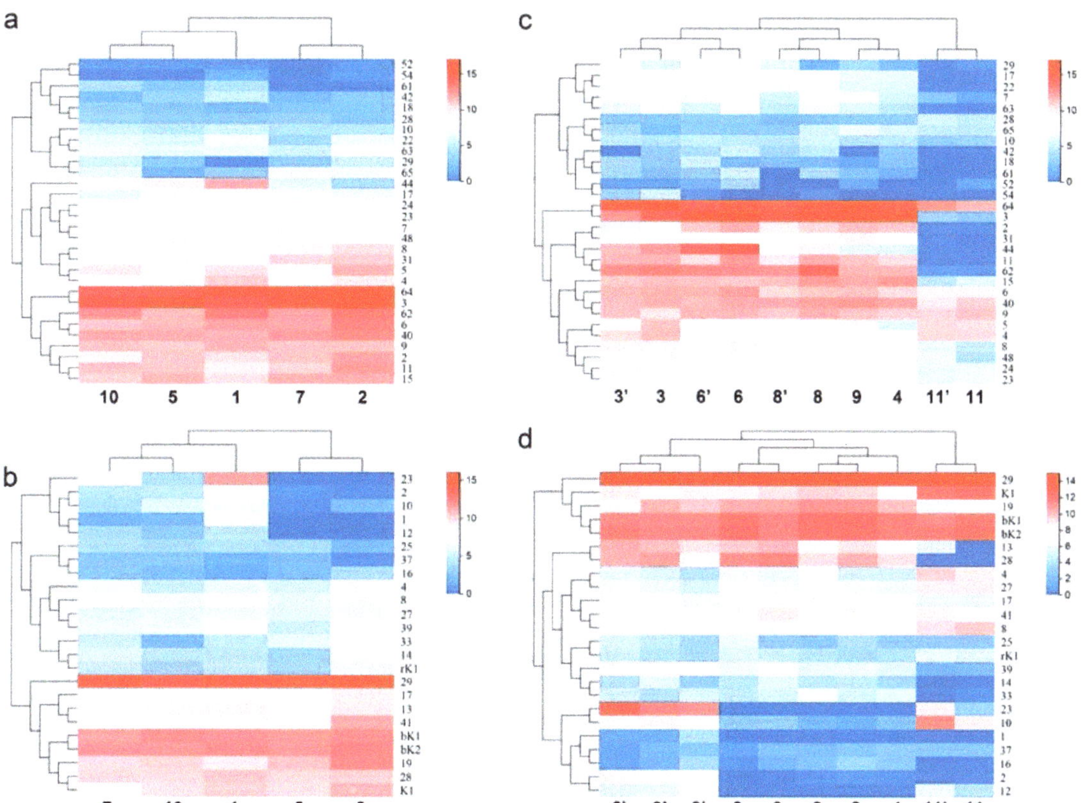

Figure 6. Population and gender differences in the expression of sodium channel toxin genes and potassium channel toxin genes of *Mesobuthus martensii* in China. (**a**,**b**) Heat map showing the difference in the gene expression of sodium channel toxin genes and potassium channel toxin genes among the populations of *M. martensii*; (**c**,**d**) heat map showing the difference in the gene expression of sodium channel toxin genes and potassium channel toxin genes between sexes of *M. martensii* and *M. eupeus*. Populations of *M. martensii* present in X axis: 1, BD, Baoding; 2, HL, Helan; LY, Luoyang (3, LYF, females from Luoyang; 3′, LYM, males from Luoyang); LZ, Lanzhou (4, LZF, females from Lanzhou); 5, SD, Suide; SZ, Shuozhou (6, SZF, females from Shuozhou; 6′, SZM, the males from Shuozhou); 7, TS, Tianshui; WN, Weinan (8, WNF, females from Weinan; 8′, WNM, males from Weinan); WZ, Wuzhong (9, WZF, females from Wuzhong); 10, YC, Yuncheng. The population of *M. eupeus* present in X axis: ME, Yinchuan (11, YCF, females of from Yinchuan; 11′, YCM, the males from Yinchuan). Numbers on Y axis represent gene names, (**a**,**b**) BmKNaTx; (**c**,**d**) BmKaKTx, bK-BmKbKTx, rK-BmKrKTx, K-BmKKTx. Please see Supplementary Information S2 for more details. The expression level was presented as log2(FPKM+1).

The expression (FPKM value) of MMa23370 (BmKNaTx44) in Helan and Tianshui was 16.08 and 89, respectively, whereas the expression in Yuncheng, Suide and Baoding was 320.37, 792.47, and 3300.75, respectively. The difference in the expression between Baoding and Helan differed by 205 times. The expression between Helan and Tianshui was close, and that between Yuncheng and Suide was close. The expression of MMa38588 (BmKNaTx65) in Helan and Tianshui was 153.82 and 188.08 and Suide was 2.89. The difference between the maximum and the minimum expression was 65 times. The expression of MMa53032 (BmKNaTx29) in Baoding, Suide, and Tianshui was 0.14, 6.26, and 7.69, respectively, and Helan and Yuncheng was 42.87 and 44.66. The expression of MMa20191 (BmKNaTx42) in Tianshui, Yuncheng, and Helan was 3.62, 3.9, and 7.06, respectively; whereas in Suide and Baoding, the expression was 11.58 and 46.72 respectively. The difference between the maximum and the minimum expression was 13 times. In addition to the intraspecific differentiation in the expression of toxin genes, the heat map of gene expression of sodium channel toxins showed that the closer the localities of populations, the higher the similarity: Helan and Tianshui were present in one subclade, while Baoding, Yuncheng and Suide were together (Figure 6a). This may suggest a link to the relatedness of populations in *M. martensii*.

2.3.2. Expression of Potassium Channel Toxin Genes in Different Populations

Potassium channel scorpion toxins act on the potassium channel protein of the cell membrane. They act on different types of potassium channels and exert various biological functions [22,23]. The difference in the expression of 24 potassium channel toxin genes from different populations showed intraspecific variations too (Figure 6b).

The expression of MMa16285 (BmKaKTx1) in Helan and Suide was 0, in Tianshui it was 2.48, 3.72 in Yuncheng, and in Baoding it was 133.1. The expression of MMa16284 (BmKaKTx2) in Helan and Suide was 0.33 and 0.66 respectively, in Tianshui it was 22.11, in Yuncheng it was 27.83, and in Baoding it was 175.59. The expression of MMa35044 (BmKaKTx10) in Helan and Suide was 4.42 and 0.37, respectively. In Tianshui it was 16.67, in Yuncheng it was 66.96, and in Baoding it was 109.7. The expression of MMa35043 (BmKaKTx12) in Helan and Suide was 0; in Yuncheng it was 7.66; in Tianshui it was 15.89; and in Baoding it was 27.89. The expression of MMa05343 (BmKaKTx23) in Helan and Suide was 0, in Yuncheng was 16.99, in Tianshui was 218.82, and in Baoding was 2490.09. The expression of MMa34788 (BmKaKTx33) in Yuncheng was 4.28, in Baoding was 29.87, in Tianshui was 37.86, in Suide was 71.39, and in Helan was 169. In addition to the intraspecific differences in the expression of toxin genes, similar to sodium channel toxin genes, the heat map of gene expression of potassium channel toxins showed that the closer the localities of populations, the higher the similarity. The difference in the gene expression of sodium channel toxin genes was sorted by regions: Helan and Suide, one arid and one sub-arid area with average annual temperature below 10 °C, present in one subclade, and Baoding, Tianshui and Yuncheng, three sub-wet areas with average annual temperature above 10 °C in another subclade (Figure 6b). Interestingly, it also seems to be related to the humidity and temperature of the localities (Figure 1).

2.3.3. Expression of Calcium Channel Toxin Genes, Chloride Channel Toxin Genes, and Defensin Genes in Different Populations

The structure and function of calcium channel scorpion toxins and chloride channel scorpion toxins have not been widely studied, except for CTX [22,23]. There were just 2–4 expressed genes available for comparison in every gene family. Similar to sodium channel and potassium channel toxin genes, differences in their gene expression and defensin genes also showed intraspecific variations (Figure 7a–c). The expression of MMa44674 (BmKClTx2) in Helan, Suide, and Tianshui was 0, in Yuncheng was 86.45, and in Baoding was 61.93. The expression of MMa15573 (BmKCaTx1) in Helan and Suide was 0, in Yuncheng was 85.4, in Tianshui was 128.08, and in Baoding was 238.5. The expression of MMa48745 (BmKCaTx2) in Helan and Suide was 0, in Yuncheng was 3.42, in Tianshui was 1.68, and in Baoding was 9.12. Six defensin genes of *M. martensii* were reported [20]. Two new defensin genes MMa09285 (BmKDfsin7) and MMa39355 (BmKDfsin8) were putative (Supplementary Information S3). The expression of MMa09285 (BmKDfsin7) in Helan and Tianshui was 0.15 and 0.51, in Yuncheng was 22.73, in Suide was 124.08, and in Baoding was 489.35. Unlike the sodium channel and potassium channel toxin genes, the heat map of gene expression of these genes did not gather according to the distribution of populations conformably; however, it suggests a relationship between clustering and humidity: Helan and Suide or Tianshui gathered in one subclade (Figure 7a,c), and Baoding and Yuncheng, or Baoding, Tianshui, and Yuncheng gathered in another subclade (Figure 7a,b). It is important to note that fewer genes may not reflect true intraspecific differentiation in *M. martensii*.

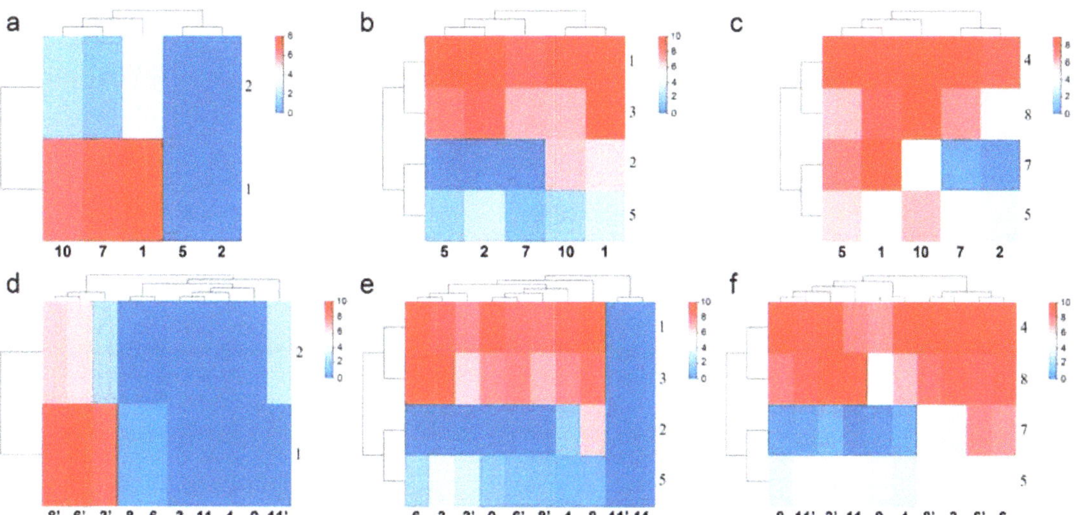

Figure 7. Populations and gender differences in the gene expression of calcium channel toxin genes, chloride channel toxin genes, and defensin genes of *M. martensii* in China. (**a–c**) Heat maps of expression differences of calcium channel toxin genes, chloride channel toxin genes, and defensin genes in different populations of *M. martensii*; (**d–f**) heat maps of expression differences of calcium channel toxin genes, chloride channel toxin genes, and defensin genes in different sexes of *M. martensii* and *M. eupeus*. Populations of *M. martensii* present in X axis: 1, BD, Baoding; 2, HL, Helan; LY, Luoyang (3, LYF, females from Luoyang; 3′, LYM, males from Luoyang); LZ, Lanzhou (4, LZF, females from Lanzhou); 5, SD, Suide; SZ, Shuozhou (6, SZF, females from Shuozhou; 6′, SZM, males from Shuozhou); 7, TS, Tianshui; WN, Weinan (8, WNF, females from Weinan; 8′, WNM, males from Weinan); WZ, Wuzhong (9, WZF, females from Wuzhong); 10, YC, Yuncheng. The population of *M. eupeus* present in X axis: ME, Yinchuan (11, YCF, females of from Yinchuan; 11′ YCM, males from Yinchuan). Numbers on Y axis represent gene names, (**a,d**) BmKCaTx; (**b,e**) BmKClTx; (**c,f**) BmKDfsin. Please see Supplementary Information S2 for more details. The expression level was presented as log2(FPKM+1).

2.4. Expression of Toxin Genes Indicated Sexual Dimorphism of Venom in Mesobuthus martensii

To reveal the difference in venoms of different sexes, we sequenced the transcriptomes of both sexes of populations from Luoyang, Shuozhou, and Weinan of *M. martensii* and *M. eupeus* from Yinchuan and the females of *M. martensii* from Wuzhong and Lanzhou. Supplementary Information S2 includes the FPKM of toxin genes of different sexes in the gene expression of *M. martensii* and *M. eupeus*.

2.4.1. Expression Differences of Sodium Channel Toxin Genes in Different Populations Are More than That in Different Sexes from the Same Population

By clustering with the differences in the expression of toxin genes of both sexes of *M. eupeus*, it was found that the difference in the expression of sodium channel toxin genes showed the following trend: interspecific difference > differences among different populations of the same species > differences between the sexes of the same population (Figure 6c).

The expression of MMa17864 (BmKNaTx5) in males from Luoyang was 413.1, and in females from Luoyang it was 2134.35, a difference that was five times larger in females than males. The expression of MMa29117 (BmKNaTx11) in males from Shuozhou was 1184.93, and in females from Shuozhou it was 4641.17, a difference that was four times larger in females than in males. In males from Weinan it was 974.72, and in females from Weinan it was 3305.2, a difference that was three times larger in females than males. Therefore, the expression of these sodium channel toxin genes of females was higher than that in males. The expression of MMa35303 (BmKNaTx18) in females from Shuozhou was 3.89, and in males from Shuozhou it was 47.85, a difference that was 12 times larger in males than females. In females from Luoyang it was 8.97, and in males from Luoyang it was 17.6, a difference that was two times larger in males than females. This is an example for the expression of sodium channel toxin genes of males was higher than that of females. Some similar examples are as follows. The expression of MMa53032 (BmKNaTx29) in females from Weinan was 2.18, and in males from Weinan it was 81.24, a difference that was 37 times larger in males as compared to females. The expression of MMa20191 (BmKNaTx42) in males from Luoyang was 0, and in females from Luoyang it was 11.43. The expression of MMa38588 (BmKNaTx65) in males from Weinan was 8.8, and in females from Weinan it was 55.85, the difference was six times larger in females than in males. In females from Luoyang, the expression was 4.42, and in males from Luoyang it was 16.73, a difference that was four times larger in males than in females. The expression of MMa14634 (BmKNaTx61) in males from Weinan was 0, and in females from Weinan it was 20.93. For males from Shuozhou it was 12.04, and for females from Shuozhou it was 52.42; the difference was four times larger in females than in males. In females from Luoyang gene expression was 6.05, and in males from Luoyang it was 16.06; the difference was three times larger in males than in females. The expression of MMa29116 (BmKNaTx63) in males from Shuozhou was 58.95, and in females from Shuozhou was 394.8, the difference was seven times larger in females than in males. In males from Weinan gene expression was 35.93, and in females from Weinan it was 144.83; the difference was four times larger in females than males.

In addition to the intraspecific differentiation in the expression of toxin genes, the heat map of gene expression of sodium channel toxins in different sexes from the different populations and species also showed that the closer the localities of populations, the higher the similarity: Lanzhou (sub-arid area) and Wuzhong (arid area) were present in one subclade, while Luoyang and Shuozhou (two sub-wet areas) were together (Figure 6c). It is important to note that although Weinan (sub-wet area) is closer to Shuozhou, the gene expression of its population is clustered with that of Wuzhong and Lanzhou. This may suggest a link to the relatedness of populations in *M. martensii*.

2.4.2. Expression Differences of Potassium Channel Toxin Genes in the Same Sexes from Different Populations of the Same Species Are Less than That in Different Sexes from the Same Population

The expression of potassium channel toxin genes also showed intraspecific sexual dimorphism. Clustering with the expression difference of potassium channel toxin genes of *M. eupeus*: interspecific difference > differences between both sexes of same populations > same-sex of different populations of same species (Figure 6d).

The expression of MMa16285 (BmKaKTx1) in females from Weinan was 0, and in males it was 15.44. There was no expression of MMa16285 (BmKaKTx1) in males and females of *M. eupeus*. The expression of MMa16284 (BmKaKTx2) in females from Luoyang was 0, in males was 112.96; in females from Weinan was 0.66, in males was 158.65; in females from Shuozhou was 0, in males was 316.69; in females, the expression was very low or no expression; in males, the expression was high. There was no expression of MMa16284 (BmKaKTx2) in males and females of *M. eupeus*. The expression of MMa35044 (BmKaKTx10) in females from Luoyang was 9.18, and in males it was 321.87, the difference was 35 times; that in females from Shuozhou was 8, and in males was 271.67, the difference was 34 times; in females from Weinan it was 2.09, in males was 120.26, the difference was 58 times; in females, the expression was low; in males, the expression was high. In *M. eupeus* females, the expression of MMa35044 (BmKaKTx10) was 370.92; in males, it was 2481.21. The expression of MMa35043 (BmKaKTx12) in females from Luoyang was 0, in males was 57.88; that in females from Shuozhou was 0, in males was 52.03; in females from Weinan was 0, in males was 137.46; the potassium channel toxin gene of female *M. martensii* was not expressed; in males, it was highly expressed. In *M. eupeus* females, the expression was 18.76; in males, it was 61.07. The expression of MMa05343 (BmKaKTx23) in females from Luoyang was 0; in males, it was 2424.3; in females from Shuozhou, it was 0; in males, it was 6707.88; in females, from Weinan it was 0; in males, it was 2095.2; the potassium channel toxin gene of female *M. martensii* was not expressed; but in males, it was highly expressed. In *M. eupeus* females, the expression of MMa05343 (BmKaKTx23) was 9.11 and in males, it was 368.02.

Although both sexes of one population gathered in different subclades, but same sex of Luoyang and Shuozhou (two close sub-wet areas) together respectively (Figure 6d). It may suggest a link to the relatedness or habitats of populations in *M. martensii*.

2.4.3. Expression Differences of Calcium Channel Toxin Genes, Chloride Channel Toxin Genes, and Defensin Genes in Sexes from Different Populations

The expression of calcium channel toxin genes suggested a population difference (Figure 7a) and sexual dimorphism (Figure 7d). Clustering with the expression difference of calcium channel toxin genes of *M. eupeus* showed male preference, all females of different populations of *M. martensii* and both sexes of *M. eupeus* showed low expression (Figure 7d). Clustering with the expression difference of toxin genes of both sexes of *M. eupeus* showed a difference in the expression of chloride channel toxin genes in different populations (Figure 7b) and different sexes (Figure 7e) of *M. martensii*; however, no difference was observed in both sexes of *M. eupeus*: interspecific difference > intraspecific differences, whereas neither of the same sex of different populations or both sexes of the same population of *M. martensii* showed more similarity (Figure 7e). All four chloride channel toxin genes showed no expression in *M. eupeus*, suggesting significant interspecific differences with *M. martensii*, whereas the two chloride channel toxin genes showed low expression in the latter. There is no logical explanation for the heat map showing the differences in the expression of defensin genes, except that both sexes from Shuozhou gathered in one subclade (Figure 7f). The expression of MMa44674 (BmKClTx2) in males from Weinan was 0; in females, it was 75.28; in males and females from Luoyang and Shuozhou, it was 0. The expression of MMa00982 (BmKClTx3) in males from Luoyang was 69.34; in females, it was 483.25, the difference was seven times. The expression of MMa15573 (BmKCaTx1) in females from Luoyang was 0, whereas in males, it was 337.25; in females, it was 0.8 from Shuozhou, whereas in males, it was 880.28; in females from Weinan, it was 0.83, whereas in males, it was 434.7; in females of *M. martensii*, the expression was extremely low or no expression; in males, it was high. In *M. eupeus*, the females did not show expression. This indicated that calcium channel toxin genes were not expressed in females. The expression of MMa48745 (BmKCaTx2) in females from Luoyang was 0, whereas in males, was 5.25; in females from Shuozhou, it was 0, whereas in males, it was 63.44; in females from Weinan, it was 0, whereas in males, it was 76.54; in females of *M. martensii*, it was not expressed; in males, it was low or high. In *M. eupeus*, the females did not express it. This indicated that calcium channel toxin genes were not expressed in females. The expression of MMa09285 (BmKDfsin7) in males from Luoyang was 1.01; in females, it was 18.12; the difference was 18 times. The expression of MMa09285 (BmKDfsin7) in males from Weinan was 27.12; in females, the expression was 0.33.

2.5. Expression of Toxin Genes in the Same Cluster of Mesobuthus martensii Genome

Nine of 17 toxin gene clusters reported in the genome sequencing of *M. martensii* showed expression [20,28]. The expression of one gene in the cluster was close in different populations and genders whereas it is not same in different members in the same populations and genders (Figure 8a–c,e,f,i). In the case of selective expression in members of the same cluster, certain members did not express in the normal state but were expressed either as silent or low in different populations or genders (Figure 8a,d,g,h). In addition, there are several cases in which the members of certain gene clusters were not expressed, and it is unclear why they are or are not expressed. (1) BmKNaTx1 was absent; and BmKNaTx3 was relatively highly expressed, whereas BmKNaTx2 was lowly expressed in *M. martensii* (Figure 8a). (2) The expression of BmKaKTx1 and BmKaKTx2 in females was very low or even not present in *M. martensii*, except the population from Baoding; however, the expression was relatively high in males (Figure 8d). (3) The expression difference of BmKNaTx4, BmKNaTx8, BmKNaTx5, and BmKNaTx7, and BmKNaTx6 in *M. martensii* was large (Figure 8h). (4) The expression of BmKNaTx9 was high relatively, whereas the expression of BmKaKTx5 and BmKaKTx3 was absent and BmKaKTx4 was low (Figure 8e). (5) The expression of and BmKaKTx8 was relatively high, whereas the expression of BmKNaTx10 was low in *M. martensii* (Figure 8b). (6) The expression of BmKNaTx15 was high relatively and BmKNaTx17 was low, whereas BmKNaTx16, BmKNaTx14, and BmKNaTx13 were not expressed in *M. martensii*; the order of this cluster is BmKNaTx16, BmKNaTx15, BmKNaTx14, BmKNaTx13, and BmKNaTx17 (Figure 8f). (7) The expression of BmKaKTx12 and BmKaKTx10 in females was very low or no expression in *M. martensii*, whereas the expression was relatively high in males. BmKaKTx13 was highly expressed relatively in different populations and genders, whereas BmKaKTx11 and BmKaKTx9 were not expressed in *M. martensii*. The order of this cluster is BmKaKTx11, BmKaKTx12, BmKaKTx10, BmKaKTx9, and BmKaKTx13 (Figure 8g). (8) The expression of BmKaKTx17 was high relatively, that of BmKrKTx1, BmKaKTx14, and BmKaKTx16 was low, and that of BmKbKTx1 was high; whereas BmKaKTx18, BmKaKTx15 were not expressed. The order of this cluster was BmKaKTx18, BmKaKTx17, BmKrKTx1, BmKaKTx14, BmKaKTx15, BmKaKTx16, and BmKbKTx1 (Figure 8i). (9) The expression of BmKNaTx24 and BmKNaTx23 was the same in same population (Figure 8c).

2.6. The Protein Evidences of Expressed Toxin Genes in Mesobuthus martensii

The protein evidences of expressed toxin genes, the multiple sequence alignment of four putative new toxin genes named following Cao et al., (2013) [20] (MMa12627 (BmKNaTx62), MMa29116 (BmKNaTx63), MMa34629 (BmKNaTx64) and MMa38588 (BmKNaTx65)), and the peptide sequences from *M. martensii* venom samples were provided (Supplementary Information S3). Twenty-one expressed genes with the MS/MS identification evidences of the *M. martensii* venom samples separated by 2-DE, SDS-PAGE, and RP-HPLCa by Xu et al., (2014) [21]. Five genes with yellow shadow have the inconclusive evidences due to the inconsistency of an amino acid residue between the multiple alignment sequences of putative toxin genes and the peptide sequences from Xu et al. (2014) [21] (Supplementary Information S3).

2.7. Validation of RNA-Seq Data by RT-qPCR Analysis

Two expressed genes namely MMa05343 (BmKaKTx23) and MMa35043 (BmKaKTx12) with $|\log2((\male\text{-FPKM}+1)/(\female\text{-FPKM}+1))| \geq 5$ from Shuozhou and Weinan populations, and eight expressed genes, namely, MMa29117 (BmKNaTx11), MMa34788 (BmKaKTx33), MMa17863 (BmKNaTx8), MMa17863-1 (BmKNaTx8-1), MMa17863-2 (BmKNaTx8-2), MMa21265 (BmKaKTx4), MMa21265-1 (BmKaKTx4-1), and MMa21265-2 (BmKaKTx4-2) from Helan (HL) and Baoding (BD) populations with $|\log2((\text{HL-FPKM}+1)/(\text{BD-FPKM}+1))| \geq 2$ were selected. Among them, -1 and -2 indicate that the gene had multiple different transcripts. The transcriptome data were verified by RT-qPCR analysis.

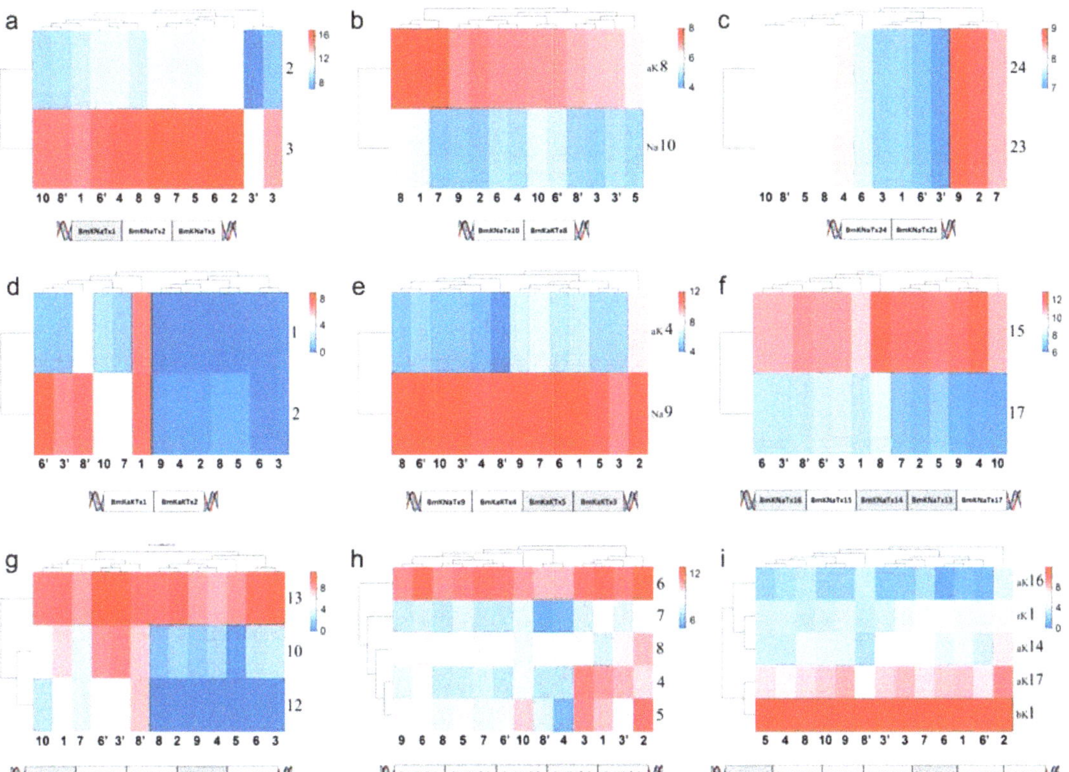

Figure 8. Expression of toxin gene cluster members in the *Mesobuthus martensii* genome, shows that the expression of one gene in the cluster was close in different populations and genders, and the members of most clusters expressed in same population and gender tended to be the different. The silent genes (with shadow in the schematic diagram of the sequence of genes on DNA) was the same in different populations and genders. The populations of *M. martensii* present in X axis: 1, BD, Baoding; 2, HL, Helan; LY, Luoyang (3, LYF, females from Luoyang; 3′, LYM, males from Luoyang); LZ, Lanzhou (4, LZF, females from Lanzhou); 5, SD, Suide; SZ, Shuozhou (6, SZF, females from Shuozhou; 6′, SZM, males from Shuozhou); 7, TS, Tianshui; WN, Weinan (8, WNF, females from Weinan; 8′, WNM, males from Weinan); WZ, Wuzhong (9, WZF, females from Wuzhong); 10, YC, Yuncheng. Population of *M. eupeus* present in X axis: ME, Yinchuan (11, MEF, females from Yinchuan; 11′, MEM, males from Yinchuan). Numbers the Y axis represents gene names, (**a**,**c**,**f**,**h**) BmKNaTx; (**d**,**g**) BmKaKTx; (**b**,**e**,**i**) Na-BmKNaTx, aK-BmKaKTx, bK-BmKbKTx, and rK-BmKrKTx. Please see Supplementary Information S2 for more details. The expression level was presented as log2(FPKM+1).

In this study, we compared the expression data obtained by RT-qPCR (blue column) and RNA-seq (red column). The ♂/♀expression level ratio of gene MMa05343 (BmKaKTx23) from Shuozhou (SZ) by RT-qPCR was 10.22, and the value in FPKM by RNA-seq was 12.71 (Figure 9a). The ♂/♀expression level ratio of gene MMa35043 (BmKaKTx12) from Shuozhou by RT-qPCR was 8.66, and the value in FPKM by RNA-seq was 5.73 (Figure 9a). The ♂/♀expression level ratio of gene MMa05343 (BmKaKTx23) from Weinan (WN) ♂/♀by RT-qPCR was 8.56, and the value in FPKM by RNA-seq was 11.03 (Figure 9b). The ♂/♀expression level ratio of gene MMa35043 (BmKaKTx12) from Weinan by RT-qPCR was 11.27, and the value in FPKM by RNA-seq was 7.11 (Figure 9b). The Helan (HL)/Baoding (BD) expression level ratio of gene MMa29117 (BmKNaTx11) by RT-qPCR was 0.87, and the value in FPKM by RNA-seq was 2.43 (Figure 9c). The HL/BD expression level ratio of gene MMa34788 (BmKaKTx33) by RT-qPCR was 1.30, and the value in FPKM by RNA-seq was 2.46. The HL/BD expression level ratio of gene MMa17863 (BmKNaTx8) by RT-qPCR was 1.68, and the value in FPKM by RNA-seq was 2.71. The HL/BD expression level ratio of gene MMa17863-1 (BmKNaTx8-1) by RT-qPCR was 1.90, while the HL/BD expression level ratio of gene MMa17863-2 (BmKNaTx8-2) by RT-qPCR was 1.26 (Figure 9c). The HL/BD expression level ratio of gene MMa21265 (BmKaKTx4) by RT-qPCR was 1.18, and the value in FPKM by RNA-seq was 2.10. The HL/BD expression level ratio of gene MMa21265-1 (BmKaKTx4-1) by RT-qPCR was 0.15, while the HL/BD expression level ratio of gene MMa21265-2 (BmKaKTx4-2) by RT-qPCR was 0.39.

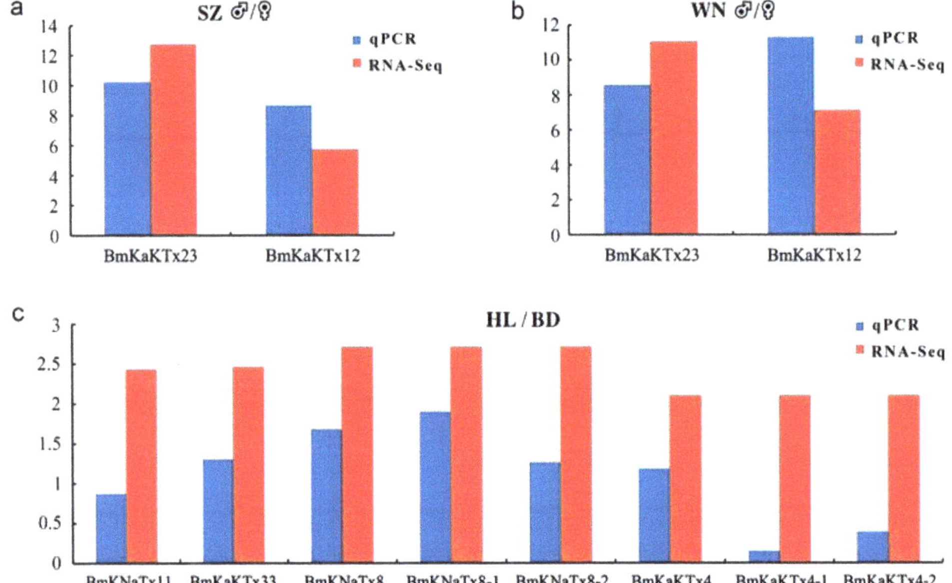

Figure 9. Comparison of gene expression data obtained by RT-qPCR and RNA-seq. (**a**) Differential expression multiple of RT-qPCR and RNA-seq in both sexes of Shuozhou population; (**b**) differential expression multiple of RT-qPCR and RNA-seq in both sexes of Weinan population; (**c**) differential expression multiple of RT-qPCR and RNA-seq in Baoding and Helan populations. BD, Baoding; HL, Helan; SZ, Shuozhou; WN, Weinan. Y asix represents the expression level of genes, obtained by RT-qPCR was presented as: $|\log2(♂\text{-RE}/♀\text{-RE})|$; obtained by RNA-seq was presented as: (**a**,**b**) value = $|\log2((♂\text{-FPKM}+1)/(♀\text{-FPKM}+1))|$, (**c**) value = $|\log2((\text{HL-FPKM}+1)/(\text{BD-FPKM}+1))|$. RE, relative expression level.

The analysis revealed that the expression trend of the selected expressed genes was basically the same as that of RNA-seq; however, the expression multiples were slightly different, indicating that the results of RNA-seq in this study had good accuracy and credibility.

3. Discussion

M. martensii is widely distributed in the vast and complex habitats of northern China. Intraspecific variation majorly contributes to their survival and adaptation. The results revealed differences in intraspecific variations in the morphology, color, toxin gene expression, and defensin gene expression in *M. martensii*. In response to the complex living environment, different populations are subjected to variable ecological factors and dual selection pressures from biological and abiotic factors.

M. martensii is an important component of traditional Chinese medicine, with a long medical history and unique functions. It has several kinds of toxins that can regulate manifold ion channels specifically and can serve as a crucial natural drug resources. In addition, its defensins play an important role in its survival and adaptation.

In recent years, intemperance in excessive collection and difficulty in artificial reproduction have led to severe destruction of natural resources and a decline in the population of *M. martensii*. Fake medicines containing other species belonging to the same genus and even different families or genera are considerably common in traditional Chinese medicinal materials. Closely related species of *M. martensii* from the same genus are widespread and similar to each other, whereas genuine medicinal materials of *M. martensii* have a large region of distribution. Intraspecific differentiation without accurate identification criteria has led to complicated medicinal material sources of Quanxie. It has severely affected accurate clinical prescription and medical research in the area of traditional Chinese medicine. Investigating its medicinal material resources and revealing its intraspecific differentiation, defensin gene and toxin gene resources will promote further in-depth research for the use of scorpions and their defensive toxins which in turn will be helpful in standardizing the identification and medical applications of Quanxie in traditional Chinese medicine.

Scorpions, as cold-blooded animals, rely on their own behavior to regulate body heat emission or absorb heat from the external environment to raise their body temperature. So the ambient temperature affects their metabolism. Therefore, it's likely that their body size (directly proportional to body weight) is also influenced by ambient temperature. In addition, it is limited by the situation of prey (richness and size, etc.). The color of their body is similar to the surrounding environment which is more conducive to the "waiting" hunting method, whereas the surrounding environment is closely related to temperature and humidity. For pectinal teeth, we speculate that the habitat in high humidity areas are more complex, whereas it is desolate and simple in low humidity areas. The size and number of pectinal teeth are related to chemical sensors. Since the airflow in complex areas is less than that in the open and desolate areas, scorpion species require a better sense of smell. However, variations in the number of large granules in the lateral sides of moveable and fixed fingers are difficult to contact with external factors. It could only be related to the size of the body as the number of large granules in the lateral sides of the moveable and fixed fingers corresponds to the number of pectinal teeth and the size of the body.

Under normal conditions, we could identify only 60 toxin genes in different populations, accounting for only about 50% of the known toxin gene members in the genome. There exist certain differences among different populations. It is not known why only these toxin genes were expressed and what is the biological significances of the difference in the expression. It is a new mystery and interesting subject. We found 21 toxin genes using MS/MS of *M. martensii* venom samples separated by 2-DE, SDS-PAGE, and RP-HPLCa by Xu et al. (2014) [21]. This suggests the biological significance of the differential expression of toxin genes.

We found that the difference in the expression of sodium channel toxin genes (Figure 6c) among different populations was conserved (populations conservation), suggesting that the expression characteristics of toxins are related to the geographical distribution and genetic relationship, whereas the expression difference in potassium channel toxin genes (Figure 6d) and calcium channel toxin genes (Figure 7d) between the two sexes was conserved (gender conservation), revealing the sexual dimorphism of toxin expression. In addition, sexual dimorphism of toxin gene expression was observed in the sexual venom gland transcriptomes of *M. eupeus* (Figure 6c,d and Figure 7d,f), suggesting that sexual dimorphism of toxin gene expression could be a common feature of species belonging to the order Scorpiones. Heat maps showed differences in the gene expression of calcium channel toxin genes, chloride channel toxin genes, and defensin genes in different sexes of *M. martensii* and *M. eupeus*, although there were too few genes (2–4) available for comparison. Hierarchical clustering did not reflect accurate results about the relationship of interspecific and intraspecific differences such as those of the toxin genes encoding for sodium channels and potassium channels. Interestingly, the expression of two calcium channel toxin genes showed preference for males and certain populations.

It was reported that 51 toxin genes encoding for sodium and potassium channels of *M. martensii* were linked into up to 17 gene clusters [20,28]. In this study, eight clusters of toxin genes were found to be not expressed under normal conditions and nine clusters were expressed. Among these, the expression of a member of one gene cluster is consistent in different populations or genders. In addition, the level of expression of each member of the same gene cluster was generally inconsistent, including certain members was not expressed. The expression characters of potassium channel toxin gene clusters and sodium channel toxin gene clusters are the same approximately; however, certain gene members vary in different sexes. This revealed selective silencing, low expression, or high expression among gene cluster members.

It is worth noting that it is not uncertain whether the expression level data of toxin genes and antimicrobial peptide gene families of different populations or sexes of *M. martensii* from different localities are significant, which may be the major weakness of this paper to accurately reveal intraspecific differentiation. The data of gene expression levels used for hierarchical clustering of the different populations or sexes did not yield significantly different results in the t-test, even if the standardization process was performed in advance. This may be due to excessive differences in expression levels among the genes in each population or each sex of one population (Supplementary Information S2). However, as we have observed, the difference of color characters among some populations is noticeable and the sexual dimorphism of morphological structures in each population is apparent. The body length of adult individuals from arid and sub-arid areas are smaller than those from sub-wet areas, and the average values of the number of pectinal teeth and the number of large granules in the lateral sides of the pedipalp chela fingers were smaller in the former. By Kruskal–Wallis H test (KW test), there were significant differences in body length, number of pectinal teeth, and number of large granules in the lateral sides of movable and fixed fingers between males or females of *M. martensii* from different localities ($p < 0.001$; Tables S1 and S2). Therefore, in terms of morphological characteristics, the populations we selected are significantly different. Due to the wide distribution range and complex habitat of this species, a more extensive sampling is clearly needed for rigorously examining intraspecific diversification and evolution in *M. martensii* [12]. Intraspecific resting metabolic rate, a key physiological trait linked with evolutionary fitness, shows that variation exists in between sexes and among populations of *M. martensii* that is closely related to the local mean temperature and mean annual days of rainfall [13]. Our findings in the expression differences of toxin gene families between genders and among populations of *M. martensii* suggest a similar case to the relevance to the environment to resting metabolic rate.

4. Materials and Methods

4.1. Collection of Mesobuthus martensii

The specimens of wild *M. martensii* were collected from the Gansu Province, Hebei Province, Henan Province, Ningxia Hui Autonomous Region, Shaanxi Province, and Shanxi Province. The localities of specimens were in the sub-wet zone (light green area), sub-arid zone (light yellow area), and arid zone (yellow area) (Figures 1, 10 and S1) and were kept in 75% alcohol. Specimens were deposited in the Museum of Hebei University (MHBU), Baoding, China. The living specimens were raised in the laboratory at room temperature (about 25 °C) temporarily. Geographical names and their abbreviations are also used to represent populations, except when introducing geographical locations and explaining distance relationships between localities.

Figure 10. *Mesobuthus martensii* from Nan yang, Henan Province, China; the small picture is its venom gland and venom.

4.2. Morphological Study

Measurements were taken using the Motic-K700 microscope (Motic China Group Co., Ltd., Xiamen, China, with accurate micrometric ocular). All materials were randomly selected from healthy adult individuals. All measurements were in millimeters. The terminology followed was described by Hjelle (1990) [29]. The measurement methods followed were those described by Sissom et al. (1990) [30]. Photographs were acquired by the macro camera Canon 650D (with a microlens). The Division map of arid and wet areas in China is based on Zhang et al. (2016) [31]. The dry humidity division of China followed the procedure described by the National Meteorological Information Center [32].

The body length, number of pectinal teeth, and number of large granules in the lateral sides of the pedipalp chela fingers of randomly selected adult individuals were measured. Kruskal–Wallis H test in SPSS24.0 (IBM, New York, NY, USA) software was used to analyze the overall differences of the characteristics of the scorpions from different localities according to the characteristic values of the scorpions.

4.3. Genetic Distance Calculation

We downloaded the COI sequence of wild *M. martensii* from 44 localities uploaded by Shi et al. (2013) in NCBI [12,33]. The coverage area of these localities included the collection sites of *M. martensii*. Next, sequence alignment was performed using ClustalX1.83 (Bioedit Company, IN, USA), and genetic distance was analyzed using MEGA6.0 (Tamura et al., 2013) [34].

4.4. Transcriptome Extraction, Sequencing, Annotation, and Gene Analysis

We used pooled scorpion venom glands for the study rather than replicates of individuals or replicates of pools. The number of adult individuals used for each pool (transcriptome) was as followed: *M. martensii*: 1-BD, 4 females and 4 males from Baoding; 2-HL, 2 females and 1 male from Helan; 3-LYF, 9 females from Luoyang; 3′-LYM, 9 males from Luoyang; 4-LZF, 7 females from Lanzhou; 5-SD, 8 females and 3 males from Suide; 6-SZF, 9 females from Shuozhou; 6′-SZM, 9 males from Shuozhou; 7-TS, 5 females and 3 males from Tianshui; 8-WNF, 9 females from Weinan; 8′-WNM, 7 males from Weinan; 9-WZF, 11 females from Wuzhong; 10-YC, 5 females and 5 males from Yuncheng. *M. eupeus*: 11-MEF, 9 females from Yinchuan; 11′-MEM, 10 males from Yinchuan. The relevant results of expression of toxin genes in the following text are mean toxin gene expression.

The transcriptomes were extracted by the Trizol method, and the quality of the extracted nucleic acid was determined. Sequencing and splicing were performed using the BGISEQ-500 analysis platform of BGI (Beijing Genomics Institute, Beijing, China), whereas its annotation was done by referring to the genome of *M. martensii* [35]. We defined expressed genes using the following criteria: FPKM $\geq$ 10 in any one population (FPKM not less than 10 in at least one population). Unless otherwise stated, the unit of expression level in our analyses is FPKM. Online BLAST annotated possible new genes and named them according to the discovery and verification information of *M. martensii* toxin genes provided by Cao et al. (2013) [20]. The heat maps were produced online by BGI. Based on the results of the detection of differential genes, a hierarchical cluster analysis was performed on concatenated differential genes, using the heatmap in R [36]. The gene cluster analysis was based on the method described by Cao et al. (2013) [20].

4.5. Proteomic Identification of Expressed Toxin Genes

The peptide sequence of expressed toxin genes was aligned with the MS/MS identification protein evidence of *M. martensii* venom samples separated by 2-DE, SDS-PAGE, and RP-HPLC [21,24] by ClustalX1.83, as well as the local blast.

4.6. RT-qPCR Analysis

RT-qPCR analysis was performed to detect the expression of 10 genes in scorpion venom gland tissues. The RT-qPCR primers were synthesized at the Invitrogen Company (Beijing, China). Scorpion GAPDH was used as the housekeeping gene. PCR amplification was conducted in an ABI 7500 real-time PCR system using the following program: 30 s at 95 °C; 40 cycles (95 °C for 5 s, 60 °C for 40 s [collecting fluorescence]). To establish the melting curve of PCR products, amplification of the product was done at 95 °C for 10 s, 60 °C for 60 s, 95 °C for 15 s. They were subsequently slowly heated from 60 °C to 99 °C (ramp rate is 0.05 °C/s). Data were analyzed using Ct values and the $2^{-\Delta\Delta CT}$ method [37].

Supplementary Materials: The following supporting information can be downloaded at: https://www.mdpi.com/article/10.3390/toxins14090630/s1, Figure S1: The *Mesobuthus martensii* populations and its relatives from sub-wet area, sub-arid area, and arid area. The populations of *M. martensii*: 1 & 1′, the females and males from Baoding; 2&2′, the females and males from Helan; 3&3′, the females and males from Luoyang; 4&4′, the females and males from Lanzhou; 5&5′, the females and males from Suide; 6&6′, the females and males from Shuozhou; 7&7′, the females and males from Tianshui; 8&8′, the females and males from Weinan; 9&9′, the females and males from Wuzhong; 10&10, the females and males from Yuncheng. The populations of *M. eupeus*: 11&11′, the females and males from Yinchuan; 12&12′, the females and males from Zhongwei (between Lanzhou and Yinchuan). Table S1: Measurement data of the male *Mesobuthus martensii* (means $\pm$ SD, **: $p < 0.001$, Kruskal-Wallis H test (KW test)) Body length (BL), number of adult individuals (NI), number of pectinal teeth (NP), number of large granules in the lateral sides of movable fingers (NM), number of large granules in the lateral sides of fixed fingers (NF). BD, Baoding; HL, Helan; LY, Luoyang; LZ, Lanzhou; SD, Suide; SZ, Shuozhou; TS, Tianshui; WN, Weinan; WZ, Wuzhong; YC, Yuncheng. Table S2: Measurement data of the female *Mesobuthus martensii* (means $\pm$ SD, **: $p < 0.001$, Kruskal-Wallis H test (KW test)) Body length (BL), number of adult individuals (NI), number of pectinal teeth (NP), number of large granules in the lateral sides of movable fingers (NM), number of large granules in the lateral sides of fixed fingers (NF). BD, Baoding; HL, Helan; LY, Luoyang; LZ, Lanzhou; SD, Suide; SZ, Shuozhou; TS, Tianshui; WN, Weinan; WZ, Wuzhong; YC, Yuncheng. Supplementary information S1: The genetic distance of COI sequences of 44 populations of *Mesobuthus martensii* in China is 0.2% to 6.2%. These 44 localities cover the representative sites of the distribution area of *M. martensii* in China [12], so it supports that the wild populations in this study of *M. martensii* belong to the same species. Supplementary information S2: The FPKM of toxin and defensin genes of different populations and sexes in the gene expression of *Mesobuthus martensii* (MM) and *Mesobuthus eupeus* (ME) in China. The populations of *M. martensii*: 1-BD, Baoding; 2-HL, Helan; LY, Luoyang (3-LYF, the females from Luoyang; 3'-LYM, the males from Luoyang); LZ, Lanzhou (4-LZF, the females from Lanzhou); 5-SD, Suide; SZ, Shuozhou (6-SZF, the females from Shuozhou; 6'-SZM, the males from Shuozhou); 7-TS, Tianshui; WN, Weinan (8-WNF, the females from Weinan; 8'-WNM, the males from Weinan); WZ, Wuzhong (9-WZF, the females from Wuzhong); 10-YC, Yuncheng. The population of *M. eupeus*: ME, Yinchuan (11-YCF, the females of from Yinchuan; 11'-YCM, the males from Yinchuan). The genes with green shadow are that have the MS/MS identification evidences of the *M. martensii* venom samples separated by 2-DE, SDS-PAGE and RP-HPLCa by Xu et al. (2014) [21]. Supplementary information S3: S3.1 The multiple sequence alignment of four putative new toxin genes (MMa12627 (BmKNaTx62), MMa29116 (BmKNaTx63), MMa34629 (BmKNaTx64) and MMa38588 (BmKNaTx65)), two predicted defensin genes (MMa09285 (BmKDfsin7) and MMa39355 (BmKDfsin8)) and their homologous genes; S3.2 The protein evidences of expressed toxin genes; S3.3 The peptide sequences with serial numbers in S3.2 from MS/MS identification of the *M. martensii* venom samples separated by 2-DE, SDS-PAGE and RP-HPLCa (the original data was from Xu et al. (2014)) [21].

Author Contributions: Z.D. designed sequencing experiments. Z.D., S.Q., S.X., C.L. and Y.L. conducted experiments on sample preparation, and RNA isolation for sequencing. Z.D. and S.Q. processed sequence data and performed bioinformatics analysis. Z.D., S.Q. and X.L. contributed to morphological data and transcriptome data analysis. Z.D. and S.Q. contributed to sequencing and experimental validation. Z.D., S.Q., X.L. and F.Z. drafted and revised the manuscript. Z.D., F.Z. and S.L. conceived the study, and directed on experiments. All authors have read and agreed to the published version of the manuscript.

Funding: This work was supported by grants to Zhiyong Di from the National Natural Sciences Foundation of China (31970403 and 31601871), the Hebei Provincial Natural Science Foundation (C2019201273), the Advanced Talents Incubation Program of the Hebei University (801260201276).

Institutional Review Board Statement: Not applicable.

Informed Consent Statement: Not applicable.

Data Availability Statement: Not applicable.

Acknowledgments: We appreciate the editors and reviewers for their valuable suggestions. The authors would like to thank Nan Wu (Hebei Agricultural University) and Guang Li (Northeast Institute of Geography and Agroecology, Chinese Academy of Sciences) for their guidance on bioinformatics analysis and statistical analysis, and Roger Dean Farley (University of California, Riverside) for his linguistic assistance during the preparation of this manuscript.

Conflicts of Interest: The authors declare no conflict of interest.

References

1. Shi, C.; Huang, Z.; Wang, L.; He, L.; Hua, Y.; Leng, L.; Zhang, D. Geographical distribution of two species of *Mesobuthus* (Scorpiones, Buthidae) in China: Insights from systematic field surveys and predictive models. *J. Arachnol.* **2007**, *35*, 215–226. [CrossRef]
2. Pan, J.H.; Yang, H.; Yu, Q.H.; Wang, F.; Qiu, Z.N.; Zhong, S.Q.; Zeng, Q.E.; Zhong, N.S. Clinical investigation of traditional Chinese medicine in treatment of 71 patients with severe acute respiratory syndrome. *Chin. J. Integr. Tradit. West Med. Intensive Crit. Care* **2003**, *4*, 204–207.
3. Yaozhiwang. Available online: https://db.yaozh.com/ (accessed on 3 June 2022).

4. Bi, M. Forty-six cases of mumps treated with scorpion and dipyridamole. *Chin. J. Intergrative Med.* **1996**, *16*, 384.

5. Zhang, Z. Discussion on heat toxin syndrome of traditional Chinese medicine from influenza A (H1N1). *Hebei J. Tradit. Chin. Med.* **2010**, *32*, 1647–1648.

6. Feng, P.; Shen, Z. Clinical analysis of 50 cases of severe hand foot mouth disease treated with integrated traditional Chinese and Western Medicine. *Lishizhen Med. Mater. Med. Res.* **2014**, *25*, 659–660.

7. Guo, J. *Record of Treating Japanese Encephalitis by Guo Keming, a Famous Doctor of Febrile Disease*; People's Medical Publishing House: Beijing, China, 2017; pp. 1–262.

8. Xu, B.; Jiang, Z.; Jiang, Y. One hundred cases of infantile pneumonia treated with comprehensive therapy of traditional Chinese Medicine. *Chin. J. Inf. Tradit. Chin. Med.* **2013**, *20*, 72–73.

9. Qi, J.; Zhu, M.; Lourenço, W.R. Redescription of *Mesobuthus martensii martensii* (Karsch, 1879) (Scorpiones: Buthidae) of China. *Rev. Ibérica Aracnol.* **2004**, *10*, 137–144.

10. Zhang, L.; Zhu, M. Morphological variation of *Mesobuthus martensii* (Karsch, 1879) (Scorpiones: Buthidae) in Northern China. *Euscorpius* **2009**, *81*, 1–18. [CrossRef]

11. Li, Y.; Fan, Q.S.; Zhang, Z.X.; Guo, P.; Ye, F.P.; Jiang, X.M.; Qiu, W.; Zhang, F.Q.; Zheng, Y.; Ao, L. Toxicity of *Buthus martensii Karsch* venom and preparation of its F(ab′)$_2$ antibody. *Chin. J. Biol.* **2014**, *27*, 647–651.

12. Shi, C.; Ji, Y.; Liu, L.; Wang, L.; Zhang, D. Impact of climate changes from Middle Miocene onwards on evolutionary diversification in Eurasia: Insights from the mesobuthid scorpions. *Mol. Ecol.* **2013**, *22*, 1700–1716. [CrossRef]

13. Wang, W.; Liu, G.-M.; Zhang, D.-X. Intraspecific variation in metabolic rate and its correlation with local environment in the Chinese scorpion *Mesobuthus martensii*. *Biol. Open* **2019**, *8*, bio041533. [CrossRef] [PubMed]

14. Gao, S.; Liang, H.; Shou, Z.; Yao, Y.; Lv, Y.; Shang, J.; Lu, W.; Jia, C.; Liu, Q.; Zhang, H.; et al. *De novo* transcriptomic and proteomic analysis and potential toxin screening of *Mesobuthus martensii* samples from four different provinces. *J. Ethnopharmacol.* **2021**, *265*, 113268. [CrossRef] [PubMed]

15. Gao, S.; Wu, F.; Chen, X.; Yang, Y.; Zhu, Y.; Xiao, L.; Shang, J.; Bao, X.; Luo, Y.; Chen, H.; et al. Sex-Biased Gene Expression of *Mesobuthus martensii* Collected from Gansu Province, China, Reveals Their Different Therapeutic Potentials. *Evid. Based Complement Alternat. Med.* **2021**, *6*, 1–15. [CrossRef] [PubMed]

16. Fet, V.; Lowe, G. Family Buthidae. In *Catalog of the Scorpions of the World (1758–1998)*; Fet, V., Sissom, W.D., Lowe, G., Braunwalder, M.E., Eds.; The New York Entomological Society: New York, NY, USA, 2000; pp. 54–286.

17. Abdel-Rahman, M.A.; Omran, M.A.A.; Abdel-Nabi, I.M.; Ueda, H.; McVean, A. Intraspecific variation in the Egyptian scorpion *Scorpio maurus palmatus* venom collected from different biotopes. *Toxicon* **2009**, *53*, 349–359. [CrossRef]

18. Zhao, R.; Ma, Y.; He, Y.; Di, Z.; Wu, Y.; Cao, Z.; Li, W. Comparative venom gland transcriptome analysis of the scorpion *Lychas mucronatus* reveals intraspecific toxic gene diversity and new venomous components. *BMC Genom.* **2010**, *11*, 452.

19. Carcamo-Noriegaa, E.N.; Olamendi-Portugal, T.; Restano-Cassulini, R.; Rowe, A.; Uribe-Romero, S.J.; Becerril, B.; Possani, L.D. Intraspecific variation of *Centruroides sculpturatus* scorpion venom from two regions of Arizona. *Arch. Biochem. Biophys.* **2018**, *638*, 52–57. [CrossRef]

20. Cao, Z.; Yu, Y.; Wu, Y.; Hao, P.; Di, Z.; He, Y.; Chen, Z.; Yang, W.; Shen, Z.; He, X.; et al. The genome of *Mesobuthus martensii* reveals a unique adaptation model of arthropod. *Nat. Commun.* **2013**, *3602*, 1–10. [CrossRef]

21. Xu, X.; Duan, Z.; Di, Z.; He, Y.; Li, J.; Li, Z.; Xie, C.; Zeng, X.; Cao, Z.; Wu, Y.; et al. Proteomic analysis of the venom from the scorpion *Mesobuthus martensii*. *J. Proteom.* **2014**, *106*, 162–180. [CrossRef]

22. Li, W.; Wu, Y.; Cao, Z.; Di, Z. Composition and classification of scorpion toxins. In *Scorpion Biology and Toxins*; Li, W., Wu, Y., Cao, Z., Di, Z., Eds.; Science Press: Beijing, China, 2016; pp. 267–316.

23. Li, W.; Wu, Y.; Cao, Z.; Di, Z. Structure and function of scorpion toxin. In *Scorpion Biology and Toxins*; Li, W., Wu, Y., Cao, Z., Di, Z., Eds.; Science Press: Beijing, China, 2016; pp. 346–438.

24. Li, W.; Wu, Y.; Cao, Z.; Di, Z. Mesobuthus martensii venom omics. In *Scorpion Biology and Toxins*; Li, W., Wu, Y., Cao, Z., Di, Z., Eds.; Science Press: Beijing, China, 2016; pp. 439–482.

25. Li, W.; Wu, Y.; Cao, Z.; Di, Z. The biological applications of scorpion toxins. In *Scorpion Biology and Toxins*; Li, W., Wu, Y., Cao, Z., Di, Z., Eds.; Science Press: Beijing, China, 2016; pp. 483–552.

26. Sun, D.; Sun, Z. Notes on the genus *Mesobuthus* (Scorpiones: Buthidae) in China, with description of a new species. *J. Arachnol.* **2011**, *39*, 59–75. [CrossRef]

27. Mirshamsi, O.; Sari, A.; Elahi, E. Phylogenetic relationships of *Mesobuthus eupeus* (C.L. Koch, 1839) inferred from *COI* sequences (Scorpiones: Buthidae). *J. Nat. Hist.* **2010**, *44*, 2851–2872. [CrossRef]

28. Li, W.; Wu, Y.; Cao, Z.; Di, Z. Molecular evolution of scorpion toxin. In *Scorpion Biology and Toxins*; Li, W., Wu, Y., Cao, Z., Di, Z., Eds.; Science Press: Beijing, China, 2016; pp. 317–345.

29. Hjelle, J.T. Anatomy and morphology. In *The Biology of Scorpions*; Polis, G.A., Ed.; Stanford University Press: Stanford, CA, USA, 1990; pp. 9–63.

30. Sissom, W.D.; Polis, G.A.; Wait, D.D. Field and laboratory methods. In *The Biology of Scorpions*; Polis, G.A., Ed.; Stanford University Press: Stanford, CA, USA, 1990; pp. 445–461.

31. Zhang, C.; Liao, Y.; Duan, J.; Song, Y.; Huang, D.; Wang, S. The Progresses of Dry-Wet Climate Divisional Research in China. *Clim. Chang. Res.* **2016**, *12*, 261–267.

32. National Meteorological Science Data Center. Available online: http://data.cma.cn/ (accessed on 3 June 2022).

33. NCBI. Available online: https://www.ncbi.nlm.nih.gov/ (accessed on 3 June 2022).
34. Tamura, K.; Stecher, G.; Peterson, D.; Filipski, A.; Kumar, S. MEGA6: Molecular Evolutionary Genetics Analysis version 6.0. *Mol. Biol. Evol.* **2013**, *30*, 2725–2729. [CrossRef]
35. Mesobuthus martensii, Whole Genome Shotgun Sequencing Project. Available online: https://www.ncbi.nlm.nih.gov/nuccore/AYEL00000000.1 (accessed on 3 June 2022).
36. Hierarchical Clustering (Hclust). Available online: https://www.rdocumentation.org/packages/stats/versions/3.6.2/topics/hclust (accessed on 3 June 2022).
37. Livak, K.; Schmittgen, T. Analysis of Relative Gene Expression Data Using Real-time Quantitative PCR and the $2^{-\Delta\Delta CT}$. *Methods* **2001**, *25*, 402–408. [CrossRef]

toxins

Article

Exploring the Pivotal Components Influencing the Side Effects Induced by an Analgesic-Antitumor Peptide from Scorpion Venom on Human Voltage-Gated Sodium Channels 1.4 and 1.5 through Computational Simulation

Fan Zhao [1], Liangyi Fang [1], Qi Wang [1], Qi Ye [1], Yanan He [1], Weizhuo Xu [2,*] and Yongbo Song [1,*]

[1] School of Life Science and Biopharmaceutics, Shenyang Pharmaceutical University, 103 Wenhua Road, Shenyang 110016, China
[2] Faculty of Functional Food and Wine, Shenyang Pharmaceutical University, 103 Wenhua Road, Shenyang 110016, China
* Correspondence: weizhuo.xu@syphu.edu.cn (W.X.); songyongbo@syphu.edu.cn (Y.S.); Tel.: +86-024-43520923 (Y.S.)

Abstract: Voltage-gated sodium channels (VGSCs, or Na_v) are important determinants of action potential generation and propagation. Efforts are underway to develop medicines targeting different channel subtypes for the treatment of related channelopathies. However, a high degree of conservation across its nine subtypes could lead to the off-target adverse effects on skeletal and cardiac muscles due to acting on primary skeletal muscle sodium channel $Na_v1.4$ and cardiac muscle sodium channel $Na_v1.5$, respectively. For a long evolutionary process, some peptide toxins from venoms have been found to be highly potent yet selective on ion channel subtypes and, therefore, hold the promising potential to be developed into therapeutic agents. In this research, all-atom molecular dynamic methods were used to elucidate the selective mechanisms of an analgesic-antitumor β-scorpion toxin (AGAP) with human $Na_v1.4$ and $Na_v1.5$ in order to unravel the primary reason for the production of its adverse reactions on the skeletal and cardiac muscles. Our results suggest that the rational distribution of residues with ring structures near position 38 and positive residues in the C-terminal on AGAP are critical factors to ensure its analgesic efficacy. Moreover, the substitution for residues with benzene is beneficial to reduce its side effects.

Keywords: voltage-gated sodium channel; $Na_v1.4$; $Na_v1.5$; analgesic-antitumor peptide; subtype selectivity; adverse drug reaction; molecular dynamics

Key Contribution: The pivotal components associated with the subtype selectivity of AGAP to human $Na_v1.4$ and $Na_v1.5$ were excavated to provide detailed information on the rational design of high-efficiency and safety peptide medicines targeting voltage-gated sodium channels.

Citation: Zhao, F.; Fang, L.; Wang, Q.; Ye, Q.; He, Y.; Xu, W.; Song, Y. Exploring the Pivotal Components Influencing the Side Effects Induced by an Analgesic-Antitumor Peptide from Scorpion Venom on Human Voltage-Gated Sodium Channels 1.4 and 1.5 through Computational Simulation. *Toxins* **2023**, *15*, 33. https://doi.org/10.3390/toxins15010033

Received: 1 December 2022
Revised: 28 December 2022
Accepted: 29 December 2022
Published: 31 December 2022

1. Introduction

Voltage-gated sodium channels (VGSCs) play important roles in membrane excitability transduction [1]. Mammals express nine subtypes of VGSCs ($Na_v1.1$–1.9) according to their different tissue distributions and functions. Modification of each subtype yields different biological responses [2,3]. Among these VGSCs, $Na_v1.4$ is mainly responsible for generating action potentials in skeletal muscles. When this subtype is activated, the action potential rapidly conveys excitement through the skeletal muscle fibers and regulates the release of Ca^{2+} from myofibrils to drive the contraction and relaxation of skeletal muscle. Clinical and electromyographical features reveal that mutations in the encoding gene *SCN4A* of $Na_v1.4$ can trigger hyperkalemic periodic paralysis and paramyotonia congenital, while sodium channel myotonias may produce gain-of-function changes [4], but loss-of-function mutations may induce hypokalamic periodic paralysis and myasthenic weakness [5]. Encoded by the gene *SCN5A*, Na$_V$1.5 is the primary segment within the intercalated disks in atrial and ventricular myocytes [6]. Gain-of-function mutations in this gene are related to the disruption of fast inactivation, persistent sodium current generation, and ventricular action potential prolongation [7]. However, loss-of-function mutations in SCN5A disrupt the membrane trafficking of the channel protein. In addition, the majority of patients carrying these mutations are diagnosed with cardiac diseases such as LQT3, Brugada syndrome, and sick sinus syndrome [8,9].

In previous comprehensive studies, by inhibiting human Na$_V$1.7 (hNa$_V$1.7) with potential analgesic effects, a kind of β-Scorpion Toxin (β-ScTx) was found and named analgesic-antitumor peptide (AGAP, Figure 1) [10,11]. Based on the genetic evidence that gain-of-function and loss-of-function mutations of this sodium channel coding gene may cause painful syndromes and pain insensitivity, Na$_V$1.7 has emerged as a promising and well-validated pain target [12–15]. Therefore, it is reasonable to believe that this VGSC subtype, which is distributed primarily in the peripheral nervous system, is an ideal target for developing non-addictive therapeutics for pain. However, it is alarming that the off-target effects of the analgesic on Na$_V$1.4 and Na$_V$1.5 could contribute to the possible side effects on the skeletal and cardiac muscles, which may be due to the relatively high conservation of different VGSC subtypes. Unrestricted VGSC blockage could cause heart failure, paralysis, and respiratory failure because it impairs the activity of Na$_V$1.4 and Na$_V$1.5, which are the primary sodium channels in the skeletal and cardiac muscles, respectively. In fact, a part of analgesic targeting hNa$_V$1.7 has to be abandoned in its clinical trials for the above reasons [16–18].

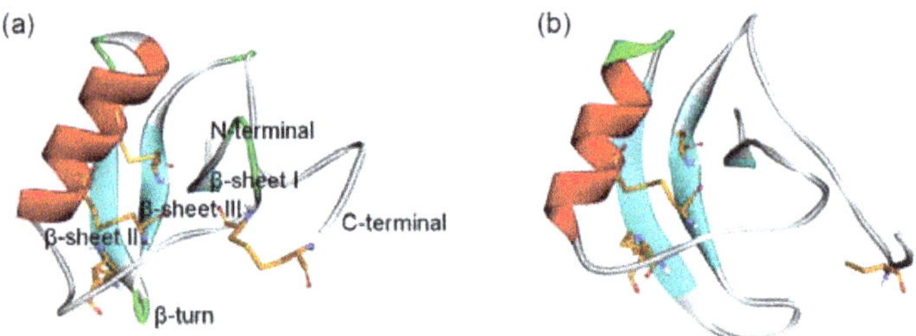

Figure 1. 3D structure of β-ScTxs. (**a**) The crystal structure of a toxin from the scorpion Centruroides noxius Hoffmann (2YC1); (**b**) The structure model of AGAP. Residues forming four disulfide bonds are orange; α-helices and β-sheets are colored red and blue, respectively. Naming notations of secondary structures are labeled in (**a**).

The usage of animal venom to develop novel medicines and serve as pharmacological instruments for monitoring voltage-gated sodium channel function has long been recognized. Among them, the scorpion toxins acting on VGSC can be divided into α- and β-Scorpion Toxins (ScTx) based on their differences in electrophysiological properties and binding sites. The α-ScTx binding on site 3 in VGSC is believed to trap the S4 segment on DIV in an inward position, which prevents it from moving normally in response to depolarization and prolongs action potentials. Meanwhile, β-ScTx can reduce the amplitude of the peak currents and potentiate the activation of the Na$^+$ channels by binding to site 4 in VGSCs (Figure 2) [19,20]. For instance, Huang et al. found that an α-like scorpion toxin, OD-1, had an agonistic effect on VGSC and served as a new excitotoxicity and seizure model to explore the underlying mechanism of a novel third-generation antiepileptic drug [21,22]. Unlike that, AGAP from Buthus martensii Karsch (Bmk) is a 66-amino acid neurotoxin that belongs to β-ScTx [23]. The outcomes of our earlier electrophysiological studies on AGAP were exactly consistent with this characteristic. In the whole-cell patch clamp tests on different VGSC subtypes, compared to the control, the peak current was significantly suppressed. The analysis of the current–voltage relationship displayed a negative shift in the voltage dependence of activation following priming depolarizations after exposure to 100 nM AGAP. In contrast, no significant action on $V_{1/2}$ values of inactivation was observed, and the duration of the recovery process stayed mainly unchanged. As a result, this peptide is defined as a kind of β-ScTx by this current–voltage relationship for AGAP-modified sodium currents [24]. Four homologous domains (DI–DIV) form the α-subunit of VGSC (Figure 2). Site 4 is formed by the extracellular loops connecting transmembrane helices S1–S2 and S3–S4 at DII. Likewise, the extracellular loops S1–S2 and S3–S4 are also the principal components of site 3, but this binding site is located on DIV [25]. After activation by a strong depolarization, i.e., the toxin trapped into the VGSC active conformation, the residues of S4 on the voltage sensing domain (VSD) of DII targeted by β-ScTx will be exposed [26]. Cumulatively, subsequent activation and repetitive action potential firing will be enhanced [27]. Therefore, β-ScTx can alert the action potential of VGSCs in the activated state. Extensive experimental and computational studies have provided further insights into the interaction mechanism between β-ScTx and VGSCs. These results identified that substitution of E779 and P782 at DII/S1-S2 or A841, N842, V843, E844, G845, and L846 at DII/S3-S4 reduced the bioactivity of Css4 (a kind of β-ScTx from Centruroides suffusus) on rNa$_v$1.2. Furthermore, the replacement of DII in this channel with the counterpart DmNa$_v$1 domain from Drosophila melanogaster inverted the insensitivity of AahIT (a kind of β-ScTx from Androctonus australis) to rNa$_v$1.2 [28,29].

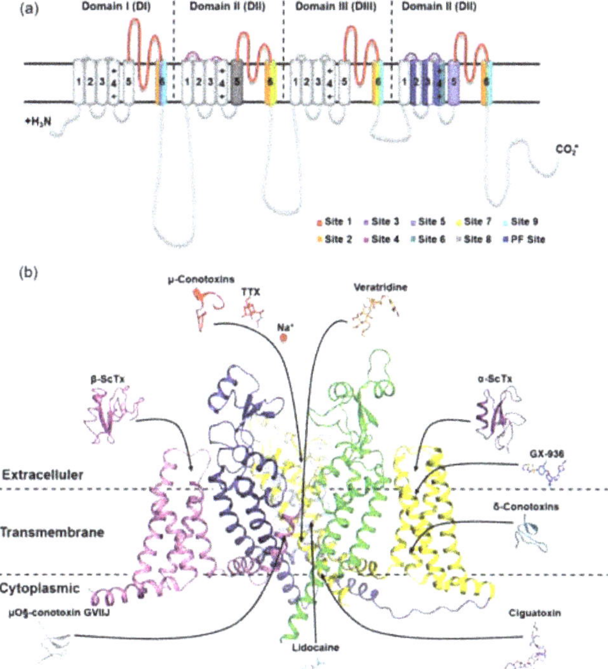

Figure 2. Location of nine receptor sites on mammalian VGSC α subunits. (**a**) Topology of mammalian VGSC α subunits indicating binding sites. Representative inhibitors of sites 1–7 and 9 and the PF site on VGSC are indicated with arrows pointing to the general location of their respective primary receptor sites and colored; (**b**) Side view of the human Na$_v$1.7 subunit complex structure (6J8G) in part with the VSD of DI and PM of DII removed for clarity. The representative inhibitors of site 8 (pyrethroids) are synthetic analogs of pyrethrins and are, therefore, not shown here. The PDB codes for peptide toxins are μ-conotoxin (1TCG), α-scorpion toxin (2ASC), β-scorpion toxins (2YC1), δ-conotoxin (1G1P), and μO§-conotoxin GVIIJ (2N8H).

With IC_{50} values of 3.25×10^{-8} M, 2.19×10^{-8} M, and 1.41×10^{-8} M, respectively, the similar biological activity of AGAP to $hNa_v1.7$, $hNa_v1.4$ and $hNa_v1.5$ is the most likely reason for the occurrence of their adverse effects on the skeletal and cardiac muscles. This possibility was confirmed in subsequent animal experiments. The measurement of heart rate, creatine kinase (CK), and lactic dehydrogenase (LDH) in mice after intravenous injection of AGAP verified the acute toxicity of this peptide to cardiac muscle and could not survive more than six days, even at a low dose of AGAP-treated group. It also led to the absence of motor function tests because no eligible mice survived in this group [30]. To address this problem, we conducted multiple cellular and molecular studies, and the results primarily identified the importance of W38 on AGAP in the specificity of different VGSC subtypes and screened two effective mutants ($AGAP^{W38G}$/$AGAP^{W38F}$). The whole-cell clamp patch test and in vivo experiments indicated that there were no significant effects on skeletal and cardiac muscles after intravenously injecting $AGAP^{W38G}$ compared with the saline-treated group [24,31]. To date, there has not been enough evidence to provide a panoramic mechanistic understanding of how AGAP interacts with different VGSC subtypes. In light of our previous results on the binding modes of AGAP and $hNa_v1.7$ [32], we herein elucidated the detailed mechanism of AGAP mutants with $hNa_v1.4$ and $hNa_v1.5$ through dynamic simulations, revealing the reason why the mutation of one single amino acid can bring about the remarkable alteration of subtype selectivity. We believe these findings are not only beneficial to avoid toxicity to the muscles and myocardium but also helpful to promote progress in developing safer and more effective treatments aimed at VGSC subtypes.

2. Results

For ease of presentation in this paper, the amino acid residues on VGSCs were indicated by three-letter abbreviations and on peptides by single-letter abbreviations. The other abbreviations were listed in the Supplementary Materials as well (Table S3).

2.1. Differences in 3D Structures of $VSD2^{hNav1.4}$ and $VSD2^{hNav1.5}$ in Comparison with $VSD2^{hNav1.7}$

MD simulations were carried out to clarify the effects of receptor structure differences on the selectivity of AGAP. The RMSDs of VSD2s on $hNa_v1.4$, $hNa_v1.5$, and $hNa_v1.7$ during MD simulations were 0.39 nm, 0.39 nm, and 0.37 nm, respectively, indicating the similar stability of these three systems (Figure 3). As the major binding site for β-ScTx, the extracellular loops connecting DII/S1–S2 and DII/S3–S4 displayed significant primary sequence alignment identity differences with the transmembrane helices (S1–S4) (Figure 4). The different residues in S1–S3 may interact with conserved negatively charged residues in S4 to form different salt bridges (Figure 5). Specifically, R114 and R117 on S4 directly engage E40 on S1 and E98 on S3 of $VSD2^{hNav1.7}$ through salt-bridge interactions so that the distances between these three helices were shortened. Similar trends were observed for R114 and E40 of $VSD2^{hNav1.5}$ but not for $VSD2^{hNav1.4}$. As a result, the binding site of β-ScTxs, the gap between the two extracellular loops on S1–S2 and S3–S4 of $VSD2^{hNav1.7}$, is much more compact than the gap of $VSD2^{hNav1.4}$ and $VSD2^{hNav1.5}$. The discrepancy between the spatial structures of the active pockets may further explain the different bonding strengths of the same types of inhibitors. However, when the IC_{50} values were identified by patch clamp to detect the different subtype selectivity of AGAP, the results were similar [24,31]. It is, therefore, still necessary to further explore the interaction mechanism of AGAP with $hNa_v1.4$, $hNa_v1.5$, and $hNa_v1.7$.

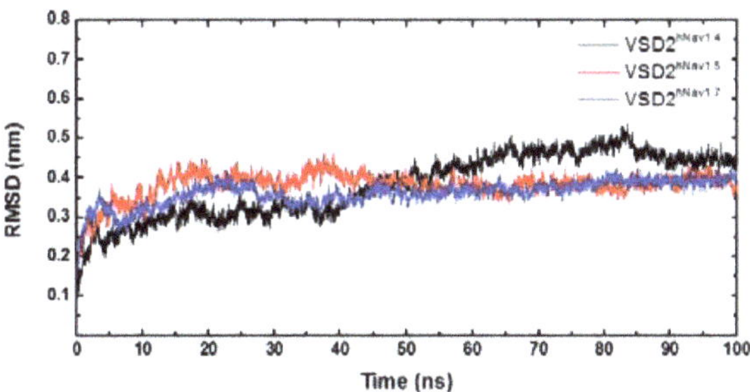

Figure 3. RMSD curves of systems of VSD2s on hNa$_v$1.4, hNa$_v$1.5, and hNa$_v$1.7.

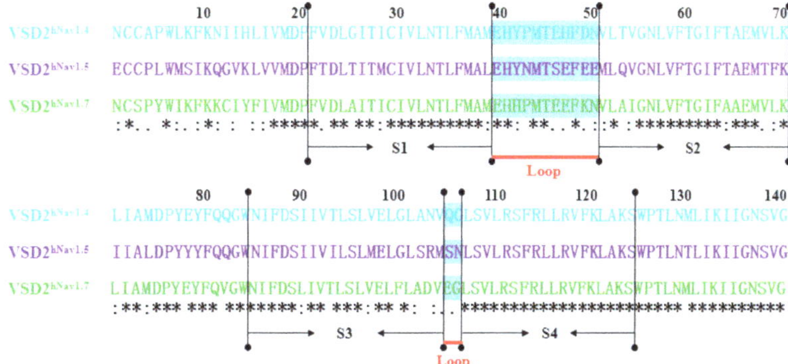

Figure 4. Sequence alignment for VSD2s on hNa$_v$1.4, hNa$_v$1.5, and hNa$_v$1.7. "*" means the residues in this location are identical in three VGSC isoforms; ":" means similar, and "." means a little similar. Dashed lines in green indicate the extracellular loops connecting S1–S2 and S3–S4 on VSD2s.

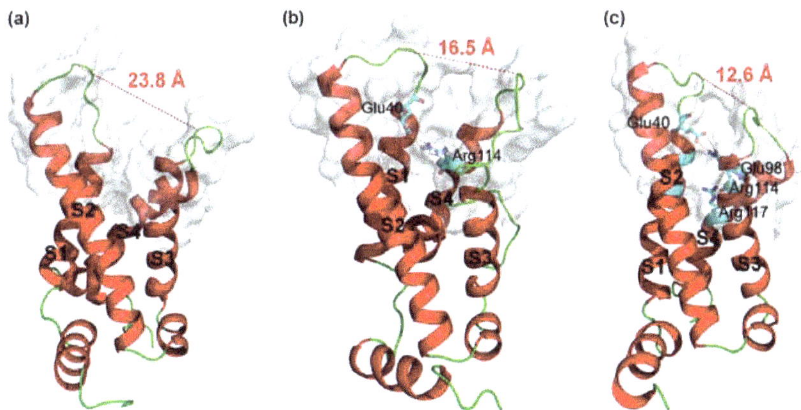

Figure 5. 3D structures of VSD2s on hNa$_v$1.4, hNa$_v$1.5, and hNa$_v$1.7. (**a**) VSD2$^{hNav1.4}$; (**b**) VSD2$^{hNav1.5}$; (**c**) VSD2$^{hNav1.7}$. Dashed lines in orange indicate salt bridges; in red are widths of the active pockets formed by loops between DII S1–2 and DII S3–4; amino acid residues formed by salt bridges are in sticks.

2.2. Analysis of the Binding Modes of AGAP and the W38G/W38F Mutant with VSD2$^{hNav1.4}$ and VSD2$^{hNav1.5}$

From 100 ns MD simulations, the static binding poses of AGAP and its two mutants AGAP$^{W38G/W38F}$ with VSD2$^{hNav1.4}$ and VSD2$^{hNav1.5}$ were obtained. Similar to hNa$_v$1.7, the major interaction regions were located on the β-turn and C-terminal in the AGAP peptides.

2.2.1. A Structural Model for the β-ScTx-hNa$_v$1.4 Complex

Six residues in the β-turn (W38, A39, V41, Y42, G43 and N44) participate in the combination of AGAP with VSD2$^{hNav1.4}$ (Figure 6a). Among these residues, W38, A39, V41, and N44 are well positioned to interact with the bond DII/S1-S2 loop, and V41, Y42, and G43 interact with the DII/S3–S4 loop. Moreover, W38 and V41 have wide ranges of interactions with VSD2$^{hNav1.4}$. Specifically, W38 contacts Met37/His41/Pro43/Leu52 in the DII/S1–S2 loop, while V41 contacts Met39/Thr53 in the DII/S1–S2 loop and Arg111/Arg114 in the DII/S3-S4 loop. Additionally, K62, C63, N64, and G65 in the AGAP C-terminal are also active in binding with VSD2$^{hNav1.4}$. They interact with Leu107/Ser108 in DII/S3–S4 except for K62, which interacts with Glu40 in the DII/S1–S2 loop. Apart from residues in these two regions, Y5, Y14, F15, and Y35 in wild type (WT) contributed to the bindings as well.

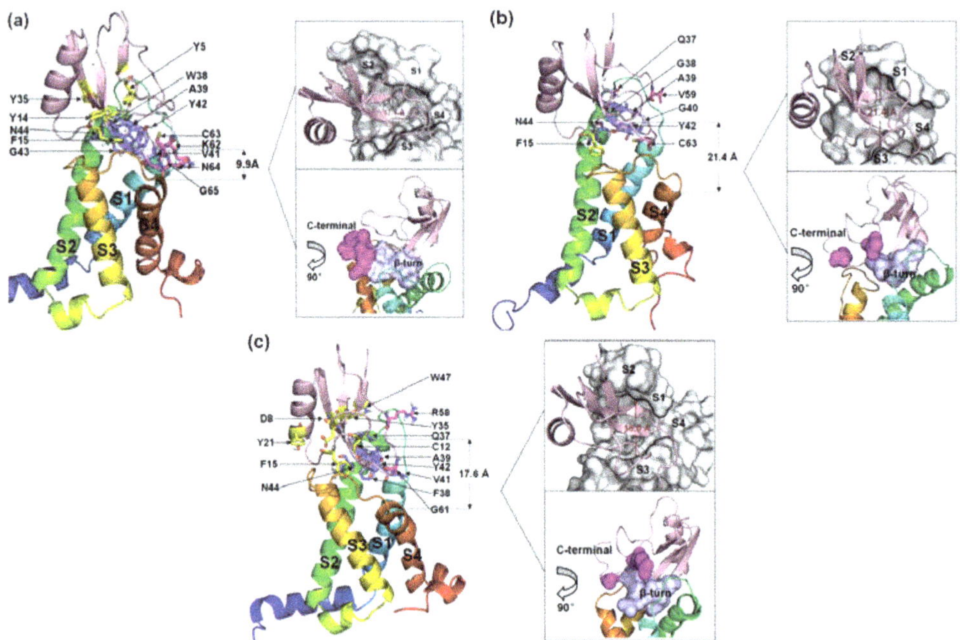

Figure 6. Final binding poses of AGAP (**a**), AGAP^{W38G} (**b**), and AGAP^{W38F} (**c**) with VSD2 on hNa$_v$1.4. The interacting residues on the AGAP peptide are represented by sticks: purple (residues 37–44), magenta (residues 58–66), and yellow (all others). The distances between the C-terminus and DII S4 were measured. The width of the active pocket on VSD2$^{hNav1.4}$ is labeled in deep red in the upper insets, which represent the top views of the complex. The interaction surfaces of the β-turn and C-terminal on AGAP/AGAPW38G/AGAPW38F with VSD2$^{hNav1.4}$ are labeled purple and magenta, respectively, in the bottom insets.

When W38 is substituted in WT by G38, the number of residues with direct contact decreases dramatically (Figure 6b). In the wild-type W38 β-turn, all residues bound to VSD2$^{hNav1.4}$ pointed toward the DII/S1–S2 loop. AGAP^{W38G} failed to continue to form multiple interactions with VSD2$^{hNav1.4}$. Meanwhile, V59 and C62 in the AGAP^{W38G} C-terminus, respectively, bind to Tyr42 and Gln105 in VSD2$^{hNav1.4}$. In addition, F15 also contributes significantly to binding by interacting with Arg114.

When W38 is substituted in WT by F38, six residues in the β-turn (Q37, F38, A39, V41, Y42, and N44) participate in the combination of AGAP with VSD2$^{hNav1.4}$ (Figure 6c). Among these residues, Q37, F38, and A39 were positioned to interact with the DII/S1–S2 loop, while V41, Y42, and N44 interacted with the DII/S3–S4 loop. Similar to the WT, V41 in AGAP can directly contact these two extracellular loops. R58 and G61 in the AGAP^{W38F} C-terminus combined with Tyr42 and Arg111, respectively, in VSD2$^{hNav1.4}$. The interactions between D8/C12/F15 with Gln105, Y21 with Pro102, Y35/W47 with Asp49 and Met44 were also important in AGAP binding on VSD2$^{hNav1.4}$.

Evidently, the binding modes of WT and AGAP^{W38F} to hNa$_v$1.4 bear more striking resemblances compared to AGAP^{W38G}, suggesting that the ring structure at position 38 has a critical effect on its combination with the DII/S3–S4 loop. In contrast, the interactions of the peptides with the DII/S1–S2 loop were more stable. Although the active pocket will naturally converge if the peptide contacts both extracellular loops, the inherent spacious active pocket characteristic of VSD2$^{hNav1.4}$ is destined to affect the affinity of AGAP.

2.2.2. A Structural Model for the β-ScTx-hNa$_v$1.5 Complex

Seven residues in the β-turn of WT (Q37, W38, A39, G40, V41, Y42, and N44) interact directly with VSD2$^{hNav1.5}$ (Figure 7a). Most of the residues were positioned to combine with the bound DII/S3–S4 loop except for Q37 and G40. Multiple interactions were observed between W38 and Glu98/Arg111/Arg114 and between N44 and Glu98/Ser102/Met104. Multiple residues in the C-terminal segments participated in the combination of the WT to DII/S3–S4 loop in VSD2$^{hNav1.5}$. Moreover, Y5 and W47 in WT also contributed to the interactions with Asn43 in the DII/S1–S2 loop. On the whole, the ligand was biased to the side of the DII/S3–S4 loop.

When W38 was substituted in WT by G38, the binding of this residue to VSD2$^{hNav1.5}$ was abolished. Unlike VSD2$^{hNav1.4}$, the number of residues in the β-turn that interacted with the receptor was equal to the number of residues in the β-turn that interacted with the receptor in the WT (Figure 7b). In contrast, almost all interactions between the AGAP C-terminus and VSD2$^{hNav1.5}$ disappeared except for the contact between C63 and Asn106. Overall, AGAP^{W38G} binds to the middle of the active pocket, with approximately equal distance to both of the extracellular loops.

When W38 was substituted in WT by F38, the role of this residue in combination with VSD2$^{hNav1.5}$ was retained (Figure 7c). The ligand bonded with the receptor was biased to the side of the DII/S1–S2 loop due to the weakening interaction between AGAP^{W38F} and the DII/S3-S4 loop. The function of the C-terminus is similar to the function of the WT. However, Y5, N19, Y35, and W47 are also important in peptide binding to hNa$_v$1.5. Among these residues, Y5 on β-sheet I and W47 on β-sheet II in the VSD are adjacent in space and close to Q37. These three residues may form a signature binding region to hNa$_v$1.5 compared to the other two VGSC subtypes.

The comparison of the binding modes of AGAP and the W38G/W38F mutant with hNa$_v$1.4, hNa$_v$1.5, and hNa$_v$1.7 shows that Q37, 38th, G40, V41, Y42, and N44 in the β-turn comprise a crucial binding region of the peptide when it interacts with VGSCs. Among these binding modes, Q37 is preferentially bound by DII/S1-S2 in VSD2$^{hNav1.5}$, while N44 always contributes significantly to the binding of the three VGSC isoforms. The ring structure at position 38 critically affects its combination with the VGSC. Alternatively, the functions of the C-terminus in combinations with hNa$_v$1.4 and hNa$_v$1.5 are roughly identical and far less powerful than those of the C-terminus in combinations with hNa$_v$1.7. Another interesting difference is that C63 in this segment is always bound with Asn106 in VSD2$^{hNav1.5}$ or the residue at position 105 in VSD2$^{hNav1.4}$ and VSD2$^{hNav1.7}$, which seems to indicate that the residues at this position in the different receptors have important implications for the subtype selectivity of the toxins. In addition, F15 and Y5/W47 are also important for the binding of VSD2$^{hNav1.4}$ and VSD2$^{hNav1.5}$, respectively.

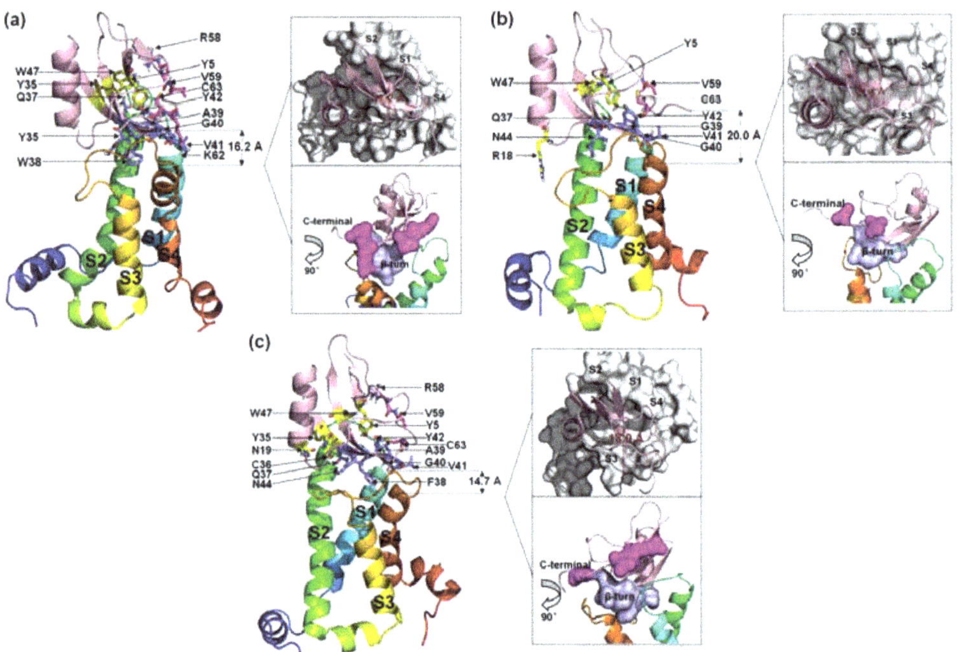

Figure 7. Final binding poses of AGAP (**a**), AGAP^{W38G} (**b**), and AGAP^{W38F} (**c**) with VSD2 on hNa$_v$1.5. The interacting residues on the AGAP peptide are represented by sticks: purple (residues 37–44), magenta (residues 58–66), and yellow (all others). The distances between the C-terminus and DII S4 were measured. The width of the active pocket on VSD2hNa$_v$1.4 is labeled deep red in the upper insets, which represent the top views of the complex. The interaction surfaces of the β-turn and C-terminal on AGAP/AGAPW38G/AGAPW38F with VSD2$^{hNav1.4}$ are labeled purple and magenta, respectively, in the bottom insets.

2.3. Analysis of Dissociation Pathways of the AGAP/AGAP^{W38G}/^{W38F} Mutant with VSD2$^{hNav1.4}$ and VSD2$^{hNav1.5}$ by SMD Simulations and PMF Calculations

2.3.1. Differences in Conformations of AGAP and the W38G/W38F Mutant with VSD2$^{hNav1.4}$ and VSD2$^{hNav1.5}$

Based on the SMD method, the specific modes and interactions of critical residues are depicted precisely by analyzing the dissociation of the peptides from the three VGSC isoforms. The results show that AGAP, AGAP^{W38G}, and AGAP^{W38F} are separated completely from VSD2$^{hNav1.4}$ after 3680 ps, 2950 ps, and 3360 ps (Figure 8a), as is VSD2$^{hNav1.5}$ after 2880 ps, 2920 ps, and 3070 ps (Figure 8b). PMF indicates that the binding free energy of AGAP, AGAP^{W38G}, and AGAP^{W38F} are 190.64 kJ·mol^{-1}, 175.68 kJ·mol^{-1}, and 150.53 kJ·mol^{-1} to hNa$_v$1.4 (Figure 9a), as well as 164.81 kJ·mol^{-1}, 146.19 kJ·mol^{-1}, and 149.48 kJ·mol^{-1} to hNa$_v$1.5 (Figure 9b). Apparently, WT has a higher affinity for hNa$_v$1.4 and hNa$_v$1.5 than the two mutants. Moreover, the same change trend is expressed between dissociation time and binding affinity. Overall, the simulation and previous patch clamp experimental results are consistent with each other. Furthermore, the proportion of electrostatic and VDW interactions in AGAP binding with hNa$_v$1.5 were found not as regular as it was with hNa$_v$1.4 and hNa$_v$1.7, which were dominated by only one single type of interaction (see Sections 2.3.2 and 2.3.3 for details).

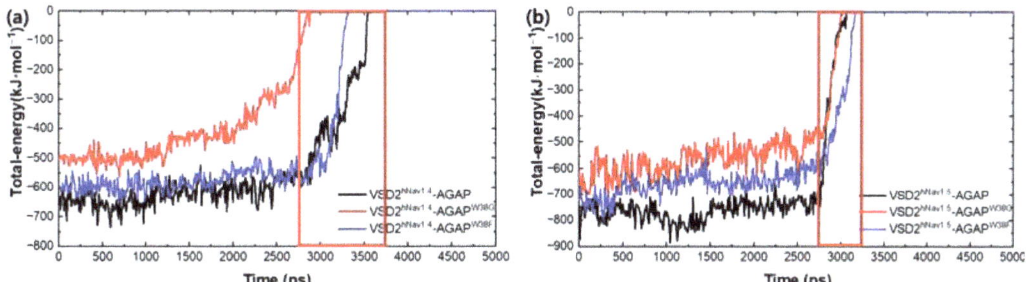

Figure 8. Total energy of AGAP and its mutants with VSD2s on hNa$_v$1.4 (**a**) and hNa$_v$1.5 (**b**).

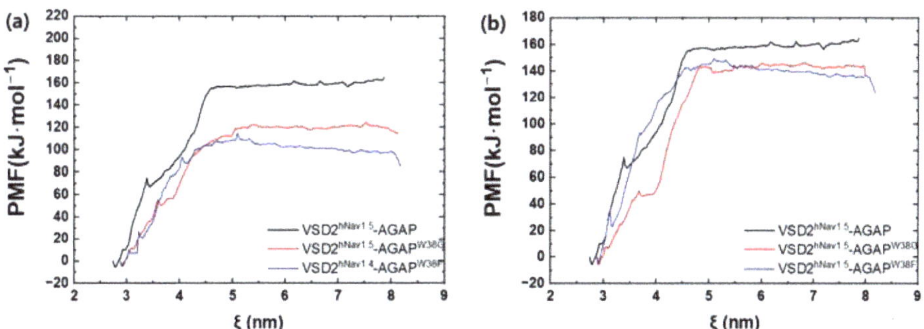

Figure 9. PMF curves of AGAP and its mutants with VSD2s on hNa$_v$1.4 (**a**) and hNa$_v$1.5 (**b**). ζ represents the reaction coordinate generated by the configurations.

According to the comparison between the conformations of the peptides about dissociation from the receptors (Figure 10) with binding modes (Figures 6 and 7), the interactions of W38 in the β-turn with DII/S1-S2 and K62 in the C-terminal with negatively charged residues in DII/S4 contribute significantly to the combination of WT to hNa$_v$1.4 and hNa$_v$1.7, but not to hNa$_v$1.5. However, mutations of position 38 can partially decrease the direct connection between G38/F38 and DII/S1–S2 but completely destroy it between the C-terminal and II/S4. Moreover, a broad binding region of β-turns to the three VGSC subtypes ensures stable and strong contact between the toxins and receptors. The interactions of the more flexible C-terminal to the VGSCs are easily reformed.

2.3.2. Specific Types of Interactions of Important Residues in the β-ScTx-hNa$_v$1.4 Complex

The decompositions of the binding free energy of ligand–residue pair interactions are employed to investigate how critical components influence the affinity and selectivity of the peptides of different VGSC subtypes. The calculation results show that the average energy contributions in van der Waals (VDW) during the dissociations of AGAP, AGAP^{W38G}, and AGAP^{W38F} from VSD2^{h}Na$_v$$^{1.4}$ are −215.77 kJ·mol^{-1}, −135.54 kJ·mol^{-1}, and −187.15 kJ·mol^{-1}, respectively, whereas in electrostatic interactions, they are −185.63 kJ·mol^{-1}, −100.74 kJ·mol^{-1}, and −178.81 kJ·mol^{-1}, respectively. It follows that VDWs bear greater responsibility than electrostatic interactions to the combinations. In contrast, our previous study indicated that the latter interaction type is the dominant factor leading to the differences in the binding free energy of the three peptides to VSD2$^{hNav1.7}$.

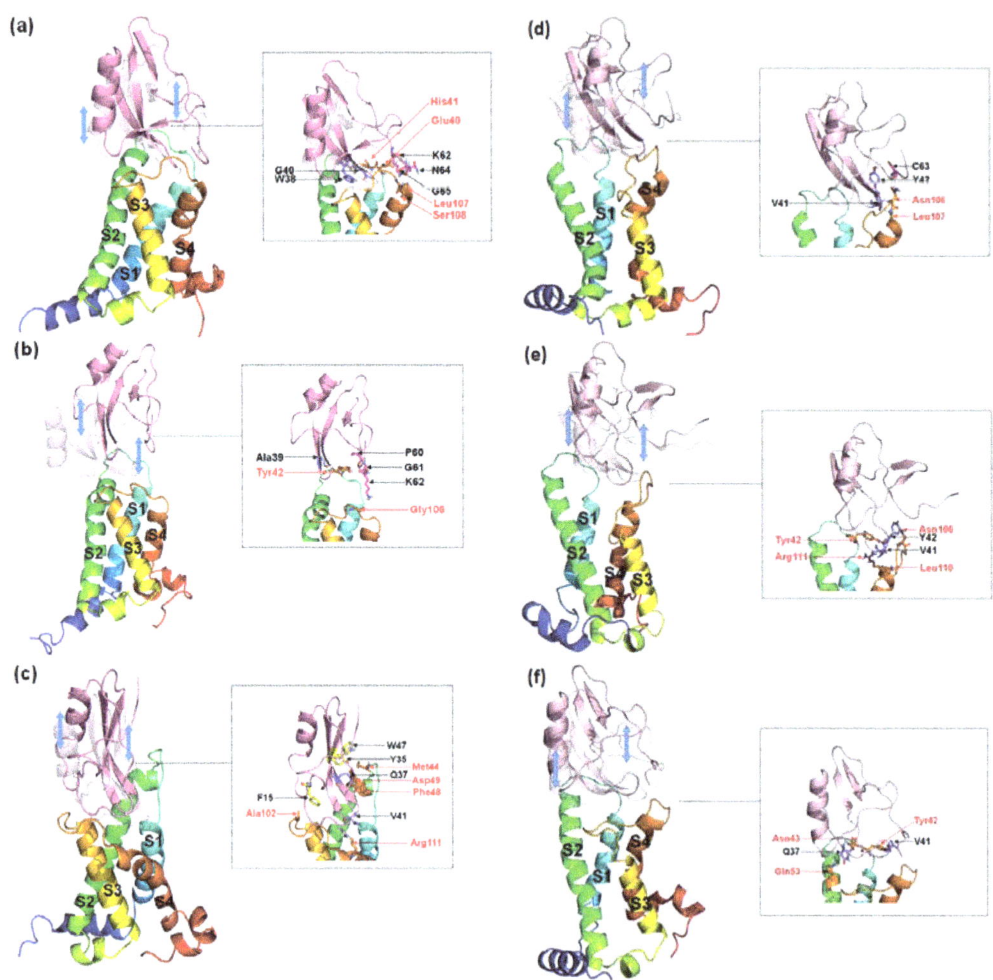

Figure 10. Representative conformations of AGAP (**a,d**), AGAP^{W38G} (**b,e**), and AGAP^{W38F} (**c,f**) with VSD2s on hNa$_v$1.4 (**a–c**) and hNa$_v$1.5 (**d–f**) during the dissociation process. These conformations reflected the state of the receptor and the ligand just prior to complete separation. The ligands in translucency are conformations before pulling. The insets depict the interaction between the receptor and the ligand when they were to be separated. Ligand residues on the β-turn are purple, and those at the C-terminal are magenta; key residues at the interface are marked black. VSD2 hNa$_v$1.5 residues are orange and are highlighted in red characters.

In particular, four residues (W/G/F38, A39, N42, and N44) in the β-turn play vital roles in VSD2$^{hNav1.4}$ trapping (Table 1, Figure S1 and Table S1). In accordance with hNa$_v$1.7, substitution W38G strongly diminished the toxin binding affinity due to steric hindrance and H-bond repulsion between this residue and DII/S1–S2. Y42 in WT accepts a π-cation contact from Arg111 in DII/S4, which in mutants is H-bonded with DII/S3–S4 for identical contribution. N44 in WT forms weak H-bonds with the two extracellular loops, which are strong with one loop in mutants. A residue located on the loop between α-helix and β-sheet I, F15, also significantly contributes by forming a π-cation contact with the highly conserved Arg114 in DII/S4. Interestingly, the important function of F15 is to work only when the toxins are bound with hNa$_V$1.4. Therefore, we deduced that this particularity is attributed to the discrepancy in the 3D structures of the three VGSC subtypes. Additionally, unlike hNa$_V$1.7, residues in the C-terminus lack powerful interactions with hNa$_V$1.4, although they are involved in the combination of the receptor and toxins.

Table 1. Average total-residue interaction of AGAP and its mutants with VSD2s on hNa$_V$1.4 during the dissociation process.

Region	Residue	Total Interaction (kJ·mol^{-1})		
		AGAP	AGAP^{W38G}	AGAP^{W38F}
β-sheet I	Y5	−0.99	-	-
	D8	-	-	−22.96
loop between	C12	-	-	−2.87
α-helix and	Y14	−2.55	-	-
β-sheet I	F15 * [1]	−23.10	−20.44	−40.96
α-helix	Y21	-	-	−2.60
β-sheet III	Y35	−3.16	-	−40.40
	Q37	-	−19.97	−48.43
	W38	−38.99	-	-
	G38	-	−16.60	-
	F38	-	-	−38.59
β-turn	A39	−10.16	−12.59	−7.65
	G40	-	−27.39	-
	V41	−37.54	-	−38.77
	Y42 *	−25.09	−18.10	−23.06
	G43	−24.60	-	-
	N44 *	−15.78	−28.25	−24.74
β-sheet II	W47	-	-	−7.25
	R58	-	-	−2.75
	V59	-	−2.47	-
	G61	-	-	−14.15
C-terminal	K62	−68.51	-	-
	C63	−17.33	−10.14	-
	N64	−19.85	-	-
	G65	−25.08	-	-

[1] The characters in "*" played important roles in all three systems.

2.3.3. Specific Types of Interactions of Important Residues in the β-ScTx-hNa$_V$1.5 Complex

The calculation results show that the average energy contributions in van der Waals (VDW) during the dissociations of AGAP, AGAP^{W38G}, and AGAP^{W38F} from VSD2$^{hNav1.5}$ are −169.03 kJ·mol^{-1}, −167.40 kJ·mol^{-1} and −218.49 kJ·mol^{-1}, respectively, whereas in electrostatic interactions, they are −267.36 kJ·mol^{-1}, −158.96 kJ·mol^{-1} and −174.17 kJ·mol^{-1}, respectively.

Specifically, seven residues (Q37, A39, G40, V41, N42, G43, and N44) in the β-turn have significant contributions to electrostatic interactions in all complexes (Table 2, Figure S2 and Table S2). Of these, N44 can always accept connection with DII/S3-S4. Similar to hNa$_V$1.4 and hNa$_V$1.7, the ring structure at position 38 still has a critical effect on the combination with hNa$_V$1.5. Notably, Y5 and W47, which are adjacent to each other and Q37 in space, always keep in direct contact with Asn43 in DII/S1–S2. Moreover, the residues in the DII/S1-S2 contact with Q37 in the peptide are also close to Asn43. Therefore, we inferred that the interaction surface formed by Y5, Q37, and W47 reacts with the new characteristics of the binding pose of the toxins with hNa$_V$1.5. Further analysis reveals that the production of the feature residues is derived from the opposite distribution of the hydrophobicity of the extracellular loop in DII/S1-S2 on hNa$_V$1.5 and hNa$_V$1.4/hNa$_V$1.7 (Figure 11). Because it is located in the middle of this loop on VSD2$^{hNav1.5}$, hydrophilic Asn43 is able to directly contact Y5 and W47 on AGAP through electrostatic interactions. Likewise, the roles of residues in the C-terminus are limited in the combination of VSD2$^{hNav1.5}$ and toxins.

Although the negatively charged R58 in this region can contact the positively charged Asn43 in DII/S1–S2, the contribution is small due to the remote distance between them.

Table 2. Average total-residue interaction of AGAP and its mutants with VSD2s on hNa$_v$1.5 during the dissociation process.

Region	Residue	Total Interaction (kJ·mol^{-1})		
		AGAP	**AGAP^{W38G}**	**AGAP^{W38F}**
β-sheet I	Y5 * [1]	−17.07	−15.63	−28.23
loop between	R18	-	−2.83	-
α-helix and β-sheet I	N19	-	-	−35.77
β-sheet III	Y35	−11.30	-	−48.22
	C36	-	-	−8.85
	Q37 *	−37.77	−44.46	−38.32
	W38	−83.59	-	-
	G38	-	-	-
	F38	-	-	−29.71
β-turn	A39 *	−10.93	−10.09	−11.77
	G40 *	−25.21	−31.28	−29.44
	V41 *	−16.80	−36.36	−25.75
	Y42 *	−23.21	−38.87	−16.02
	N44 *	−64.25	−24.25	−45.47
β-sheet II	W47 *	−31.40	−23.19	−19.55
	R58	−19.23	-	−8.82
C-terminal	V59	−5.30	−2.39	−2.30
	K62	−33.93	-	-
	C63	−21.27	−14.85	−1.11

[1] The characters in "*" played important roles in all three systems.

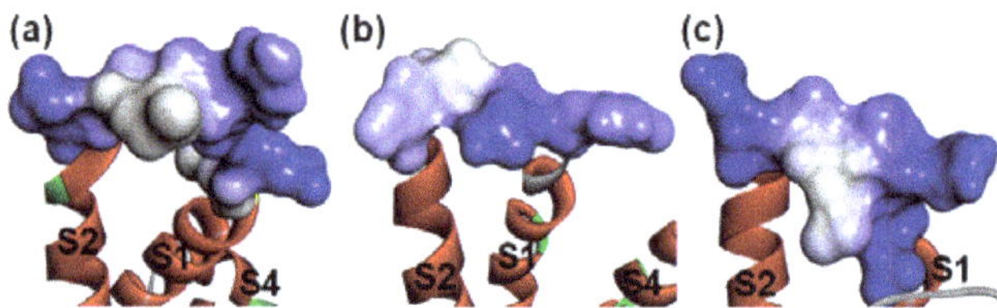

Figure 11. Hydrophobicity of the loop between S1-S2 on VSD2s on VGSCs. (**a**) hNa$_v$1.4; (**b**) hNa$_v$1.5; (**c**) hNa$_v$1.7. The default color spectrum used is blue–white–brown. Surfaces in blue correspond to hydrophilic residues, whereas surfaces in brown correspond to hydrophobic residues. Residues at positions 40–46 on VSD2s are shown on the surface.

3. Discussion

Combined with our previous study [32] and this research, two major factors are highly related to the selectivity of AGAP and its mutants to hNa$_v$1.4, hNa$_v$1.5, and hNa$_v$1.7. First, there are progressive dissimilarities in the 3D structures of AGAP bound, in which VSD2$^{hNav1.7}$ is the narrowest, VSD2$^{hNav1.5}$ is wider, and VSD2$^{hNav1.4}$ is the widest. These differences in the binding poses established the feature residues of AGAP binding. Second, the affinity of WT AGAP to VGSCs is always higher than the affinity of the two mutants, according to the binding energy calculated by simulations and the IC$_{50}$ values detected by experiments. This evidence fully demonstrates that the residue at position 38 on AGAP is one of the dominant factors affecting the selectivity of AGAP to the three VGSC subtypes. Moreover, the results indicate that the ring structure of this position is presented as the hinge structure, which provides a significant contribution when it contacts the channels through VDW interactions. Similarly, K62 is also the pivotal residue in the AGAP C-terminal, which may offer an impressive energy contribution to the negatively charged residue on VGSCs when salt bridges form. Benefiting from the narrow active pocket of

AGAP binding to VSD2$^{hNav1.7}$, K62 can contact negatively charged Asp103 and Glu105 in the DII/S3-S4 loop through powerful electrostatic interactions.

Additionally, a significant correlation was found between W38 and K62. For intermolecular interactions, these two residues displayed a synergistic effect to determine the selectivity of AGAP to different VGSC subtypes. For instance, although the affinity of W38 and F38 is similar to the affinity of hNa$_V$1.4, the binding free energy of WT AGAP is much higher than that of AGAP^{W38F} because of the formation of a salt bridge between K62 on WT and Glu40 on VSD2$^{hNav1.4}$. In contrast to AGAP^{W38F}, the contribution of position 38 on AGAP^{W38G} is minimal, while the contribution of K62 is still considerable when the toxin contacts VSD2$^{hNav1.7}$. Finally, the affinities of the two mutants to hNa$_V$1.7 are equivalent. For intramolecular interactions, the function of K62 is affected by the substitution of residue at position 38 as well. The electrostatic interactions between K62/C63 and the negatively charged residues in the DII/S3–S4 loop on VSD2$^{hNav1.7}$ were discovered to arise from the existence of the internal reaction chain on WT (C63-C12-D8-N11-R58-Y42), which limits the swing of the C-terminus. On AGAP^{W38F}, this chain is broken because of the abolition of the interaction between Y42 and R58 by A39. In WT and AGAP^{W38G}, the group of A39 interacting with Y42 is occupied by the combination with the receptor. Likewise, the formation of the salt bridge between K62 and the negatively charged residue in the DII/S1-S2 loop on VSD2$^{hNav1.4}$ comes from a similar internal chain on WT (C63-C12-D8-N11-G61-Y42). After the mutation, the electrostatic interaction between Y42 and G61 is distributed because the reaction group in the former is occupied by A13 and Y14. The reason is the swerve of the benzene on Y42 because of the interactions between it and Gln105/Arg111 on VSD2$^{hNav1.4}$. In the meantime, the shortage of the internal reaction chain to restrict the direction of the C-terminal on AGAP results in the modest energy contribution of this region when it is bound with hNa$_V$1.5.

However, there is still a sensitive difference in the affinity of the residue at position 38 to hNa$_V$1.5 on WT, AGAP^{W38G,} and AGAP^{W38F}. This result highlights that a single factor is not enough to determine the selectivity of AGAP to the three VGSC subtypes.

Additionally, some limitations of this study should not be omitted. Firstly, following the change of membrane potential, there are three different statuses: open, inactive, and closed state. Herein, the structures in an open state were selected to explore the interactions between AGAP and different VGSC subtypes. However, the combination of toxins and sodium channels should be dynamically regulated. For β-ScTx, a voltage sensor trapping model is universally accepted. In this model, the toxin first attaches to its receptor site on VGSC in its inactive state, leading to a concentration-dependent decrease of the peak current. As the channel is activated by a strong depolarization, a new binding site on the channel to β-ScTx is exposed due to the change of conformation of VGSC. Finally, the tightly bound toxin traps the activated conformation of the sodium channel in a process independent of unimolecular concentration [27,33]. This could possibly be the reason for parts of the studies exploring β-ScTx with activated state VGSC through computational and/or experimental approaches [34,35]. For this work, as mentioned above, the outcoming of the patch clamp test manifested that there was little effect of AGAP on the inactivation process [24]. In this way, AGAP appears to suppress the sodium channels, mainly in the open state. Therefore, the interactions between this peptide and activated VGSC were explored in our study. Nevertheless, it does not mean that the complete dynamic process of the toxin binding to the channels is fully considered. These variables should be taken into account in our further studies. Secondly, AGAP draws our attention because of its analgesic effect targeting hNa$_V$1.7. Furthermore, its potential skeletal and cardiac muscle toxicity, as well as the similar biological activity to hNa$_V$1.7, hNa$_V$1.4, and hNa$_V$1.5, led us to pay more attention to the interactions between AGAP and these three VGSC subtypes. After intravenous injection of AGAP, central nervous system diseases such as epilepsy and migraine were not observed in mice, so the combinations of AGAP with the VGSC subtypes mainly distributed in this region (Na$_V$1.1, Na$_V$1.2, Na$_V$1.3, and Na$_V$1.6) were not examined in this study. On the other hand, Na$_V$1.7, Na$_V$1.8, and Na$_V$1.9 are primarily expressed in the peripheral nervous system, and all have the potential to be a non-addictive analgesic target. Our previous studies by whole-cell clamp patch indicated that AGAP could only suppress the activity of $hNa_v1.8$ with a 25% reduction in peak current while $hNa_v1.7$ with a reduction in 68% [24]. Limited by the experimental materials, the inhibiting effect of AGAP on $hNa_v1.9$ were not discussed in our research. In future investigations, the mechanism of this toxin and other VGSC subtypes should be complemented through both computational simulation and experimental methods. Thirdly, the rational design scheme of AGAP proposed in this paper still needs to be tested with additional experiments. In the sequel of our study, the new mutants will be obtained by genetic engineering. Their actual effects on different VGSC subtypes should be illustrated through whole-cell clamp patch or animal experiments, such as forced swimming test, rota-rod test, and mouse-twisting model.

4. Conclusions

Overall, the in-depth simulation analysis revealed several significant commonalities and differences in AGAP binding on the three VGSC subtypes. In general, retaining the ring structure of the amino acid residue at or near position 38, as well as increasing the rational distribution of basic residues in the C-terminal on AGAP, are indicated to be advantageous for improving the affinity of $hNa_v1.7$, which was disclosed in our previous studies [32]. In contrast, disruption of the ring structure of F15 and Y5/W47 on AGAP effectively reduced its binding activity for $hNa_v1.4$ and $hNa_v1.5$.

To elucidate these findings more specifically, from the perspective of AGAP: (i) β-turn is the essential region of AGAP to combine with the VGSC because the majority of the binding residues (at position 37–44) are located within it; (ii) the conserved N44 in the β-turn is always H-bonded to the DII/S3-S4 on the three VGSCs with a prominent energy contribution; (iii) the deficiency of the carbonyl in the R group at position 105 on the DII/S3-S4 loop of $hNa_v1.4$ and $hNa_v1.5$ may dramatically decrease its electrostatic contacts with Y42 in the β-turn or K62 in the C-terminal on the AGAP; (iv) the unique residues on AGAP may function vitally when combined with different VGSC subtypes, for example, F15 for $hNa_v1.4$ and Y5, Q37 and W47 for $hNa_v1.5$. From the perspective of three VGSC subtypes: (i) The charged residue at position 49 on VSD2 always accepts a π-cation contact with the residues bearing ring structure (Y35, W38, and Y42) in the β-turn of AGAP; (ii) The highly conserved negatively charged residues in DII/S4 (Arg111 and/or Arg114) participate in the combination with the peptides by forming H-bond or π-cation interactions with G40 and/or V41 in the AGAP β-turns. This response is in accordance with the voltage sensor trapping model, whereby the activated conformation of VSD2, in which DII/S4 moves outward, is trapped by β-ScTxs through strong binding with it. Further research should be undertaken to verify the above results based on experimental methods such as animal toxicity test, western blotting, blood assays, and clamp patch.

We believe that the enlargement of the study of the interaction mechanism of AGAP with $hNa_v1.4$, $hNa_v1.5$, and $hNa_v1.7$ will shed more light on elaborating the selectivity of scorpion toxins to different VGSC subtypes. Furthermore, this research provides more abundant theoretical knowledge and references for developing selective medicines targeting VGSCs.

5. Materials and Methods

5.1. Homology Modeling and Molecular Docking

The theoretical model of $hNa_v1.4$ was obtained from the Protein Data Bank (PDB ID: 6AGF). Five hundred conformations of $hNa_v1.5$ and AGAP were built through Modeler 9.9 [36] due to the protein sequences obtained from UniProt (accession numbers were Q15858 and Q95P69, respectively) [37]. Crystal structures of cardiac sodium channel from *Rattus norvegicus* (PDD ID: 6UZ0) and α-Tx11 from *Buthus martensii* (PDB: 2KBH) were selected as the templates for homology modeling because of their high sequence identity with $hNa_v1.5$ and AGAP, respectively. Models with the least discrete optimized protein energy (DOPE) score were validated by Ramachandran plots and profile-3D and chosen as the best ones. Two mutants of AGAP (AGAP^{W38G}/AGAP^{W38F}) were obtained by modeler site-directed mutagenesis. To accurately predict the three-dimensional structures of the target sequences in a real physiological environment, molecular dynamics (MD) simulations were applied. The parameters used here are described in Section 5.2.

The optimized structures of AGAP and its mutants were docked into the binding sites on VSD2$^{hNav1.4}$ and VSD2$^{hNav1.5}$ by ZDOCK, which is suitable for studying the interactions between biomacromolecules [38]. The active pockets were restricted in the extracellular region of VSD2. The matching algorithms are used to carry out the process of docking. In accordance with the RMSD cutoff, 2000 poses were divided into 60 clusters with an angular step size of $6°$. The small angular step size ensures that the most possible binding modes can be considered in the docking. The cluster containing the largest number of docking poses usually figures out the binding site that the ligand is most likely to combine. Furthermore, the configurations with the lowest binding energy in the largest cluster were screened out for further simulations after excluding those in direct conflict with the position of the membrane and published research results about the binding site of β-ScTx with VGSC [20,28,29]. However, the molecular docking was carried out in a vacuum, which is insufficient to reflect the realistic conformations of AGAP with VGSCs. Therefore, molecular dynamic simulations were needed to optimize these conformations.

5.2. Molecular Dynamics

The prediction structures and docking complexes were carried out on 100 ns time scale molecular dynamics simulations utilizing the GROMACS 2018 package [39]. A 1-palmitoyl-2-oleoyl-sn-glycero-3-phosphocholine (POPC) bilayer model was used with a united-atom force field [40] to describe the phospholipid bilayer of hNa$_v$1.4 and hNa$_v$1.5, and GROMOS-53a6 force-field parameters were assigned to the other parts in the systems. InflateGro methodology [41] was performed to accurately embed these two channels into the lipids. The SPC water model [42] was introduced as the solvation, and counterions were added to neutralize the systems. The steepest descent algorithm and conjugate gradient algorithm were used to minimize the energy and remove the bad contacts first. Subsequently, the simulation conditions were heated to 310 K using a modified Berendsen thermostat [43] with non-hydrogen solute atoms restrained. Simulation in the NPT ensemble follows this desired temperature and 1 atm constant pressure for 1 ns. The equilibration systems were subjected to a 100 ns MD simulation with no constraints applied. Moreover, the particle mesh Ewald (PME) method [44] and LINear Constraint Solver (LINCS) [45] were performed to assess the long-range electrostatic interactions and redress the lengths of all the bonds. The MD trajectory and snapshots were saved every 10 ps and analyzed by the tools of the Gromacs package, PyMol [46], and VMD [47]. The equilibrated trajectory was extracted to perform cluster analysis using the Gromos clustering algorithm with a tolerance of 0.15 nm for root–mean–square deviation (RMSD) [48].

5.3. Steered Molecular Dynamics and PMF Calculations

Steered molecular dynamics (SMD) simulations were carried out on the equilibrated model after MD [49]. With a biasing force constant of 500 kJ·mol^{-1}·nm^{-2} and a pulling velocity of 0.001 nm·ps^{-1}, a force pulled AGAP/AGAP^{W38G}/AGAP^{W38F} away from the binding surface of VSD2$^{hNav1.4}$ and VSD2$^{hNav1.5}$ along the z-dimension in the 5 ns simulation runs at 310 K and 1 atm. Herein, with respect to the atoms of VSD2s, the β-ScTxs were taken to be static. Snapshots of the model were captured every 10 ns. The umbrella sampling method, weighted histogram analysis method [50], and PMF (potential of mean force) curves for this process were calculated to describe the binding free energy between the ligand and receptor.

Supplementary Materials: The following supporting information can be downloaded at: https://www.mdpi.com/article/10.3390/toxins15010033/s1, Figure S1: VDW interaction and electrostatic interaction of residues on AGAP and its mutants that directly contact with VSD2s on hNa$_v$1.4; Figure S2: VDW interaction and electrostatic interaction of residues on AGAP and its mutants that directly contact with VSD2s on hNa$_v$1.5; Table S1: Residues involved in the formations of hydrogen bonds in SMD simulation of AGAP and its mutants with VSD2s on hNa$_v$1.4; Table S2: Residues involved in the formations of hydrogen bonds in SMD simulation of AGAP and its mutants with VSD2s on hNa$_v$1.5; Table S3: Abbreviations in the paper.

Author Contributions: Conceptualization, F.Z. and Y.S.; software, F.Z. and L.F.; validation, W.X. and Y.S.; formal analysis, F.Z., L.F., Q.W. and Q.Y.; investigation, Q.Y. and Y.H.; resources, F.Z., W.X. and Y.S.; data curation, F.Z., L.F., Q.W. and Q.Y.; writing—original draft preparation, F.Z. and L.F.; writing—review and editing, F.Z., W.X. and Y.S.; visualization, L.F. and Q.W.; supervision, W.X. and Y.S.; project administration, W.X. and Y.S. All authors have read and agreed to the published version of the manuscript.

Funding: This research was funded by the National Natural Science Foundation of China [81973227], the Basic Scientific Research Project of Colleges and University of Liaoning Provincial Department of Education [LJKZ0936, LJKMZ20221359], and Liaoning Scientific Provincial Nature Fund Guidance Plan [2019-ZD-0462].

Institutional Review Board Statement: Not applicable.

Informed Consent Statement: Not applicable.

Data Availability Statement: The datasets used and analyzed during the current study are available from the corresponding author upon reasonable request.

Conflicts of Interest: The authors declare no conflict of interest.

References

1. Catterall, W.A. Cellular and molecular biology of voltage-gated sodium channels. *Physiol. Rev.* **1992**, *72* (Suppl. S4), 15–48. [CrossRef] [PubMed]
2. Saito, Y.A.; Strege, P.R.; Tester, D.J.; Locke, G.R., III; Talley, N.J.; Bernard, C.E.; Rae, J.L.; Makielski, J.C.; Ackerman, M.J.; Farrugia, G. Sodium channel mutation in irritable bowel syndrome: Evidence for an ion channelopathy. *Am. J. Physiol. Gastrointest. Liver Physiol.* **2009**, *296*, G211–G218. [CrossRef]
3. Puntmann, V.O.; Taylor, P.C.; Barr, A.; Schnackenburg, B.; Jahnke, C.; Paetsch, I. Towards understanding the phenotypes of myocardial involvement in the presence of self-limiting and sustained systemic inflammation: A magnetic resonance imaging study. *Rheumatology* **2010**, *49*, 528–535. [CrossRef] [PubMed]
4. Ke, Q.; Ye, J.; Tang, S.; Wang, J.; Luo, B.; Ji, F.; Zhang, X.; Yu, Y.; Cheng, X.; Li, Y. N1366S mutation of human skeletal muscle sodium channel causes paramyotonia congenita. *J. Physiol.* **2017**, *595*, 6837–6850. [CrossRef] [PubMed]
5. Cannon, S.C. Channelopathies of skeletal muscle excitability. *Compr. Physiol.* **2015**, *5*, 761–790. [PubMed]
6. Martin, B.; Gabris, B.; Barakat, A.F.; Henry, B.L.; Giannini, M.; Reddy, R.P.; Wang, X.; Romero, G.; Salama, G. Relaxin reverses maladaptive remodeling of the aged heart through Wnt-signaling. *Sci. Rep.* **2019**, *9*, 18545. [CrossRef] [PubMed]
7. Miller, D.; Wang, L.; Zhong, J. Sodium channels, cardiac arrhythmia, and therapeutic strategy. *Adv. Pharmacol.* **2014**, *70*, 367–392. [PubMed]
8. Liu, M.; Yang, K.C.; Dudley, S.C. Cardiac Sodium Channel Mutations: Why so Many Phenotypes? *Curr. Top Membr.* **2016**, *78*, 513–559. [CrossRef] [PubMed]
9. Remme, C.A.; Bezzina, C.R. Sodium channel (dys)function and cardiac arrhythmias. *Cardiovasc. Ther.* **2010**, *28*, 287–294. [CrossRef] [PubMed]
10. Liu, Y.F.; Ma, R.L.; Wang, S.L.; Duan, Z.Y.; Zhang, J.H.; Wu, L.J.; Wu, C.F. Expression of an antitumor-analgesic peptide from the venom of Chinese scorpion *Buthus martensii* karsch in *Escherichia coli*. *Protein Expr. Purif.* **2003**, *27*, 253–258. [CrossRef] [PubMed]
11. Mao, Q.H.; Ruan, J.P.; Cai, X.T.; Lu, W.G.; Ye, J.; Yang, J.; Yang, Y.; Sun, X.Y.; Cao, J.L.; Cao, P. Antinociceptive effects of analgesic-antitumor peptide (AGAP), a neurotoxin from the scorpion *Buthus martensii* Karsch, on formalin-induced inflammatory pain through a mitogen-activated protein kinases-dependent mechanism in mice. *PLoS ONE* **2013**, *8*, e78239. [CrossRef] [PubMed]
12. Yang, Y.; Wang, Y.; Li, S.; Xu, Z.; Li, H.; Ma, L.; Fan, J.; Bu, D.; Liu, B.; Fan, Z.; et al. Mutations in SCN9A, encoding a sodium channel alpha subunit, in patients with primary erythermalgia. *J. Med. Genet.* **2004**, *41*, 171–174. [CrossRef] [PubMed]
13. Cox, J.J.; Reimann, F.; Nicholas, A.K.; Thornton, G.; Roberts, E.; Springell, K.; Karbani, G.; Jafri, H.; Mannan, J.; Raashid, Y.; et al. An SCN9A channelopathy causes congenital inability to experience pain. *Nature* **2006**, *444*, 894–898. [CrossRef] [PubMed]
14. Goodwin, G.; Mcmahon, S.B. The physiological function of different voltage-gated sodium channels in pain. *Nat. Rev. Neurosci.* **2021**, *22*, 263–274. [CrossRef]

15. Chew, L.A.; Bellampalli, S.S.; Dustrude, E.T.; Khanna, R. Mining the $Na_V1.7$ interactome: Opportunities for chronic pain therapeutics. *Biochem. Pharm.* **2019**, *163*, 9–20. [CrossRef] [PubMed]

16. Loussouarn, G.; Sternberg, D.; Nicole, S.; Marionneau, C.; Le Bouffant, F.; Toumaniantz, G.; Barc, J.; Malak, O.A.; Fressart, V.; Péréon, Y.; et al. Physiological and Pathophysiological Insights of $Na_V1.4$ and $Na_V1.5$ Comparison. *Front. Pharm.* **2016**, *14*, 314.

17. Flinspach, M.; Xu, Q.; Piekarz, A.D.; Fellows, R.; Hagan, R.; Gibbs, A.; Liu, Y.; Neff, R.A.; Freedman, J.; Eckert, W.A.; et al. Insensitivity to pain induced by a potent selective closed-state Nav1.7 inhibitor. *Sci. Rep.* **2017**, *3*, 39662. [CrossRef] [PubMed]

18. Hagen, N.A.; Lapointe, B.; Ong-Lam, M.; Dubuc, B.; Walde, D.; Gagnon, B.; Love, R.; Goel, R.; Hawley, P.; Ngoc, A.H.; et al. A multicentre open-label safety and efficacy study of tetrodotoxin for cancer pain. *Curr. Oncol.* **2011**, *18*, e109–e116. [CrossRef] [PubMed]

19. Israel, M.R.; Tay, B.; Deuis, J.R.; Vetter, I. Sodium Channels and Venom Peptide Pharmacology. *Adv. Pharmacol.* **2017**, *79*, 67–116. [PubMed]

20. De Lera Ruiz, M.; Kraus, R.L. Voltage-Gated Sodium Channels: Structure, Function, Pharmacology, and Clinical Indications. *J. Med. Chem.* **2015**, *58*, 7093–7118. [CrossRef]

21. Lai, M.C.; Wu, S.N.; Huang, C.W. The Specific Effects of OD-1, a Peptide Activator, on Voltage-Gated Sodium Current and Seizure Susceptibility. *Int. J. Mol. Sci.* **2020**, *21*, 8254. [CrossRef]

22. Hung, T.Y.; Wu, S.N.; Huang, C.W. The Integrated Effects of Brivaracetam, a Selective Analog of Levetiracetam, on Ionic Currents and Neuronal Excitability. *Biomedicines* **2021**, *9*, 369. [CrossRef] [PubMed]

23. Couraud, F.; Jover, E.; Dubois, J.M.; Rochat, H. Two types of scorpion receptor sites, one related to the activation, the other to the inactivation of the action potential sodium channel. *Toxicon* **1982**, *20*, 9–16. [CrossRef]

24. Xu, Y.; Meng, X.; Hou, X.; Sun, J.; Kong, X.; Sun, Y.; Liu, Z.; Ma, Y.; Niu, Y.; Song, Y.; et al. A mutant of the *Buthus martensii* Karsch antitumor-analgesic peptide exhibits reduced inhibition to hNa(v)1.4 and hNa(v)1.5 channels while retaining analgesic activity. *J. Biol. Chem.* **2017**, *292*, 18270–18280. [CrossRef] [PubMed]

25. Cestèle, S.; Catterall, W.A. Molecular mechanisms of neurotoxin action on voltage-gated sodium channels. *Biochimie* **2000**, *82*, 883–892. [CrossRef] [PubMed]

26. Cestèle, S.; Qu, Y.; Rogers, J.C.; Rochat, H.; Scheuer, T.; Catterall, W.A. Voltage sensor-trapping: Enhanced activation of sodium channels by beta-scorpion toxin bound to the S3-S4 loop in domain II. *Neuron* **1998**, *21*, 919–931. [CrossRef] [PubMed]

27. Pedraza Escalona, M.; Possani, L.D. Scorpion beta-toxins and voltage-gated sodium channels: Interactions and effects. *Front. Biosci.* **2013**, *18*, 572–587. [CrossRef] [PubMed]

28. Gurevitz, M. Mapping of scorpion toxin receptor sites at voltage-gated sodium channels. *Toxicon* **2012**, *60*, 502–511. [CrossRef] [PubMed]

29. Zhang, J.Z.; Yarov-Yarovoy, V.; Scheuer, T.; Karbat, I.; Cohen, L.; Gordon, D.; Gurevitz, M.; Catterall, W.A. Structure-function map of the receptor site for β-scorpion toxins in domain II of voltage-gated sodium channels. *J. Biol. Chem.* **2011**, *286*, 33641–33651. [CrossRef] [PubMed]

30. Xu, Y.; Sun, J.; Yu, Y.; Kong, X.; Meng, X.; Liu, Y.; Cui, Y.; Su, Y.; Zhao, M.; Zhang, J. Trp: A conserved aromatic residue crucial to the interaction of a scorpion peptide with sodium channels. *J. Biochem.* **2020**, *168*, 633–641. [CrossRef]

31. Xu, Y.; Sun, J.; Liu, H.; Sun, J.; Yu, Y.; Su, Y.; Cui, Y.; Zhao, M.; Zhang, J. Scorpion Toxins Targeting Voltage-gated Sodium Channels Associated with Pain. *Curr. Pharm. Biotechnol.* **2018**, *19*, 848–855. [CrossRef] [PubMed]

32. Zhao, F.; Wang, J.L.; Ming, H.Y.; Zhang, Y.N.; Dun, Y.Q.; Zhang, J.H.; Song, Y.B. Insights into the binding mode and functional components of the analgesic-antitumour peptide from *Buthus martensii* Karsch to human voltage-gated sodium channel 1.7 based on dynamic simulation analysis. *J. Biomol. Struct. Dyn.* **2020**, *38*, 1868–1879. [CrossRef] [PubMed]

33. Zhorov, B.S.; Du, Y.; Song, W.; Luo, N.; Gordon, D.; Gurevitz, M.; Dong, K. Mapping the interaction surface of scorpion β-toxins with an insect sodium channel. *Biochem. J.* **2021**, *478*, 2843–2869. [CrossRef]

34. Chen, R.; Chung, S.H. Conserved functional surface of antimammalian scorpion β-toxins. *J. Phys. Chem. B* **2012**, *116*, 4796–4800. [CrossRef] [PubMed]

35. Zhu, S.; Gao, B.; Peigneur, S.; Tytgat, J. How a Scorpion Toxin Selectively Captures a Prey Sodium Channel: The Molecular and Evolutionary Basis Uncovered. *Mol. Biol. Evol.* **2020**, *37*, 3149–3164. [CrossRef] [PubMed]

36. Webb, B.; Sali, A. Comparative Protein Structure Modeling Using MODELLER. *Curr. Protoc. Protein Sci.* **2016**, *54*, 5.6.1–5.6.37.

37. Consortium, U.P. UniProt: A hub for protein information. *Nucleic Acids Res.* **2015**, *43*, D204–D212. [CrossRef]

38. Chen, R.; Li, L.; Weng, Z. ZDOCK: An initial-stage protein-docking algorithm. *Proteins Struct. Funct. Bioinform.* **2010**, *52*, 80–87. [CrossRef]

39. Abraham, M.J.; Murtola, T.; Schulz, R.; Páll, S.; Smith, J.C.; Hess, B.; Lindahl, E. GROMACS: High performance molecular simulations through multi-level parallelism from laptops to supercomputers. *Softwarex* **2015**, *1*, 19–25. [CrossRef]

40. Berger, O.; Edholm, O.; Jahnig, F. Molecular dynamics simulations of a fluid bilayer of dipalmitoylphosphatidylcholine at full hydration, constant pressure, and constant temperature. *Biophys. J.* **1997**, *72*, 2002–2013. [CrossRef]

41. Kandt, C.; Ash, W.L.; Tieleman, D.P. Setting up and running molecular dynamics simulations of membrane proteins. *Methods* **2007**, *41*, 475–488. [CrossRef] [PubMed]

42. Amira, S.; Spangberg, D.; Hermansson, K. Derivation and evaluation of a flexible SPC model for liquid water. *Chem. Phys.* **2004**, *303*, 327–334. [CrossRef]

43. Lemak, A.S.; Balabaev, N.K. On The Berendsen Thermostat. *Mol. Simul.* **1994**, *13*, 177–187. [CrossRef]

44. Petersen, H.G. Accuracy and efficiency of the particle mesh Ewald method. *J. Chem. Phys.* **1995**, *103*, 3668–3679. [CrossRef]
45. Hess, B.; Bekker, H.; Berendsen, H.J.C.; Fraaije, J.G. LINCS: A linear constraint solver for molecular simulations. *J. Chem. Theory Comput.* **1997**, *4*, 1463–1472. [CrossRef]
46. Makarewicz, T.; Kazmierkiewicz, R. Molecular dynamics simulation by GROMACS using GUI plugin for PyMOL. *J. Chem. Inf. Model.* **2013**, *53*, 1229–1234. [CrossRef] [PubMed]
47. Humphrey, W.; Dalke, A.; Schulten, K. VMD: Visual molecular dynamics. *J. Mol. Graph.* **1996**, *14*, 33–38. [CrossRef]
48. Daura, X.; Gademann, K.; Jaun, B.; Seebach, D.; Van Gunsteren, W.F.; Mark, A.E. Peptide Folding: When Simulation Meets Experiment. *Angew. Chem. Int. Ed.* **1999**, *38*, 236–240. [CrossRef]
49. Do, P.C.; Lee, E.H.; Le, L. Steered Molecular Dynamics Simulation in Rational Drug Design. *J. Chem. Inf. Model.* **2018**, *58*, 1473–1482. [CrossRef]
50. Kumar, S.; Bouzida, D.; Swendsen, R.H.; Kollman, P.A.; Rosenberg, J.M. The weighted histogram analysis method for free-energy calculations on biomolecules. I. The method. *J. Comput. Chem.* **1992**, *13*, 1011–1021. [CrossRef]

Article

Smp24, a Scorpion-Venom Peptide, Exhibits Potent Antitumor Effects against Hepatoma HepG2 Cells via Multi-Mechanisms In Vivo and In Vitro

Tienthanh Nguyen [1,2,†], Ruiyin Guo [1,2,†], Jinwei Chai [2], Jiena Wu [2], Junfang Liu [1], Xin Chen [1], Mohamed A. Abdel-Rahman [3], Hu Xia [1,*] and Xueqing Xu [2,*]

[1] Department of Pulmonary and Critical Care Medicine, Zhujiang Hospital, Southern Medical University, Guangzhou 510280, China
[2] Guangdong Provincial Key Laboratory of New Drug Screening, School of Pharmaceutical Sciences, Southern Medical University, Guangzhou 510515, China
[3] Zoology Department, Faculty of Science, Suez Canal University, Ismailia 41522, Egypt
* Correspondence: xiahu@smu.edu.cn (H.X.); xu2003@smu.edu.cn (X.X.); Tel.: +86-20-61648537 (X.X.)
† These authors contributed equally to this work.

Abstract: Scorpion-venom-derived peptides have become a promising anticancer agent due to their cytotoxicity against tumor cells via multiple mechanisms. The suppressive effect of the cationic antimicrobial peptide Smp24, which is derived from the venom of *Scorpio Maurus palmatus*, on the proliferation of the hepatoma cell line HepG2 has been reported earlier. However, its mode of action against HepG2 hepatoma cells remains unclear. In the current research, Smp24 was discovered to suppress the viability of HepG2 cells while having a minor effect on normal LO2 cells. Moreover, endocytosis and pore formation were demonstrated to be involved in the uptake of Smp24 into HepG2 cells, which subsequently interacted with the mitochondrial membrane and caused the decrease in its potential, cytoskeleton reorganization, ROS accumulation, mitochondrial dysfunction, and alteration of apoptosis- and autophagy-related signaling pathways. The protecting activity of Smp24 in the HepG2 xenograft mice model was also demonstrated. Therefore, our data suggest that the antitumor effect of Smp24 is closely related to the induction of cell apoptosis, cycle arrest, and autophagy via cell membrane disruption and mitochondrial dysfunction, suggesting a potential alternative in hepatocellular carcinoma treatment.

Keywords: antitumor peptide; membrane disruption; mitochondrial dysfunction; apoptosis; cell cycle arrest; autophagy

Key Contribution: The antitumor mechanism of Smp24 against HepG2 cells was demonstrated to be associated with the disruption of cell membrane and mitochondrial dysfunction, leading to the suppression of cell viability by induction of cell apoptosis, cycle arrest, and autophagy.

Citation: Nguyen, T.; Guo, R.; Chai, J.; Wu, J.; Liu, J.; Chen, X.; Abdel-Rahman, M.A.; Xia, H.; Xu, X. Smp24, a Scorpion-Venom Peptide, Exhibits Potent Antitumor Effects against Hepatoma HepG2 Cells via Multi-Mechanisms In Vivo and In Vitro. *Toxins* **2022**, *14*, 717. https://doi.org/10.3390/toxins14100717

Received: 14 September 2022
Accepted: 18 October 2022
Published: 21 October 2022

1. Introduction

Scorpions and their venom have a long history of serving people as traditional medicine against multiple conditions. Specially, scorpion-derived peptides are proved to become a promising remedy against severe diseases such as cancer and immune-related disease [1,2]. The active peptides from scorpions generally belong to disulfide-bridged peptides (DBPs) and non-disulfide-bridged peptides (NDBPs). As the important components of scorpion venom, DBPs are associated with neurotoxicity, while NDBPs are multiple functional cationic peptides possessing antibacterial, antiviral, anticancer, and immune-modulatory activities [3]. These peptides may serve as potential candidates for drug development with a broad range of applications [4]. Due to the need for novel anti-cancer drugs with low cytotoxicity and treatment resistance, antimicrobial peptides (AMPs) can represent a potential therapy by selectively targeting tumor cells through binding the negative-charged phosphatidylserine and syndecans on their surface, causing cell death by multiple mechanisms [5]. For instance, MSP-4, an AMP identified from the Nile tilapia (*Oreochromis niloticus*), has been described to decrease the viability of the human osteosarcoma cell line MG-63 through induction of cell cycle arrest and apoptosis by activating a Fas/FasL- and mitochondria-mediated pathway [6]. The cationic AMP reported from the hemocyanin of *Litopenaeus vannamei*, LvHemB1, inhibits the viability of various cancer cell lines while not affecting the normal liver cell lines [7]. Moreover, LvHemB1 exerts its antitumor effect by causing mitochondrial membrane potential loss, increasing reactive oxygen species (ROS) production and apoptotic proteins. We previously reported Smp24 (IWSFLIKAATKLLPSLFGGGKKDS), a cationic AMP identified from the venom of *Scorpio Maurus palmatus* with a wide antimicrobial spectrum against various bacteria and fungi [8]. The cytotoxicity of Smp24 against two acute leukemia cell lines (KG1-a and CCRF-CEM), as well as four lung cancer cell lines (A549, H3122, PC-9, and H460) and nontumor cell lines (HRECs, CD34$^+$, MCR-5, and HaCaT) has been described in further studies [9–11]. Smp24 also has cytotoxic effects against HepG2 hepatoma cells, as presented in ATP release assays; nevertheless, its mode of action against HepG2 hepatoma cells remains unclear. In this study, we explored the cytotoxicity of Smp24 and its impacts on the cell membrane, cell cycle distribution, apoptosis, and autophagy of HepG2 cells. Our results reveal that the antitumor activity of Smp24 is related to membrane disruption and mitochondrial dysfunction, inducing cell cycle arrest, apoptosis, and autophagy.

2. Results

2.1. Smp24 Inhibits the Proliferation of Hepatoma Cells

The cytotoxicity of Smp24 against both human hepatoma cell HepG2 and normal hepatic cell LO2 was evaluated. While compared with the minor effects on LO2 cells, Smp24 presented a remarkable inhibitory effect on the proliferation of HepG2 (Figure 1A). In detail, after 24 h incubation, the IC$_{50}$ values of Smp24 against HepG2 and LO2 cells were approximately 5.524 µM and 16.68 µM, respectively. Moreover, as exhibited in Figure 1B, Smp24 decreased the viability of HepG2 hepatoma cells in concentration- and time-dependent manners. Consistently, in comparison with control group, the fluorescence of EdU-labeled cells was gradually decreased with the increasing Smp24 concentrations (Figure 1C). The morphological changes of HepG2 cells were visible after 24 h incubation with Smp24 (Figure 1D). Cells treated with Smp24 changed to a spherical shape with the appearance of cellular debris and floating cells, while it had an epithelial-like morphology with smooth surfaces in the control group. All these data indicated that Smp24 suppresses the viability of HepG2 cells in a dosage-dependent manner.

2.2. Smp24 Enters into HepG2 Cells through Endocytosis and Pore Formation

The surface charge was calculated due to the fact that the interaction between cancer cells and cationic peptide plays a crucial role in its inhibitory effect. As presented in Figure 2A, the zeta potential of HepG2 cells incubated with Smp24 (2.5, 5, and 10 µM) was increased from −14.67 mV to −8.11 mV in a concentration-dependent manner. Therefore, Smp24 could react with the HepG2 cell surface or cross the cell membrane, subsequently culminating in alterations of cell surface charge. To further evaluate the potential internalization activity of Smp24, the fluorescence of FITC-labeled Smp24 was observed. Smp24 was able to enter the HepG2 cells after 6 h of incubation, followed by the penetration into the nucleus in 24 h, as presented by the increase in fluorescence while compared with the control group (Figure 2B). Moreover, flow cytometry suggested that Smp24 also dose-dependently accomplished its internalization activity (Figure 2C). As a crucial component involved in the interaction between membrane and cell-penetrating peptide [12], the impact of heparan sulfate on the cellular internalization of Smp24 was assessed. Pretreatment with heparan sulfate (5, 10, and 20 µg/mL) led to the decreasing fluorescence of FITC-labeled Smp24 by approximately 1.56%, 48.12%, and 66.82%, respectively. Furthermore, treatment with ammonium chloride significantly reduced the cellular uptake of Smp24. The internalization effect of Smp24 in different temperatures was further analyzed. The cellular internalization of Smp24 at 37 °C was approximately 45.89% higher than that at 4 °C, showing that Smp24 penetrated HepG2 cell membrane in an energy-dependent and thermosensitive manner. These findings suggested the internalization of Smp24 into HepG2 cells via pore formation and endocytosis.

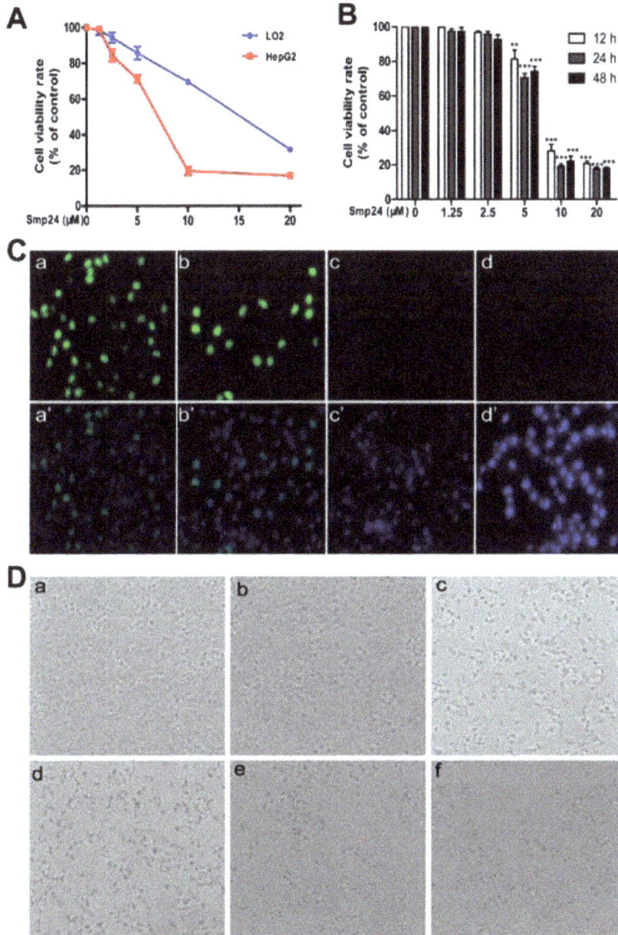

Figure 1. Smp24 suppressed the proliferation of HepG2 cells. (**A**) Effect of Smp24 on the viability of HepG2 and LO2 cells. After treatment with Smp24 (0–20 μM) for 24 h, MTT assays were used to determine cell viabilities. (**B**) Cytotoxicity of different concentrations of Smp24 against HepG2 cells after 12, 24, and 48 h of treatment and their IC_{50} values were detected by MTT assay. (**C**) EdU cell proliferation assay. Photographs were captured at 400× magnification. Panels a–d present cells stained with EdU while panels a'–d' present the merge images of cells stained by EdU and DAPI. Panels a and a': control cells; panels b–d and b'–d': cells treated with 2.5, 5, and 10 μM Smp24 for 24 h, respectively. (**D**) Changes in cell morphology in the presence of Smp24. Photographs were taken at 100× magnification. Panel a: control group; panels b–f: cells treated with 1.25, 2.5, 5, 10, and 20 μM of Smp24 for 24 h, respectively. The data are shown as the mean ± SEM (n = 3). ** $p < 0.01$ and *** $p < 0.001$ are regarded to be statistically significant in comparison with the control group.

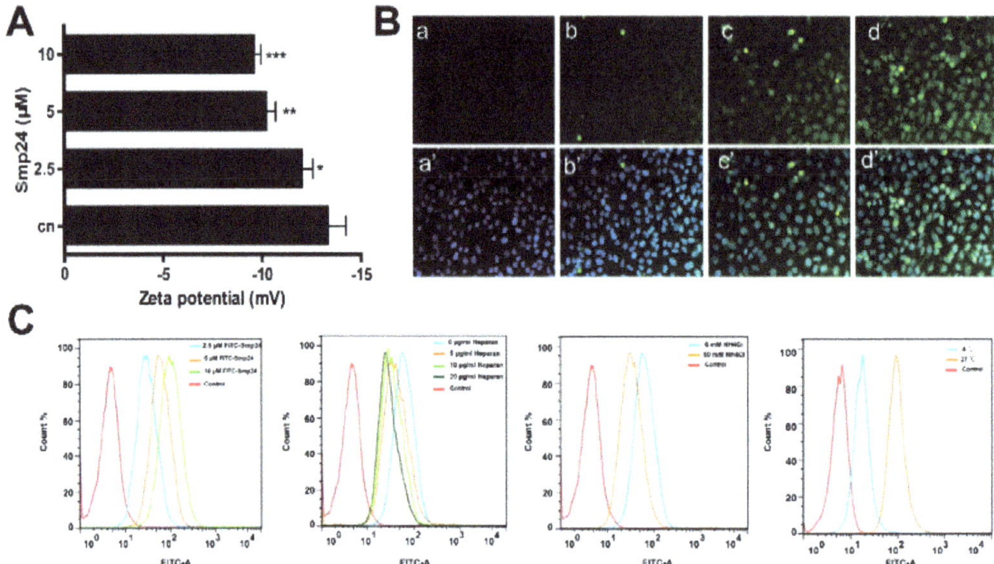

Figure 2. Uptake mechanism of Smp24 into HepG2 cells. (**A**) Zeta potential assay. Changes of cell membrane surface potential after Smp24 (0, 2.5, 5, and 10 μM) was co-incubated with HepG2 cells for 5 min with PBS as the control; (**B**) fluorescence microscope detection of Smp24 entering HepG2 cells. The upper panel depicts FITC-labeled Smp24-stained cells, the lower panel depicts cells stained by FITC-labeled Smp24 and DAPI, panels a and a': the control groups, panels b–d and b'–d': the 6, 12, and 24 h Smp24 administration groups, respectively; photographs were taken at 400× magnification. (**C**) Flow cytometry measurement of Smp24 entering HepG2 cells in the absence and presence of ammonium chloride, heparan sulfate (5, 10, and 20 μg/mL) or different temperatures for 1 h. The data are shown as the means ± SEM (*n* = 3). * *p* < 0.05, ** *p* < 0.01, and *** *p* < 0.001 are regarded to be statistically significant in comparison with control group without Smp24 treatment.

2.3. Smp24 Changes Cell Cytoskeleton Conformation

Actin-forming microfilaments in the cytoskeleton are involved in many cellular processes, such as cell signaling, division, motility, cytokinesis, the building-up and maintenance of cell junctions, and shape [13]. Accordingly, filamentous (F)-actin was stained with rhodamine–phalloidin to identify the effect of Smp24 on the cytoskeleton. In comparison with the control cells, which had flat microfilament bundles and sharp boundaries, treatment with Smp24 resulted in the reorganization of intracellular actin filaments together with the demolition of F-actin, leading to the structural dissociation of HepG2 cells. Furthermore, this phenomenon was aggravated by increasing Smp24 concentration (Figure 3). The results revealed the effect of Smp24 on the cytoskeleton in HepG2 cells by affecting the F-actin network.

2.4. Smp24 Destroys the Cellular and Mitochondrial Membrane of HepG2 Cells

The LDH-release assay is a common cell death/cytotoxicity assay used to assess plasma membrane damage to a cell population. As shown in Figure 4A, LDH release from HepG2 cells was increased with treatment of Smp24 in concentration- and time-dependent manners. SEM imaging was carried out to confirm the impact of Smp24 on the cell membrane of HepG2. Compared with the well-defined cell membrane of control cells, Smp24 treatment led to visible changes in morphology, with unclear boundaries of cell membrane together with leakage of cellular contents (Figure 4B). Consequently, the leaked calcein from HepG2 cells was measured to evaluate the impact of Smp24 on cell membrane permeability. As showed by the changing fluorescence intensity in Figure 2C, the calcein leakage from HepG2 cells was gradually increased by the administration of Smp24. Thus, the membrane integrity of HepG2 cells could be disrupted by Smp24. Furtherly, $CoCl_2$ is a calcein fluorescence quencher in cytoplasm rather than in mitochondria. The calcein leakage fluorescence in the presence of $CoCl_2$ was concentration-dependently decreased by Smp24 in comparison with the control group. Additionally, there was no statistically noteworthy change in the fluorescence intensity of cells incubated with both NAC and CsA. These findings indicate that Smp24 affects both the mitochondrial and cellular membranes of HepG2 cells in a nonspecific manner.

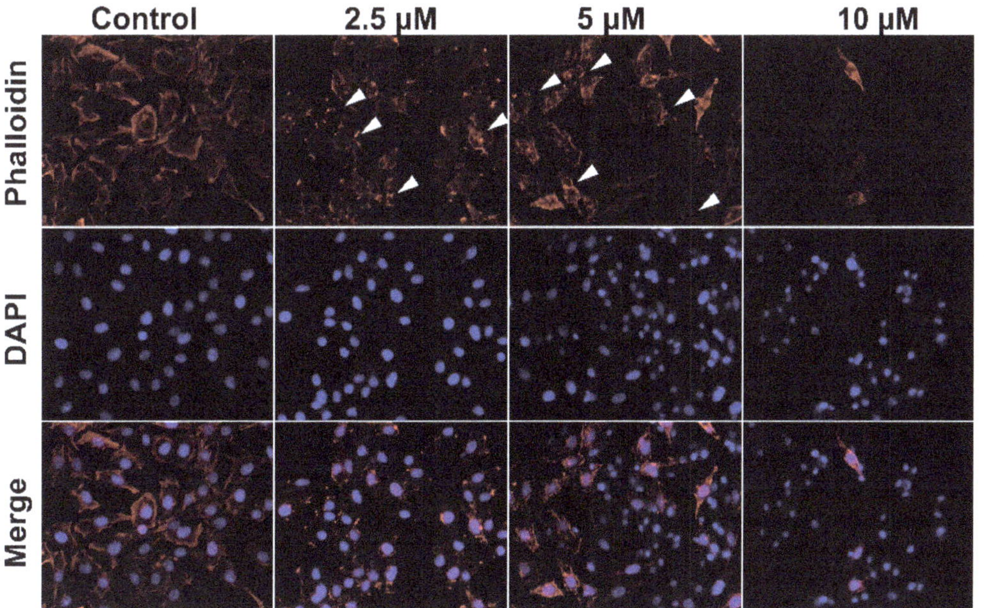

Figure 3. The changes in the cytoskeleton of HepG2 cells in the presence of Smp24. After being treated with Smp24 (2.5, 5, and 10 µM) for 24 h, HepG2 cells were subsequently dyed by rhodamine–phalloidin and DAPI and were examined under fluorescence microscopy with 400× magnification. White arrows indicate disorganized microfilament bundles. Scale bar, 20 µm.

2.5. Smp24 Declines Mitochondrial Membrane Potential but Promotes ROS Production

The disruption of the mitochondrial membrane may result in the dysfunction of mitochondria, loss of mitochondrial membrane potential, as well as a high level of ROS [14], which finally cause metabolic cell death. Thus, the changes in membrane potential of Smp24-treated cells were assessed with JC-1 staining, which formed red fluorescence at a high mitochondrial membrane potential while forming green fluorescence at a low one. In comparison with the control group, Smp24 concentration-dependently increased the green fluorescence while declining the red fluorescence (Figure 5A), indicating that Smp24 could decrease the mitochondrial membrane potential.

DCFH-DA can be hydrolyzed by intracellular esterase to produce DCFH which does not permeate cell membranes and can be oxidized to generate fluorescent DCF. Consistent with the result in the JC-1 staining assay, Smp24 dose-dependently induced ROS production in HepG2 cells (Figure 5B), compared with the control cells. In addition, co-treatment with NAC notably ameliorated all the abnormal phenomena caused by Smp24. All these findings revealed that Smp24 is responsible for the loss of mitochondrial membrane potential and increase in ROS production, implying a damaged mitochondrial membrane and function.

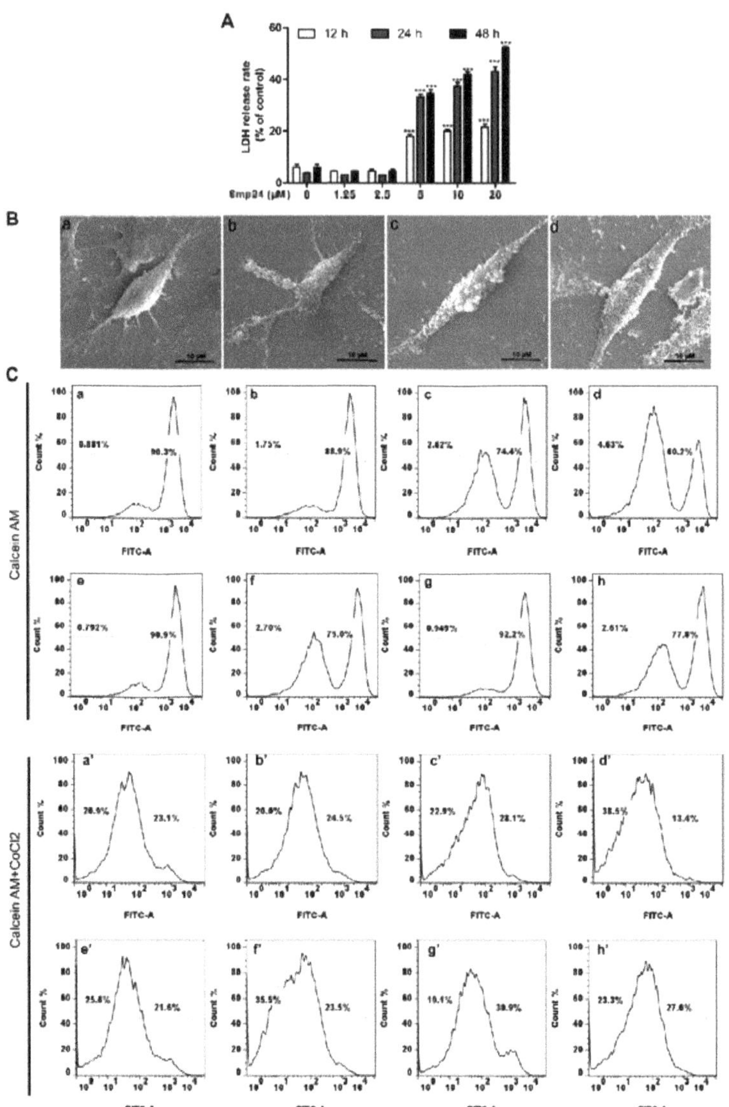

Figure 4. Cellular and mitochondrial membrane damage stimulated by Smp24 in HepG2 cells. (**A**) Levels of LDH release from Smp24-treated HepG2 cells for 12, 24, and 48 h. (**B**) The morphological structure changes of HepG2 cells under SEM. Panels a–d: HepG2 cells in the presence of 0, 2.5, 5, and 10 µM of Smp24, respectively. Scale bar: 10 µm. (**C**) Flow cytometry analysis of calcein fluorescence with (**top panels**) and without $CoCl_2$ (**bottom panels**). After pre-incubation with NAC (2 mM) or CsA (1 µM), cells were subsequently treated by 5 µM Smp24 for 24 h, followed by incubation with 1 µM calcein AM or 1 µM calcein AM + 1 mM $CoCl_2$ at 37 °C for 30 min. Panels a–d and Panels a′–d′: HepG2 cells sequentially treated with 0, 2.5, 5, and 10 µM of Smp24, respectively; panels e and e′: HepG2 cells treated by 2 mM NAC; panels f and f′: HepG2 cells treated by 2 mM NAC + 5 µM Smp24; panels g and g′: HepG2 cells treated by 1 µM CsA; panels h and h′: HepG2 cells treated by 1 µM CsA + 5 µM Smp24. The data are displayed as the means ± SEM (n = 3). *** $p < 0.001$ are considered statistically significant in comparison with the control group without peptide.

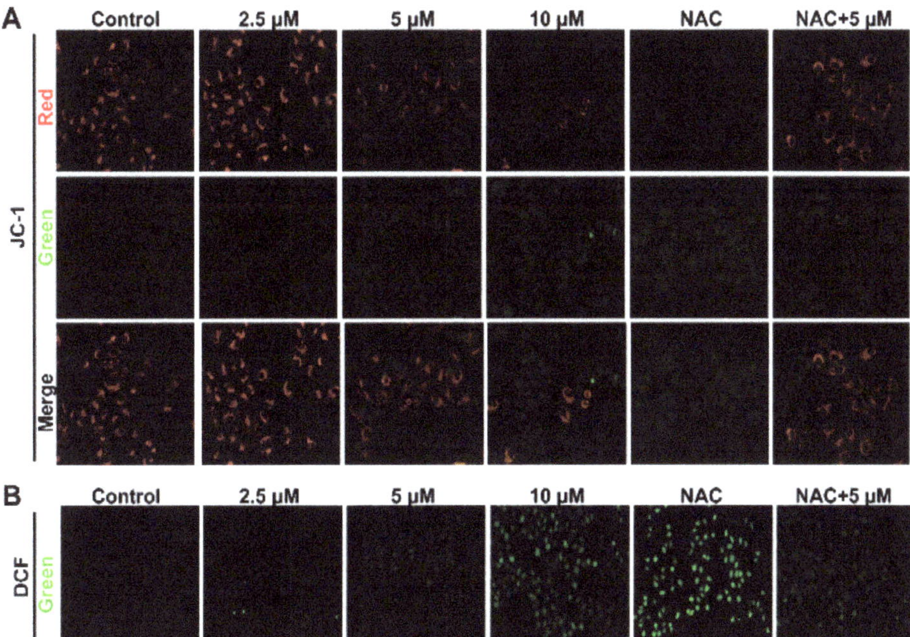

Figure 5. Impacts of Smp24 on mitochondrial membrane potential and ROS accumulation of HepG2 cells. (**A**) Mitochondrial membrane potential changes in HepG2 cells induced by Smp24. Cells were incubated with 0–10 μM Smp24, 2 mM NAC, or with 2 mM NAC + 5 μM Smp24 for 12 h, as described in Material and Methods, and then fluorescent images were taken under fluorescence microscope at 400× magnification. (**B**) ROS levels in HepG2 cells. HepG2 cells were incubated with 0–10 μM Smp24 for 12 h, followed by DCFH-DA staining at 37 °C for 30 min before detection by fluorescence microscope.

2.6. Smp24 Activates Mitochondrion-Mediated Intrinsic Pathway and Induces Apoptosis in HepG2 Cells

The variations in mitochondrial outer membrane permeability contribute to the apoptotic cascade in cell death pathways [15]. Thus, HepG2 cell apoptosis induced by Smp24 was investigated using DAPI and annexin V-FITC/PI staining assays. As presented in Figure 6A, typical apoptotic features, such as chromatin condensation and the appearance of apoptotic bodies, were obvious in HepG2 cells incubated with Smp24. The apoptotic cell rate measured using an annexin V-FITC/PI staining assay was also significantly increased from 8.64% to 46.8% after exposure to Smp24 (Figure 6B). Additionally, co-treatment with antioxidant NAC remarkably reduced the pro-apoptotic effect of Smp24, in which the apoptotic cell proportion decreased from 28.76% to 18.80%.

Due to their important role in the process of apoptosis, the mitochondria/cytochrome C mediated apoptotic pathways were examined to further explore the underlining mechanism of the apoptosis-induced activity of Smp24 in HepG2 cells [16,17]. Under our condition, treatment with Smp24 concentration-dependently upregulated the expression of cytochrome C (Figure 6C,D). What is more, Smp24 significantly increased the levels of cleaved caspase-3, caspase-9, and PARP, while its precursors were declined in a concentration-dependent manner. Simultaneously, compared with normal cells, Smp24-treated HepG2 cells possessed a remarkable increase in the expression of Bax while possessing a significant decrease in that of Bcl-1.

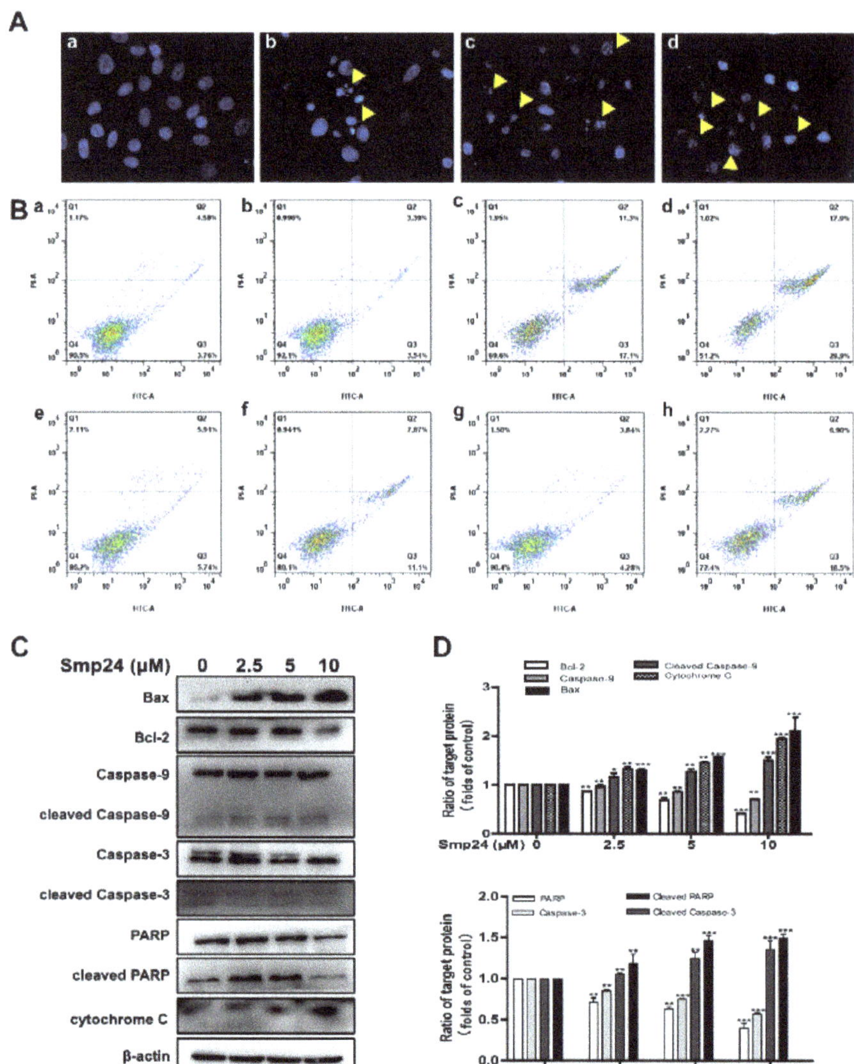

Figure 6. Apoptosis of HepG2 cells induced by Smp24. (**A**) DAPI staining images of apoptotic cells. HepG2 cells were incubated with 2.5, 5, and 10 μM Smp24 for 24 h, followed by DAPI staining. The apoptotic bodies were marked by yellow triangles. Panels a–d: HepG2 cells sequentially treated by 0, 2.5, 5, and 10 μM of Smp24. (**B**) Flow cytometry of apoptotic cells. HepG2 cells were incubated with 2.5, 5, and 10 μM of Smp24, and subsequently dyed with annexin V-FITC/PI. Panels a–d: HepG2 cells sequentially incubated with 0, 2.5, 5, and 10 μM of Smp24; panels e–f: HepG2 cells sequentially incubated with 2 mM NAC and 2 mM NAC + 5 μM Smp24; panels g–h: HepG2 cells sequentially incubated with 1 μM CsA and 1 μM CsA + 5 μM Smp24, respectively. (**C**) Representative Western blots of the apoptotic pathway induced by Smp24 in HepG2 cells. (**D**) Quantification analysis of band densities in (**C**). Bars mean the relative expression of the target protein to the control. Image J software (64-bit, National Institutes of Health, MD, USA) was applied to analyze the band, and the data are displayed as mean ± SEM ($n = 3$). ** $p < 0.01$ and *** $p < 0.001$ are regarded to be statistically significant in comparison with the control group.

2.7. Smp24 Regulates the Expression of G2/M Phase-Related Proteins and Induces Cell Cycle Arrest

Mitochondrial damage is known to be related to the occurrence of cell-cycle accumulation, leading to apoptosis in cancer cells [18]. Therefore, flow cytometry was conducted to investigate the cell cycle change induced by Smp24 in HepG2 cells. After 24 h of treatment, the proportion of HepG2 cells arrested in the S phase was increased by Smp24 (2.5, 5, and 10 μM) from 12.87% to approximately 15.95%, 18.85%, and 20.85%, respectively, while those in the G2/M phase were also raised from 6.83% to 7.83%, 9.83%, and 14.83%. Moreover, the proportion of cells in phase G0/G1 was obviously dropped by Smp24 in a concentration-dependent manner (Figure 7A). In line with the changes in cell cycle, treatment with Smp24 upregulated the expression of cyclin A, cyclin B, p21$^{\text{Wafl/Cip1}}$, as well as p53, while dose-dependently depressed the expression of CDK2 and cyclin E (Figure 7B,C).

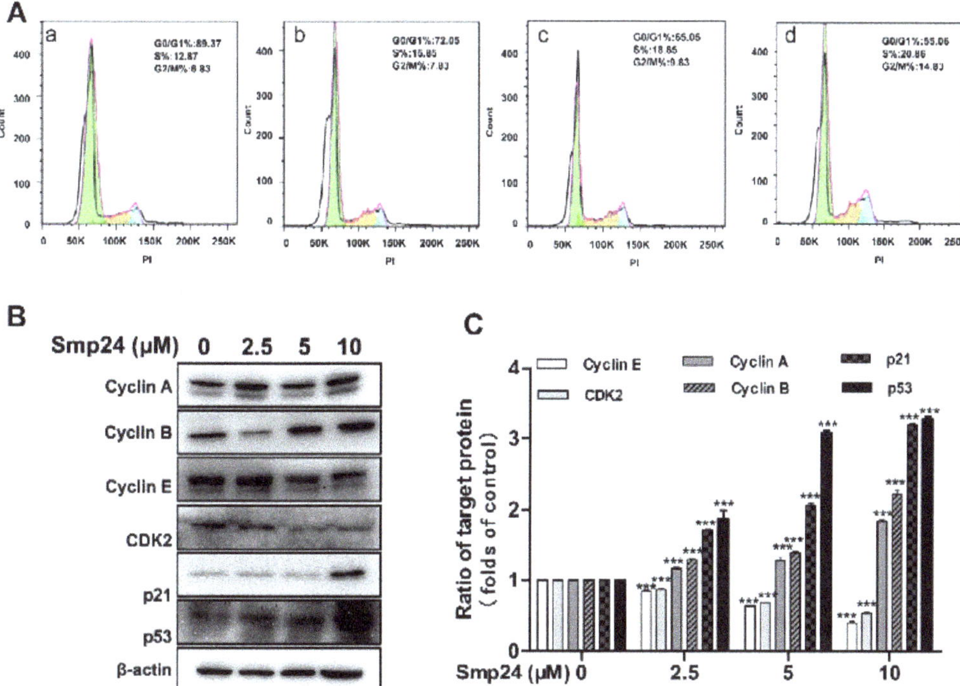

Figure 7. Cycle arrest of HepG2 cells induced by Smp24. (**A**) Cell cycle analysis by flow cytometry. Panels a–d: 0, 2.5, 5, and 10 μM Smp24-treated groups, respectively. (**B**) Western blot images of HepG2 cells treated by Smp24 (2.5, 5, and 10 μM) for 24 h. (**C**) Statistical analysis of (**B**). Bars mean the relative expression of the target protein to the control. Image J software was applied to analyze the band, and the data are shown as mean ± SEM (*n* = 3). *** *p* < 0.001 is regarded to be statistically significant in comparison with the control group.

2.8. Smp24 Regulates Autophagy-Related Signaling Pathways

Autophagy is a catabolic pathway essential for organismal homeostasis; together with apoptosis, it might be considered as a potential cancer cellular control [19,20]. Hence, the effect of Smp24 on autophagy-regulated signaling pathways was explored. As displayed in Figure 8A, when compared to the control cells, Smp24 markedly reduced the contents of phosphorylated mTOR, PI3K, as well as AKT, in a dose-dependent manner. Furthermore, treatment with Smp24 dose-dependently suppressed the expression of microtubule-associated protein LC3A/B-I and p62 while enhancing the levels of autophagosome formation marker LC3A/B-II [21].

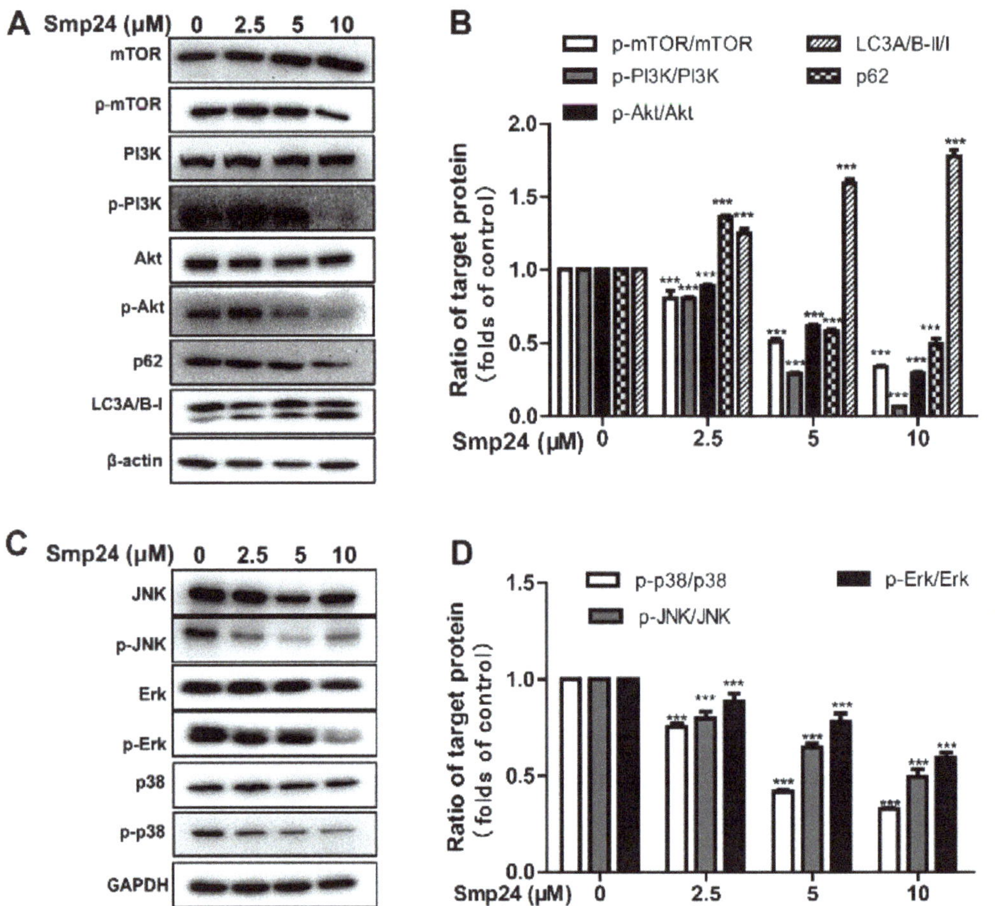

Figure 8. Autophagy and signaling pathways regulated by Smp24. (**A,C**) Representative Western blot images of proteins belonging to PI3K/Akt/mTOR and MAPK signaling pathways. HepG2 cells were exposed to 2.5, 5, and 10 μM Smp24 for 24 h, with PBS as the negative control. (**B,D**) Quantification analysis of band densities in (**A,C**). Bars mean the relative expression of the target protein to the control group. Data are shown as mean ± SEM (n = 3). *** $p < 0.001$ is regarded to be statistically significant in comparison with the control group.

The MAPK pathway may become a promising target in cancer therapy owing to the important role in multiple cell process, especially in autophagy, apoptosis, and cell cycle arrest [22,23]. Under our conditions, Smp24 significantly decreased the expression of phosphorylated-JNK, ERK, and p38, while it did not affect its total protein expression (Figure 8C,D). These results suggested that Smp24 induced autophagy in HepG2 cells by regulating autophagy-regulated signaling pathways.

2.9. Smp24 Shows Antitumor Effects In Vivo

The in vivo antihepatoma effect of Smp24 was evaluated in HepG2 xenograft mice. As presented in Figure 9B–D, Smp24 significantly reduced tumors in both weight and volume by approximately 55.4% and 56.3%, respectively, in comparison with the control group at the end of experiment. Furthermore, Smp24 did not cause notable changes in body weight and crucial organs including liver, heart, lung, spleen, and kidney in the xenograft mice (Figure 9E,F). Consistent with the above results, histopathological examination indicated the antitumor effect of Smp24 by the appearance of abnormal changes in cancer cells, such as nuclear condensation, cell shrinkage, and inflammatory cell infiltration (Figure 9G). Furthermore, in line with the upregulated expression induced by Smp24 in vitro, the expression of cleaved caspase-3 was also verified in tumor tissue by immunohistochemical staining assay (Figure 9H). These findings reveal the antitumor effect of Smp24 in vivo.

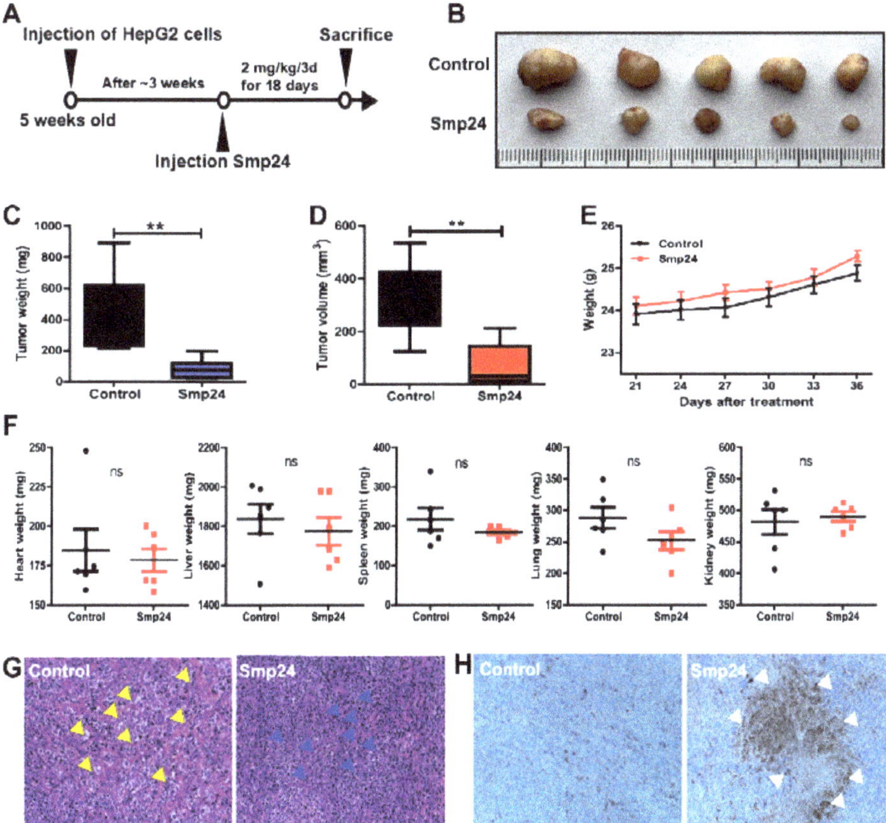

Figure 9. Effect of Smp24 on xenograft mice. Once the volume of tumors increased to approximately 50 mm^3, physiological saline or Smp24 (2 mg/kg) was injected near the tumor site every three days for a total of six times. The tumor volumes and the body weight of mice was also recorded daily. The tumors and important organs including liver, heart, lung, spleen, and kidney at the end of experiment were separated for analysis. (**A**) Schedule for in vivo experiment. (**B**) Images of tumors removed from mice. Scale unit: centimeter. (**C,D**) Tumor weight and volume at the end of experiments. (**E**) Body weight changes of nude mice after Smp24 injection. (**F**) Organ weight changes at the end of experiments. (**G**) HE staining images of tumor tissue. (**H**) IHC images of cleaved caspase-3 in tumor tissue from xenograft mice. Photographs were taken under light microscope at 200× magnification. Yellow arrow: enlarged tumor cells. Blue arrow: shrinking cells. White arrow: positive apoptotic staining (brown areas). The data are expressed as mean ± SEM ($n = 5$). ns: no significance, ** $p < 0.01$ is regarded to be statistically significant in comparison with control group.

3. Discussion

Despite their multidrug resistance and toxicity towards normal tissue, radiotherapy and chemotherapy remain the most common non-surgical remedies in cancer treatment. However, new treatments with less side effects are urgently needed. AMPs become promising anticancer agents due to their targeting of cell membranes instead of specific receptors, making them less likely to induce drug resistance in tumor cells [24]. The cytotoxicity of Smp24 against HepG2 cells have been reported in a previous study [8]; however, its underlying mechanism remains elusive. In the present study, the antitumor activity of Smp24 against HepG2 has been confirmed both in vitro and in vivo. We also revealed that Smp24 significantly suppresses the proliferation of HepG2 cells, while having a slight effect on that of the normal hepatic cell LO2, suggesting a highly selective activity against tumor cells (Figure 1A). It is noteworthy that the IC_{50} values of Smp24 against HepG2 cells was slightly higher than the results in our previous study against A549 cells [10,11]. Therefore, there are some common mechanisms against two kinds of tumor by Smp24.

Cationic AMPs commonly possess cancer-selective toxicity due to the fact that fundamental differences exist in cell membranes between normal cells and cancer cells [25]. The normal mammalian cell membrane outer leaflet is primarily composed of neutral zwitterionic phospholipids, whereas the cancer cell membrane consists of exclusive anionic constituents such as O-glycosylated mucins [26], phosphatidylserine [27], sialylated gangliosides [28], and heparan sulfate [29]. The presence of positively charged residues in AMPs enhances electrostatic binding to negative charges induced by anionic constituents in cancer cell membranes, causing changes in surface charge [30]. In agreement, Smp24 enhances the zeta potential of HepG2 cells, while its cellular internalization is suppressed in the presence of heparan sulfate (Figure 2A,C).

Antitumor AMPs generally possess a consistent membranolytic mode of action as pore formation, which cause cancer cell membrane disruption or permeation, leading to cell lysis and death [31,32]. In this study, evidence from LDH staining, SEM analysis, and calcein AM assay has revealed that Smp24 is likely a pore-forming peptide due to the occurrence of disruption and increasing permeability of cancer cell membranes (Figures 2 and 4). However, further studies are required to observe the pore formed by Smp24. Additionally, the responsibility of endocytosis for the uptake of Smp24 by HepG2 cells is also demonstrated, which is similar with melittin, a cationic AMP that enters cancer cells via the endocytosis-dependent pathway [33].

The membranolytic effect of AMPs is not only defined on the cell membrane, but also on the mitochondrial membrane, causing the decrease in its potential, altering the permeability of the mitochondrial membrane, increasing ROS production, and eventually mitochondrial dysfunction [5]. For instance, myristol-CM4, a new synthetic analog of AMP CM4, exerts its antitumor capacity against breast cancer cells by targeting mitochondria, which leads to a release of pro-apoptotic factors such as cytochrome C and induces mitochondria-dependent apoptosis [34]. Consistently, the damaged mitochondrial membrane caused by Smp24 was presented by the decline in its potential, and an increase in ROS accumulation, as well as calcein leakage from mitochondria (Figures 4 and 5). Consequently, Smp24 enhances apoptosis in HepG2 cells via activating the intrinsic mitochondrial apoptotic pathway (Figure 6). These data reveal that the antitumor mechanism of Smp24 is attributable to target mitochondria and induce mitochondria-mediated apoptosis.

The cytoskeleton is known to contribute to cancer progress by inducing cell proliferation and activating oncogenes, leading to tumorigenesis [35]. As an important component in cytoskeletal structures, the actin skeleton is involved in multiple mitochondrial functions, including mitochondrial dynamics, trafficking and autophagy, mitochondrial biogenesis, and metabolism [36]. Notably, the disruption of actin dynamics might cause the decrease in mitochondrial membrane potential, increase the ROS accumulation, induce the intrinsic apoptosis pathway, and finally lead to cell death [37]. In line with this, the reorganization of F-actin in the HepG2 cell cytoskeleton was observed in the presence of Smp24 (Figure 3). However, contrary to its effects on A549 cells [10,11], Smp24 could not inhibit the mobility of HepG2 cells in the wound-healing and transwell assays (data not shown). Therefore, there are some discrepancies in its mechanism against two cell lines and further research is necessary.

Mitochondria are crucial in cellular metabolism and their dysfunction might affect cell proliferation, causing cell cycle accumulation and apoptosis [38]. In agreement, mitochondria damage caused by Smp24 causes cell cycle suspension in the S and G2/M phases in HepG2 cells, as presented in flow cytometry analysis (Figure 7). p53 is a transcription factor which exerts a critical effect both in the G1/S and G2/M checkpoint of cell cycle and can be regulated by the transcriptional activation of p21$^{Waft1/Clip1}$ [39,40]. Concurrently, p21 $^{Waft1/Clip1}$ also causes G1/S phase cell cycle arrest by suppressing the kinase activity of CDK2/cyclin A, CDK2/cyclin E, and CDK1/cyclin A, while the inhibition of G2/M transition is related to the binding with CDK1/cyclin B1 [41]. Consistently, Western blotting reveals that treatment with Smp24 dose-dependently suppresses the expression of cyclin E and CDK2, but upregulates the expression of cyclin A, cyclin B, p21$^{Waft1/Clip1}$, and p53, inducing cell cycle accumulation at the S and G2/M phases (Figure 7).

Both autophagy and apoptosis are of pivotal importance in cellular homeostasis, and there is abundant evidence proving the co-regulation of these pathways. For example, overexpression of the autophagy effector Beclin 1 may lead to the release of Bak/Bax from Bcl-2 to enhance apoptosis, whereas an excessive Beclin 1-dependent autophagy occurs in the absence of Bcl-2 [42,43]. Several peptides have been identified to possess antitumor activity through both the autophagy and apoptosis pathways, such as CTLEW [44] and FK-16 [45]. Furthermore, the PI3K/Akt/mTOR pathways are regarded as promising targets for cancer therapy due to their important role in the process of cell proliferation, growth, survival, and mobility [46]. Consequently, multiple candidate anticancer agents suppress tumor growth by inhibiting PI3K/Akt/mTOR pathways and then inducing autophagy, as well as apoptosis of cancer cells [47,48]. In line with these, Smp24 suppressed the phosphorylation of PI3K/Akt/mTOR and increased the autophagic flux of LC3A/B-II/I, as well as the degradation of p62 (Figure 8A,B). It also is acknowledged that the MAPK signaling pathway is associated with the autophagy and apoptosis of cancer cells [22,23,49] and some compounds suppressing both the PI3K/Akt/mTOR and MAPK pathways induce autophagy and apoptosis of hepatocellular carcinoma cells [50]. Similarly, Smp24 suppresses the MAPK signaling pathways (Figure 8C,D). It is also noteworthy that Smp43, another AMP derived from *Scorpio Maurus palmatus*, and quercetin, a member of the flavonoid family, induce autophagy and apoptosis by activation of the MAPK signaling pathway [51,52]. The discrepancy in the mechanism might be associated with the cell response, which relies on the nature, strength, and duration of the MAPK pathways activated [53]. Thus, further research is needed to explore the detailed mechanism of how Smp24 induces both apoptosis and autophagy by inhibiting MAPK signaling pathways.

Despite the effectiveness and selectivity against tumor cells in vitro experiments, the use of peptide-based anticancer agents in clinical trials is still a challenge due to the degradation by proteases in vivo [54]. Our animal experiments revealed that 2 mg/kg Smp24 treatment leads to a significant decrease in tumor weight and volume, while it does not cause notable changes in body weight and important organs, which is consistent with the previous study [10,11], indicating the considerable stability and high selectivity of Smp24 against tumor tissues in vivo (Figure 9). Thus, Smp24 is an ideal antitumor candidate molecule.

4. Conclusions

In summary, the present study reveals the underlying mechanism of Smp24 against HepG2 cell proliferation. Smp24 enters HepG2 cells via pore formation and endocytosis, resulting in mitochondrial dysfunctions and membrane defects, consequently causing cell necrosis, cycle arrest, apoptosis, and autophagy. The anti-hepatoma activity is also verified in xenograft mice. Thus, our results provide evidence of the antitumor mechanisms of AMPs and suggest that Smp24 might be a promising candidate in hepatocellular carcinoma therapy.

5. Materials and Methods

5.1. Animals and Ethics Statement

Six-week-old BALB/c nude mice (18–20 g) were bought from the Laboratory Animal Center of Southern Medical University. Mice were randomly divided into the control and Smp24 (2 mg/kg)-treated groups and were separately housed in groups of five in a SPF mini-barrier system at the central animal facility of Southern Medical University. Animals lived under controlled 21 ± 2 °C room temperature, 60% humidity with a 12 h light-dark cycle. All experiments with animals were given the approval by the Animal Ethics Committee of Southern Medical University (Guangzhou, China) with ethical approval number: L2019226.

5.2. Chemicals and Cell Culture

Phosphate-buffered saline (PBS), fetal bovine serum (FBS), Dulbecco's modified Eagle's medium (DMEM), and trypsin were all bought from Gibco (Grand Island, NY, USA). HepG2 and LO2 cell lines were purchased from the American Type Culture Collection (Manassas, VA, USA). Cells were grown in RPMI-1640 medium including 1% penicillin-streptomycin and 10% FBS in a 37 °C incubator supplied with 5% CO_2. cytochrome C, Bax, Bcl-2, p53, p21, cleaved PARP, PARP, cleaved caspase-3, caspase-3, cleaved caspase-9, caspase-9, CDK2, cyclin A, cyclin B, cyclin E, p-JNK, JNK, p-Erk, Erk, p-p38, p38, p-Akt, Akt, p-mTOR, mTOR, p-FAK, FAK, p-PI3K, PI3K, LC3A/B, p62, GAPDH, β-actin, and all secondary antibodies were purchased from Cell Signaling Technology (Beverly, MA, USA). DAPI, cyclosporin A (CsA, an inhibitor of permeability transition), N-acetyl-L-cysteine (NAC), ROS assay kit, LDH-release assay kit, cell cycle and apoptosis analysis kit, and mitochondrial membrane potential assay kit for JC-1 were bought from Beyotime Institute of Biotechnology (Shanghai, China). Smp24 and FITC-labeled Smp24 were obtained as previously described by us [9].

5.3. Cell Viability and Proliferation Analysis

Effect of Smp24 on cellular viability was determined using MTT assays as previously described by us [10,11]. In brief, HepG2 and LO2 cells (1×10^4 cells/well) were exposed to a sequence of concentrations of Smp24 (1.25–20 μM) in 96-well plates for 12, 24, and 48 h, respectively. After that, cells were incubated with 10 μL MTT (5 mg/mL) at 37 °C for 4 h in the dark. The cell medium was subsequently discarded and replaced by 200 μL DMSO before the determination of optical density was conducted using a microplate reader (Tecan Company, Männedorf, Switzerland) at an absorbance of 490 nm. Considering that the MTT assay may be biased, with disruption of mitochondrial dehydrogenase system resulting in mitochondrial dysfunction or apoptosis, the cell proliferation was measured with the BeyoClick™ EdU cell proliferation kit with Alexa Fluor 488 (Beyotime Institute of Biotechnology, Shanghai, China) in accordance with the manufacturer's manual. Briefly, HepG2 cells (2×10^5 cells/well) were treated with a sequence of concentrations of Smp24 (0, 1.25, 2.5, 5, and 10 μM) in 6-well plates for 24 h. Subsequently, the cells were treated with EdU for 2 h, followed by staining with Alexa Fluor 488 under protection from light for 30 min. The fluorescence intensity was analyzed using flow cytometry (Becton Dickinson Company, Bedford, MA, USA) with 495 nm excitation and 519 nm mission wavelengths. The experiments were repeated in triplicate.

5.4. Peptide Internalization Measurement

Flow cytometry and fluorescence microscope (Axio Observer, Zeiss, Oberkochen, Germany) were used to measure the internalization of FITC-labeled Smp24. In brief, HepG2 cells (1×10^5 cells/well) were grown overnight on 24-well plate and then co-incubated with 2.5, 5, and 10 μM of FITC-labeled Smp24 for 1 h and 6 h at 37 °C. After washing with PBS, the fluorescence intensity was determined with flow cytometry. For the location analysis of Smp24 in cells, after being treated with 5 μM FITC-labeled Smp24 at 37 °C for different times (6, 12, and 24 h), the cells were washed with PBS, fixed with 4% paraformaldehyde (PFA) for 30 min, stained with DAPI for 10 min, and finally observed under fluorescence microscope at 400× magnification. Approximately three single-plane images each well were captured.

To ascertain whether heparan sulfate affects the internalization of Smp24, FITC-labeled Smp24 at 5 μM was mixed with heparan sulfate at 5, 10, or 20 μg/mL in RPMI-1640 medium for 30 min, and then were incubated together with HepG2 cells at the density of 1×10^5 cells/well for 1 h. The cell fluorescence intensity was measured with flow cytometry.

To ascertain whether the internalization of Smp24 depends on the cellular energy state, HepG2 cells were pre-incubated for 30 min at 37 °C and 4 °C, and then with 5 μM FITC-labeled Smp24 for another 1 h. After that, the cells were washed with PBS and measured with flow cytometry. In another set of experiments to further confirm the effects of the cellular energy state on its internalization, HepG2 cells were pre-incubated with 50 mM NH_4Cl for 30 min before being treated with 5 μM FITC-labeled Smp24 for another 1 h and then measured with the flow cytometry. All experiments were repeated at least three times.

5.5. Membrane Integrity Assay

Calcein AM staining was conducted to evaluate the membrane integrity of HepG2 cells in accordance with the manufacturer's manual (Beyotime Institute of Biotechnology, Shanghai, China). In short, HepG2 cells (1×10^5 cells/well) were grown overnight in a 12-well plate and subsequently incubated with a sequence of concentrations of Smp24 for 24 h. Cells were then digested by trypsin, washed with PBS for three times, and treated with 1 μM calcein AM for 30 min at 37 °C. The cells were finally analyzed with flow cytometry to detect the levels of calcein AM after being washed again with PBS. In another set of experiments, cells were pre-incubated with 1 μM CsA before treatment with calcein AM or were incubated with $CoCl_2$ after treatment with calcein AM to determine the integrity of mitochondrial membrane. Zeta potential analysis was carried out to further confirm the membrane integrity. In short, 1×10^5 HepG2 cells were re-suspended in PBS and mixed with Smp24 (0, 2.5, 5, 10 μM) for 10 min at room temperature. Folded capillary cell (DTS1070) and Zetasizer system (Nano ZS; Malvern Instruments Ltd., Worcestershire, UK) were applied to measure zeta (ξ) potential of the aboveHepG2 cells. All experiments were repeated at least three times.

5.6. Cell Morphology Observation

In the present study, 2×10^5 HepG2 cells were cultivated on 6-well plate overnight and then exposed to a sequence of concentrations of Smp24 (0–20 μM) for 24 h. The cellular morphology was then observed with an inverted phase contrast microscope (100× magnification). Approximately 3 images each well were captured.

5.7. LDH-Release Assay

The LDH assay was accomplished in accordance with the manufacturer's manual. Briefly, after incubation with a gradient of concentrations of Smp24 (0–20 μM) for 12, 24, and 48 h, respectively, each well containing HepG2 cells was supplemented with 10 μL of LDH-release solution and incubated for 1 h. After being added to a new plate, the supernatants in each well were then incubated with 60 μL of substrate solution in the dark for 30 min. A microplate reader (Tecan Company, Männedorf, Switzerland) was used to determine the absorbance value of the mixture at 490 nm. The LDH release rate (%) = (absorbance of sample − absorbance of control)/(absorbance of max LDH activity) × 100. All experiments were carried out at least three times.

5.8. Scanning Electron Microscopy Determination

After 24 h of growth in a 12-well plate on the glass coverslips, HepG2 cells (1.2×10^5 cells/well) were co-cultured with a gradient of concentrations of Smp24 (0–10 μM) for another

24 h with PBS as a negative control. Subsequently, the cells were fixed with 4% glutaric dialdehyde and 2.5% glutaric dialdehyde at 22 °C for 2 h and 4 °C for 8 h, respectively. A series of gradient ethanol/water solutions was then used to dehydrate the samples. After that, cells were coated by gold and observed with Phenom ProX instrument (Phenom World, Eindhoven, The Netherlands) at 15 kV. The experiments were detected in triplicate.

5.9. Fluorescence Microscopy Analysis

Fluorescence microscopy analysis was used to observe the F-actin reorganization, intracellular ROS accumulation, and mitochondrial membrane potential. To observe the changes in F-actin, after overnight incubation in 24-well plates, HepG2 cells (7×10^4 cells/well) were cultivated with Smp24 at the concentration of 0–10 μM for another 24 h. After that, cells were fixed with 4% PFA, dyed with rhodamine–phalloidin for 30 min, washed with PBS for three times, and dyed again with DAPI for another 10 min. Subsequently, fluorescence microscope at $400\times$ magnification was used to observe the changes in F-actin.

To determine intracellular ROS levels, HepG2 cells were co-cultured with Smp24 (0, 2.5, 5, and 10 μM) for 12 h and then stained with 10 μM 2,7-dichlorodihydro-fluoresceindiacetate (DCFH-DA) for 30 min at 37 °C under protection from light. The supernatant was removed, and cells were washed three times with serum-free cell culture medium to adequately remove DCFH-DA that did not enter the cells. An inverted fluorescence microscopy at $200\times$ magnification was applied to clarify the changes in cells. NAC-treated cells were considered as a positive control.

To detect the impact of Smp24 on the mitochondrial membrane potential, HepG2 cells were co-cultured with a gradient of concentrations of Smp24 for 12 h and then were dyed with JC-1 at 37 °C for 30 min in accordance with the kit instruction manual. The supernatant was removed, and cells were washed twice with JC-1 staining buffer ($1\times$) and supplemented with 2 mL of cell culture medium, which may contain serum and phenol red. The changing mitochondrial membrane potential was observed using fluorescence microscopy at $400\times$ magnification, with NAC as a positive control. Approximately 3 random images each well were obtained.

5.10. Cell Cycle and Apoptosis Measurement

To define whether Smp24 affects cell cycle and apoptosis in HepG2 cells, flow cytometry was used. HepG2 cells (2×10^5 cells/well) were seeded in 6-well plates overnight and then co-cultured with a gradient of concentrations of Smp24 or 2 mM NAC for 24 h. After that, the cells were harvested with trypsin digestion, washed with cold PBS three times, fixed with 70% ethanol at 4 °C for 12 h, and finally dyed with PI for 30 min at 37 °C to measure cell cycle. For the apoptosis measurement, the cells were dyed with annexin V-FITC/PI instead of only PI at 22 °C for 15 min before being analyzed with flow cytometry. All experiments were conducted in triplicate.

5.11. Western Blot Analysis

HepG2 cells (2×10^5 cells/well) were incubated with Smp24 (0, 2.5, 5, and 10 μM) in 6-well plates for 24 h. After that, the cells were harvested by trypsin digestion and then lysed with RIPA lysis solution including 1% protease and phosphatase inhibitors (FDbio Science Biotech Co., Ltd., Hangzhou, China) at 4 °C for 15 min. Subsequently, the supernatant was separated by SDS-PAGE and then transferred onto PVDF membrane (Millipore, Billerica, MA, USA). After being incubated with right primary antibodies (1:1000) for 12 h at 4 °C and then with corresponding secondary antibodies (1:2000) for 1 h at 25 °C, a hypersensitive ECL chemiluminescence agent was used to visualize blot bands which were analyzed with Image J software (64-bit, National Institutes of Health, MD, USA) (n = 3 replicates).

5.12. Animal Experiments

The right flank of each mouse was subcutaneously injected with HepG2 cells (5×10^6 cells/mouse) with viability over 90%. The changes in tumor size were examined every day by palpation and determined using caliper in double planes. The volume of tumors was calculated according to the following equation: volume (mm^3) = (smallest diameter)^2 × (largest diameter)/2. Once the volume of tumors increased to approximately 50 mm^3, physiological saline or Smp24 (2 mg/kg) was injected once every three days for a total of six times near the tumor site. Including the tumor volumes, the body weight of mice was also recorded daily. At the end of experiment, the tumors and important organs including liver, heart, lung, spleen, and kidney were separated, weighed, and photographed. After being embedded in paraffin, HE staining reagent was applied to assess the morphological changes of tumor cells in vivo. To evaluate the cell apoptosis levels of tumor tissue, immunohistochemical staining was carried out to examine the contents of cleaved caspase-3 in tumor tissues. The results were obtained from five mice each group.

5.13. Statistical Analysis

All data were expressed as mean ± SEM and analyzed using GraphPad Prism 5.0 (GraphPad Software, Inc., La Jolla, CA, USA) by one-way ANOVA with Bonferroni's multiple comparison. Statistical significances were shown as * $p < 0.05$, ** $p < 0.01$, and *** $p < 0.001$.

Author Contributions: Data curation, T.N., R.G., J.C., J.W. and J.L.; Formal analysis, T.N., R.G., J.C., J.W. and J.L.; Project administration, H.X. and X.X.; Resources, X.C. and M.A.A.-R.; Supervision, H.X. and X.X.; Validation, H.X. and X.X.; Writing—original draft, T.N., R.G., J.C., J.W. and J.L.; Writing— review and editing, M.A.A.-R., H.X. and X.X. All authors have read and agreed to the published version of the manuscript.

Funding: This research was funded by the Chinese National Natural Science Foundation (31861143050, 31772476, 31911530077, 82070038) and in part by the Academy of Scientific Research and Technology (ASRT, Egypt; China–Egypt Scientific and Technological Cooperation Program).

Institutional Review Board Statement: All protocols and procedures involving live animals were approved by the Animal Care and Use Ethics of Southern Medical University (ethical approval date: 12 November 2019).

Informed Consent Statement: Not applicable.

Data Availability Statement: The datasets used and analyzed during the current study are available from the corresponding author on reasonable request.

Conflicts of Interest: The authors declare no conflict of interest.

References

1. Li, B.; Lyu, P.; Xi, X.; Ge, L.; Mahadevappa, R.; Shaw, C.; Kwok, H.F. Triggering of cancer cell cycle arrest by a novel scorpion venom-derived peptide-Gonearrestide. *J. Cell. Mol. Med.* **2018**, *22*, 4460–4473. [CrossRef]
2. Shen, B.; Cao, Z.; Li, W.; Sabatier, J.M.; Wu, Y. Treating autoimmune disorders with venom-derived peptides. *Expert Opin. Biol. Ther.* **2017**, *17*, 1065–1075. [CrossRef] [PubMed]
3. Almaaytah, A.; Albalas, Q. Scorpion venom peptides with no disulfide bridges: A review. *Peptides* **2014**, *51*, 35–45. [CrossRef] [PubMed]
4. Hmed, B.; Serriayt, H.T.; Mounir, Z.K. Scorpion peptides: Potential use for new drug development. *J. Toxicol.* **2013**, *2013*, 958797. [CrossRef] [PubMed]
5. Tornesello, A.L.; Borrelli, A.; Buonaguro, L.; Buonaguro, F.M.; Tornesello, M.L. Antimicrobial Peptides as Anticancer Agents: Functional Properties and Biological Activities. *Molecules* **2020**, *25*, 2850. [CrossRef]
6. Kuo, H.M.; Tseng, C.C.; Chen, N.F.; Tai, M.H.; Hung, H.C.; Feng, C.W.; Cheng, S.Y.; Huang, S.Y.; Jean, Y.H.; Wen, Z.H. MSP-4, an Antimicrobial Peptide, Induces Apoptosis via Activation of Extrinsic Fas/FasL- and Intrinsic Mitochondria-Mediated Pathways in One Osteosarcoma Cell Line. *Mar. Drugs* **2018**, *16*, 8. [CrossRef]
7. Liu, S.; Aweya, J.J.; Zheng, L.; Zheng, Z.; Huang, H.; Wang, F.; Yao, D.; Ou, T.; Zhang, Y. LvHemB1, a novel cationic antimicrobial peptide derived from the hemocyanin of *Litopenaeus vannamei*, induces cancer cell death by targeting mitochondrial voltage-dependent anion channel 1. *Cell Biol. Toxicol.* **2022**, *38*, 87–110. [CrossRef]
8. Harrison, P.L.; Abdel-Rahman, M.A.; Strong, P.N.; Tawfik, M.M.; Miller, K. Characterisation of three alpha-helical antimicrobial peptides from the venom of *Scorpio maurus palmatus*. *Toxicon* **2016**, *117*, 30–36. [CrossRef]
9. Elrayess, R.A.; Mohallal, M.E.; Mobarak, Y.M.; Ebaid, H.M.; Haywood-Small, S.; Miller, K.; Strong, P.N.; Abdel-Rahman, M.A. Scorpion Venom Antimicrobial Peptides Induce Caspase-1 Dependant Pyroptotic Cell Death. *Front. Pharmacol.* **2021**, *12*, 788874. [CrossRef]

10. Guo, R.; Chen, X.; Nguyen, T.; Chai, J.; Gao, Y.; Wu, J.; Li, J.; Abdel-Rahman, M.A.; Chen, X.; Xu, X. The Strong Anti-Tumor Effect of Smp24 in Lung Adenocarcinoma A549 Cells Depends on Its Induction of Mitochondrial Dysfunctions and ROS Accumulation. *Toxins* **2022**, *14*, 590. [CrossRef]

11. Guo, R.; Liu, J.; Chai, J.; Gao, Y.; Abdel-Rahman, M.A.; Xu, X. Scorpion Peptide Smp24 Exhibits a Potent Antitumor Effect on Human Lung Cancer Cells by Damaging the Membrane and Cytoskeleton In Vivo and In Vitro. *Toxins* **2022**, *14*, 483. [CrossRef] [PubMed]

12. Poon, G.M.; Gariépy, J. Cell-surface proteoglycans as molecular portals for cationic peptide and polymer entry into cells. *Biochem. Soc. Trans.* **2007**, *35*, 788–793. [CrossRef] [PubMed]

13. Doherty, G.J.; McMahon, H.T. Mediation, modulation, and consequences of membrane-cytoskeleton interactions. *Ann. Rev. Biophys.* **2008**, *37*, 65–95. [CrossRef] [PubMed]

14. Reddy, D.; Kumavath, R.; Ghosh, P.; Barh, D. Mitochondria as a therapeutic target for aging and neurodegenerative diseases. *Curr. Alzheimer Res.* **2011**, *8*, 393–409. [CrossRef]

15. Chipuk, J.E.; Bouchier-Hayes, L.; Green, D.R. Mitochondrial outer membrane permeabilization during apoptosis: The innocent bystander scenario. *Cell Death Differ.* **2006**, *13*, 1396–1402. [CrossRef]

16. Jiang, S.; Cai, J.; Wallace, D.C.; Jones, D.P. Cytochrome c-mediated apoptosis in cells lacking mitochondrial DNA. Signaling pathway involving release and caspase 3 activation is conserved. *J. Biol. Chem.* **1999**, *274*, 29905–29911. [CrossRef]

17. Xiong, S.; Mu, T.; Wang, G.; Jiang, X. Mitochondria-mediated apoptosis in mammals. *Protein Cell* **2014**, *5*, 737–749. [CrossRef]

18. Hao, X.; Bu, W.; Lv, G.; Xu, L.; Hou, D.; Wang, J.; Liu, X.; Yang, T.; Zhang, X.; Liu, Q.; et al. Disrupted mitochondrial homeostasis coupled with mitotic arrest generates antineoplastic oxidative stress. *Oncogene* **2022**, *41*, 427–443. [CrossRef]

19. Mulcahy, L.J.; Thorburn, A. Autophagy in cancer: Moving from understanding mechanism to improving therapy responses in patients. *Cell Death Differ* **2020**, *27*, 843–857. [CrossRef]

20. Su, M.; Mei, Y.; Sinha, S. Role of the Crosstalk between Autophagy and Apoptosis in Cancer. *J. Oncol.* **2013**, *2013*, 102735. [CrossRef]

21. Koukourakis, M.I.; Kalamida, D.; Giatromanolaki, A.; Zois, C.E.; Sivridis, E.; Pouliliou, S.; Mitrakas, A.; Gatter, K.C.; Harris, A.L. Autophagosome Proteins LC3A, LC3B and LC3C Have Distinct Subcellular Distribution Kinetics and Expression in Cancer Cell Lines. *PLoS ONE* **2015**, *10*, e0137675. [CrossRef] [PubMed]

22. Chen, C.; Gao, H.; Su, X. Autophagy-related signaling pathways are involved in cancer. *Exp. Ther. Med.* **2021**, *22*, 710. [CrossRef] [PubMed]

23. Santarpia, L.; Lippman, S.M.; El-Naggar, A.K. Targeting the MAPK-RAS-RAF signaling pathway in cancer therapy. *Expert Opin. Ther. Targets* **2012**, *16*, 103–119. [CrossRef]

24. Kunda, N.K. Antimicrobial peptides as novel therapeutics for non-small cell lung cancer. *Drug Discov. Today* **2020**, *25*, 238–247. [CrossRef] [PubMed]

25. Schweizer, F. Cationic amphiphilic peptides with cancer-selective toxicity. *Eur. J. Pharmacol.* **2009**, *62*, 190–194. [CrossRef]

26. Yoon, W.H.; Park, H.D.; Lim, K.; Hwang, B.D. Effect of O-glycosylated mucin on invasion and metastasis of HM7 human colon cancer cells. *Biochem. Biophys. Res. Commun.* **1996**, *222*, 694–699. [CrossRef] [PubMed]

27. Wang, C.; Chen, Y.W.; Zhang, L.; Gong, X.G.; Zhou, Y.; Shang, D.J. Melanoma cell surface-expressed phosphatidylserine as a therapeutic target for cationic anticancer peptide, temporin-1CEa. *J. Drug Target.* **2016**, *24*, 548–556. [CrossRef]

28. Lee, H.S.; Park, C.B.; Kim, J.M.; Jang, S.A.; Park, I.Y.; Kim, M.S.; Cho, J.H.; Kim, S.C. Mechanism of anticancer activity of buforin IIb, a histone H2A-derived peptide. *Cancer Lett.* **2008**, *271*, 47–55. [CrossRef]

29. Fadnes, B.; Rekdal, O.; Uhlin-Hansen, L. The anticancer activity of lytic peptides is inhibited by heparan sulfate on the surface of the tumor cells. *BMC Cancer* **2009**, *9*, 183. [CrossRef]

30. Hoskin, D.W.; Ramamoorthy, A. Studies on anticancer activities of antimicrobial peptides. *Biochim. Biophys. Acta* **2008**, *1778*, 357–375. [CrossRef]

31. Gaspar, D.; Veiga, A.S.; Castanho, M.A. From antimicrobial to anticancer peptides. A review. *Front. Microbiol.* **2013**, *4*, 294. [CrossRef] [PubMed]

32. Benfield, A.H.; Henriques, S.T. Mode-of-Action of Antimicrobial Peptides: Membrane Disruption vs. Intracellular Mechanisms. *Front. Med. Technol.* **2020**, *2*, 610997. [CrossRef] [PubMed]

33. Kohno, M.; Horibe, T.; Ohara, K.; Ito, S.; Kawakami, K. The membrane-lytic peptides K8L9 and melittin enter cancer cells via receptor endocytosis following subcytotoxic exposure. *Chem. Biol.* **2014**, *21*, 1522–1532. [CrossRef] [PubMed]

34. Li, C.; Liu, H.; Yang, Y.; Xu, X.; Lv, T.; Zhang, H.; Liu, K.; Zhang, S.; Chen, Y. N-myristoylation of Antimicrobial Peptide CM4 Enhances Its Anticancer Activity by Interacting with Cell Membrane and Targeting Mitochondria in Breast Cancer Cells. *Front. Pharmacol.* **2018**, *9*, 1297. [CrossRef]

35. Gaspar, P.; Holder, M.V.; Aerne, B.L.; Janody, F.; Tapon, N. Zyxin antagonizes the FERM protein expanded to couple F-actin and Yorkie-dependent organ growth. *Curr. Biol.* **2015**, *25*, 679–689. [CrossRef]

36. Illescas, M.; Peñas, A.; Arenas, J.; Martín, M.A.; Ugalde, C. Regulation of Mitochondrial Function by the Actin Cytoskeleton. *Front. Cell Dev. Biol.* **2021**, *9*, 95838. [CrossRef]

37. Gourlay, C.W.; Ayscough, K.R. The actin cytoskeleton: A key regulator of apoptosis and ageing? *Nat. Rev. Mol. Cell Biol.* **2005**, *6*, 583–589. [CrossRef]

38. Antico, A.V.; Elguero, M.E.; Poderoso, J.J.; Carreras, M.C. Mitochondrial regulation of cell cycle and proliferation. *Antioxid. Redox Signal.* **2012**, *16*, 1150–1180. [CrossRef]

39. El-Deiry, W.S.; Tokino, T.; Velculescu, V.E.; Levy, D.B.; Parsons, R.; Trent, J.M.; Lin, D.; Mercer, W.E.; Kinzler, K.W.; Vogelstein, B. WAF1, a potential mediator of p53 tumor suppression. *Cell* **1993**, *75*, 817–825. [CrossRef]
40. Martín-Caballero, J.; Flores, J.M.; García-Palencia, P.; Serrano, M. Tumor susceptibility of p21(Waf1/Cip1)-deficient mice. *Cancer Res.* **2001**, *61*, 6234–6238.
41. Abbas, T.; Dutta, A. p21 in cancer: Intricate networks and multiple activities. *Nat. Rev. Cancer* **2009**, *9*, 400–414. [CrossRef] [PubMed]
42. Sun, Y.; Liu, J.H.; Jin, L.; Lin, S.M.; Yang, Y.; Sui, Y.X.; Shi, H. Over-expression of the Beclin1 gene upregulates chemosensity to anti-cancer drugs by enhancing therapy-induced apoptosis in cervix squamous carcinoma CaSki cells. *Cancer Lett.* **2010**, *294*, 204–210. [CrossRef] [PubMed]
43. Akar, U.; Chaves-Reyez, A.; Barria, M.; Tari, A.; Sanguino, A.; Kondo, Y.; Kondo, S.; Arun, B.; Lopez-Berestein, G.; Ozpolat, B. Silencing of Bcl-2 expression by small interfering RNA induces autophagic cell death in MCF-7 breast cancer cells. *Autophagy* **2008**, *4*, 669–679. [CrossRef] [PubMed]
44. Ma, S.; Huang, D.; Zhai, M.; Yang, L.; Peng, S.; Chen, C.; Feng, X.; Weng, Q.; Zhang, B.; Xu, M. Isolation of a novel bio-peptide from walnut residual protein inducing apoptosis and autophagy on cancer cells. *BMC Complement. Altern. Med.* **2015**, *15*, 413. [CrossRef]
45. Ren, S.X.; Shen, J.; Cheng, A.S.; Lu, L.; Chan, R.L.; Li, Z.J.; Wang, X.J.; Wong, C.C.; Zhang, L.; Ng, S.S.; et al. FK-16 derived from the anticancer peptide LL-37 induces caspase-independent apoptosis and autophagic cell death in colon cancer cells. *PLoS ONE* **2013**, *8*, e63641. [CrossRef] [PubMed]
46. Alzahrani, A.S. PI3K/Akt/mTOR inhibitors in cancer: At the bench and bedside. *Semin. Cancer Biol.* **2019**, *59*, 125–132. [CrossRef]
47. Yang, J.; Pi, C.; Wang, G. Inhibition of PI3K/Akt/mTOR pathway by apigenin induces apoptosis and autophagy in hepatocellular carcinoma cells. *Biomed. Pharmacother.* **2018**, *103*, 699–707. [CrossRef]
48. Zhou, L.; Li, S.; Sun, J. Ginkgolic acid induces apoptosis and autophagy of endometrial carcinoma cells via inhibiting PI3K/Akt/mTOR pathway in vivo and in vitro. *Hum. Exp. Toxicol.* **2021**, *40*, 2156–2164. [CrossRef]
49. Su, Z.; Yang, Z.; Xu, Y.; Chen, Y.; Yu, Q. Apoptosis, autophagy, necroptosis, and cancer metastasis. *Mol. Cancer* **2015**, *14*, 48. [CrossRef]
50. Reddy, D.; Kumavath, R.; Ghosh, P.; Barh, D. Lanatoside C Induces G2/M Cell Cycle Arrest and Suppresses Cancer Cell Growth by Attenuating MAPK, Wnt, JAK-STAT, and PI3K/AKT/mTOR Signaling Pathways. *Biomolecules* **2019**, *9*, 792. [CrossRef]
51. Chai, J.; Yang, W.; Gao, Y.; Guo, R.; Peng, Q.; Abdel-Rahman, M.A.; Xu, X. Antitumor Effects of Scorpion Peptide Smp43 through Mitochondrial Dysfunction and Membrane Disruption on Hepatocellular Carcinoma. *J. Nat. Prod.* **2021**, *84*, 3147–3160. [CrossRef] [PubMed]
52. Ji, Y.; Li, L.; Ma, Y.X.; Li, W.T.; Li, L.; Zhu, H.Z.; Wu, M.H.; Zhou, J.R. Quercetin inhibits growth of hepatocellular carcinoma by apoptosis induction in part via autophagy stimulation in mice. *J. Nutr. Biochem.* **2019**, *69*, 108–119. [CrossRef] [PubMed]
53. Marshall, C.J. Specificity of receptor tyrosine kinase signaling: Transient versus sustained extracellular signal-regulated kinase activation. *Cell* **1995**, *80*, 179–185. [CrossRef]
54. Li, C.M.; Haratipour, P.; Lingeman, R.G.; Perry, J.; Gu, L.; Hickey, R.J.; Malkas, L.H. Novel Peptide Therapeutic Approaches for Cancer Treatment. *Cells* **2021**, *10*, 2908. [CrossRef]

Article

The Strong Anti-Tumor Effect of Smp24 in Lung Adenocarcinoma A549 Cells Depends on Its Induction of Mitochondrial Dysfunctions and ROS Accumulation

Ruiyin Guo [1,2,†], Xuewen Chen [1,†], Tienthanh Nguyen [1,2], Jinwei Chai [2], Yahua Gao [2], Jiena Wu [2], Jinqiao Li [2], Mohamed A. Abdel-Rahman [3], Xin Chen [1,*] and Xueqing Xu [2,*]

[1] Department of Pulmonary and Critical Care Medicine, Zhujiang Hospital, Southern Medical University, Guangzhou 510280, China
[2] Guangdong Provincial Key Laboratory of New Drug Screening, School of Pharmaceutical Sciences, Southern Medical University, Guangzhou 510515, China
[3] Zoology Department, Faculty of Science, Suez Canal University, Ismailia 41522, Egypt
* Correspondence: chen_xin1020@163.com (X.C.); xu2003@smu.edu.cn (X.X.); Tel.: +86-20-62782296 (X.C.); +86-20-61648537 (X.X.); Fax: +86-20-61643010 (X.C.); +86-20-61648655 (X.X.)
† These authors contributed equally to this work.

Abstract: Non-small cell lung cancer (NSCLC) is the leading cause of death in lung cancer due to its aggressiveness and rapid migration. The potent antitumor effect of Smp24, an antimicrobial peptide derived from Egyptian scorpion *Scorpio maurus palmatus* via damaging the membrane and cytoskeleton have been reported earlier. However, its effects on mitochondrial functions and ROS accumulation in human lung cancer cells remain unknown. In the current study, we discovered that Smp24 can interact with the cell membrane and be internalized into A549 cells via endocytosis, followed by targeting mitochondria and affect mitochondrial function, which significantly causes ROS overproduction, altering mitochondrial membrane potential and the expression of cell cycle distribution-related proteins, mitochondrial apoptotic pathway, MAPK, as well as PI3K/Akt/mTOR/FAK signaling pathways. In summary, the antitumor effect of Smp24 against A549 cells is related to the induction of apoptosis, autophagy plus cell cycle arrest via mitochondrial dysfunction, and ROS accumulation. Accordingly, our findings shed light on the anticancer mechanism of Smp24, which may contribute to its further development as a potential agent in the treatment of lung cancer cells.

Keywords: scorpion venom; antimicrobial peptides; *Scorpio maurus palmatus*; Smp24; A549; apoptosis; autophagy; necrosis; cell cycle arrest

Key Contribution: Smp24 significantly exerts an antitumor effect via induction of apoptosis, autophagy, necrosis plus cell cycle arrest due to its induction of mitochondrial dysfunction and reactive oxygen species overproduction.

Citation: Guo, R.; Chen, X.; Nguyen, T.; Chai, J.; Gao, Y.; Wu, J.; Li, J.; Abdel-Rahman, M.A.; Chen, X.; Xu, X. The Strong Anti-Tumor Effect of Smp24 in Lung Adenocarcinoma A549 Cells Depends on Its Induction of Mitochondrial Dysfunctions and ROS Accumulation. *Toxins* **2022**, *14*, 590. https://doi.org/10.3390/toxins14090590

Received: 27 July 2022
Accepted: 25 August 2022
Published: 27 August 2022

1. Introduction

Despite the effectiveness of traditional therapy in cancer treatment, such as chemotherapy, radiotherapy, and immune therapy, lung adenocarcinoma remains the major cause of cancer death due to its aggressiveness and rapid migration [1]. Besides, the cytotoxicity and drug resistance caused by those traditional therapy lead to the urgent need of alternative treatment with low cytotoxicity and treatment resistance.

In addition to the importance in the innate immune defense, antimicrobial peptides (AMPs) have emerged as promising drug candidates against multiple diseases, especially in anticancer therapy. Most of AMPs suppress the tumor cells by their direct interaction with the cell membrane rather than specific receptors, which prevents the drug resistance caused by other traditional treatment methods [2,3]. Notably, in addition to the membrane disruption of cancer cells, AMPs have been proven to exert an antitumor effect via targeting different cellular structures, interfering intrinsic pathways, which consequently results in abnormal changes in multiple events, such as apoptosis, autophagy, and cell cycle distribution [2,4]. Mitochondria are double-membrane-bound cell organelles found in most eukaryotic organisms that generate most of the chemical energy adenosine triphosphate (ATP). In addition to the production of ATP, the mitochondria have a major role in calcium ion storage and cellular proliferation regulation, such as apoptosis and cell cycle distribution regulation, making it a promising target in cancer treatments. There are abundant AMPs that have been reported to target mitochondria, leading to its dysfunction, which consequently alters the mitochondrial membrane potential and ROS production, inducing cancer cell apoptosis and cell cycle arrest. For instance, melittin, an amphiphilic alpha-helical peptide derived from honeybee venom (*Apis mellifera*), suppresses the proliferation of human gastric cancer cells via activating the mitochondrial pathway, thereby inducing the apoptosis process [5]. Previous study revealed the dedication of both necrosis and apoptosis in the antitumor mechanism of Brevinin-1RL1 by inducing extrinsic and mitochondria intrinsic apoptosis [6]. Moreover, the apoptosis of osteosarcoma MG63 cells is induced by MSP-4 via activation of Fas/FasL—and mitochondria-mediated pathway [7]. Interestingly, both pardaxin, an AMP isolated from *Pardachirus marmoratus*, and our previously reported peptide Smp43 share a similar mode of action in antitumor via inducing apoptosis and cell cycle arrest after disruption of mitochondrial membrane, both Smp43 and pardaxin also induce autophagy of hepatoma cells and ovarian cancer cells, respectively [8,9]. In previous study, we demonstrated the potent antitumor effect of Smp24 (IWSFLIKAATKLLP-SLFGGGKKDS), another venom-derived AMP of *Scorpio maurus palmatus*, toward human non-small-cell lung cancer cell (NSCLC) A549. Notably, Smp24 caused mitochondrial damage which was represented by the release of calcein AM in flow cytometry analysis [10]. However, whether Smp24 has the above effects in human non-small-cell lung cancer cell remain elusive. In the current study, we investigated the interaction between Smp24 and mitochondria as well as its effect on apoptosis, cell cycle distribution, and autophagy. Our findings revealed that treatment with Smp24 led to the disruption of the mitochondrial membrane and accumulation of ROS, inducing apoptosis, cell cycle arrest, and autophagy of A549 cells. This discovery exposes the antitumor mechanism of Smp24 against NSCLC A549, suggesting its application in lung carcinoma therapy.

2. Results

2.1. Smp24 Suppresses the Proliferation of Human Lung Cancer Cells

As shown in Figure 1A, consistent with the previous report [10], Smp24 significantly suppressed the proliferation of A549 with the IC_{50} value at approximately 4.06 µM, and it exhibited less inhibitory potency toward normal cells, MRC-5, as evidenced from higher IC_{50} of approximately 14.68 ± 0.79 µM. Furthermore, a concentration- and time-dependent cytotoxicity toward A549 cells was induced by Smp24. Furthermore, the survival rate of Smp24-treated A549 cells was significantly increased by the incubation with various tested inhibitors (Figure 1B) in comparison with the control group without inhibitor treatment. The above findings indicated cytotoxicity and cytostatic effect of Smp24 against A549 cells.

2.2. Smp24 Is Internalized into A549 Cells via Endocytosis

Smp24 has been reported to be a membrane lytic peptide [11,12] and the changes in cell surface electrostatics under different conditions reflect cellular phenomena, such as adhesion and interaction with peptides [13]. Hence, we examined the interaction between tumor cells and Smp24 by measuring the membrane potential. After incubation of Smp24 (0, 1.25, 2.5, and 5 µM) with A549 cells, an increase in zeta potential from −12.88 to −9.61 mV was observed (Figure 2A), reflecting that the interaction of cationic Smp24 peptide caused an increase in the net charge of the cell membrane surfaces. Furthermore, we investigated whether Smp24 could be internalized into A549 cells. As presented in Figure 2B, the increasing fluorescence in the cells treated with FITC-labeled Smp24 indicated the internalization of Smp24 into A549 cells in a concentration- and a time-dependent manner, which is consistent with the cell-penetrating property reported in our previous study [10]. Anionic heparan sulfate is an essential component of the cancer cell membrane and the extracellular matrix, which contributes a crucial role in the interaction between the cell membrane and a cell-penetrating peptide [14]. Hence, we further investigated whether the cellular endocytosis of Smp24 was affected by heparan sulfate. As presented in Figure 2C, the decrease in internalization of 5 μM FITC-labeled Smp24 was approximately 1.43%, 46.12%, and 68.82% after 1 h pretreatment with 5, 10, and 20 μg/mL heparan sulfate, respectively, while compared with the control group. Thereafter, we examined the role of energy in the translocation of Smp24. As the common endocytic inhibitors, ammonium chloride alters the pH of acidic endocytic vesicles, while sodium azide can directly eliminate ATP production within the cell membrane [15]. As a result, pretreatment of A549 cells with 40 μM NaN_3 and 50 mM NH_4Cl for 1 h obviously inhibited the cellular uptake of Smp24 (Figure 2D,E). As a further confirmation assay, we tested the effect of temperature on the cellular endocytosis (Figure 2F). After 1 h of incubation at 37 °C, the cellular uptake of FITC-labeled Smp24 was approximately 42.89% higher than that at 4 °C, suggesting an energy-dependent and thermo-sensitive endocytosis of Smp24 across A549 cell membrane. Notably, compared to the results of 1 h incubation, more Smp24 was internalized into cells after co-incubation for 6 h, as evidenced by the increasing fluorescence (Figure S1A). What's more, pre-incubation of A549 with heparan sulfate and endocytic inhibitors for 6 h could not decrease the cellular uptake of Smp24 (Figure S1B–D). Furthermore, the cellular uptake of Smp24 at 4 °C and 37 °C after 6 h of incubation was approximately 6.51% and 96.07%, respectively (Figure S1E).

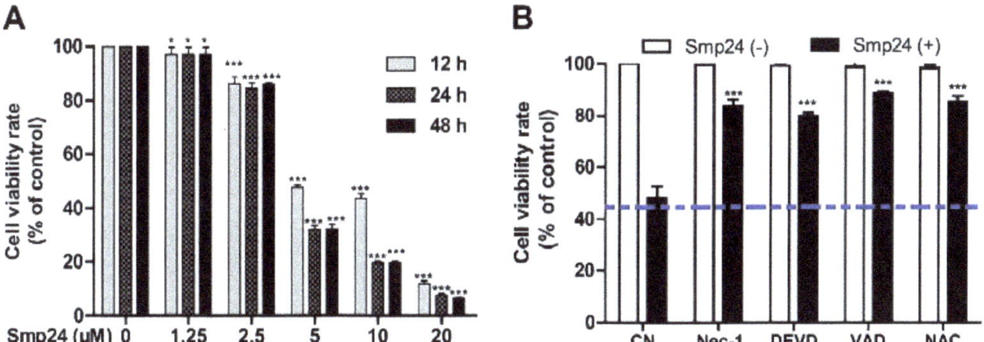

Figure 1. Effect of Smp24 on proliferation of A549 cells. (**A**) Viability of A549 cells treated with different concentrations of Smp24 for 12, 24, and 48 h. Negative control: cells treated with equivalent solvent for corresponding time. (**B**) Effects of inhibitors on the viability of Smp24-treated A549 cells. A549 cells were treated with 5 μM Smp24 for 12 h after being pre-incubated with the inhibitors 40 μM necrostatin-1 (Nec-1), 40 μM Z-DEVD-FMK (DEVD), 40 μM Z-VAD-FMK (VAD), and 2 mM NAC for 30 min. Negative control: cells were treated with above condition without Smp24. Bule dashed line represents the mean value of cell viability rate after treatment with Smp24 for 24 h. Data are normalized to control and presented as mean ± SEM (n = 3). * $p < 0.05$ and *** $p < 0.001$ are considered statistically significant when compared with the control group.

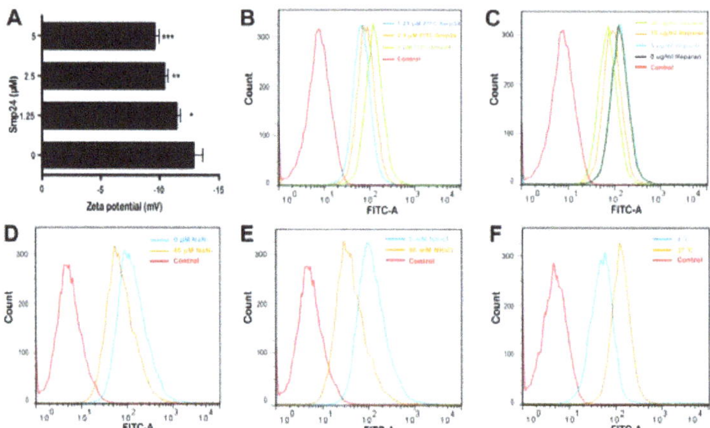

Figure 2. Internalization of Smp24 into A549 cells. (**A**) Effect of Smp24 on the zeta potential of A549 cells. Negative control: cells treated with equivalent solvent. (**B**) The internalization of FITC-labeled Smp24 into A549 cells after 1 h of treatment. (**C–F**) Effect of heparan sulfate, NaN_3, NH_4Cl, and temperature on the intracellular uptake of FITC-labeled Smp24 for 1 h, respectively. Negative control: cells treated with equivalent free-FITC solution. The values are presented as mean $\pm$ SEM ($n = 3$). * $p < 0.05$, ** $p < 0.01$, and *** $p < 0.001$ are considered statistically significant compared to the control group.

2.3. Smp24 Promotes ROS Production and Mitochondrial Membrane Potential Decrease

Based on the critical impact of cellular ROS in cell proliferation regulation, we examined whether Smp24 induced ROS production in human lung cancer cells. As presented in Figure 3A, the cell fluorescence intensity was significantly increased in a concentration-dependent manner in Smp24-treated A549 cells, while compared with the control group. In addition, the co-treatment with antioxidant NAC significantly decreased the ROS contents induced by Smp24 (Figure 3A). The obtained results demonstrated that Smp24 elevates the ROS levels in A549 cells.

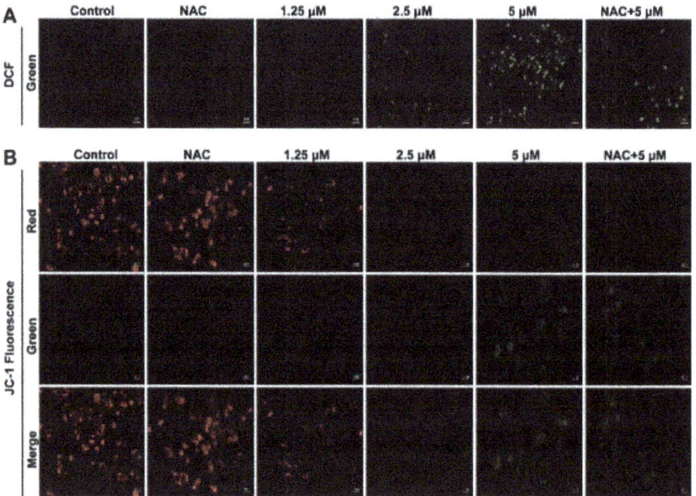

Figure 3. Influence of Smp24 on ROS production and mitochondrial membrane potential of A549 cells. (**A**) Characteristic fluorescence photographs of A549 cells stained with DCFH-DA. Scale bar, 50 μm. (**B**) Representative JC-1 fluorescence photographs of A549 cells. Scale bar, 20 μm. Negative control: cells treated with equivalent solvent for corresponding time.

ROS production is closely associated with mitochondrial membrane potential. Hence, the influence of Smp24 on mitochondrial membrane potential of A549 cells was determined using JC-1 staining. In the control group, red fluorescence was significantly observed, while a remarkable transformation from red to green staining in a concentration-dependent manner was marked in the Smp24-treated cells. What's more, NAC could attenuate the green fluorescence intensities in the cells stimulated by 5 μM Smp24 (Figure 3B). All these results demonstrated that treatment with Smp24 results in the depolarization of the mitochondrial membrane potential and inducing ROS accumulation in A549 cells.

2.4. Smp24 Induces Mitochondrion-Mediated Apoptosis in A549 Cells

ROS production is vital in the apoptosis process in cancer cells [16]. As shown by DAPI staining in Figure 4A, significant reduction in A549 cells size were observed after 24 h of Smp24 treatment, with typical apoptotic features, such as losing nuclear integrity, shrunken nuclei, formation of apoptotic bodies and chromatin condensation, indicating that the growth inhibition of A549 cells induced by Smp24 was associated with apoptosis. Consistently, the proportion of Annexin V-FITC/PI-positive apoptotic cells were remarkably increased by approximately 5.64% to 36.76% when cells were treated with Smp24 (0–5 μM) (Figure 4B, panels a–e). Besides, the presence of NAC significantly attenuated Smp24-induced apoptosis from approximately 36.76% to 19.80% (Figure 4B, panel f).

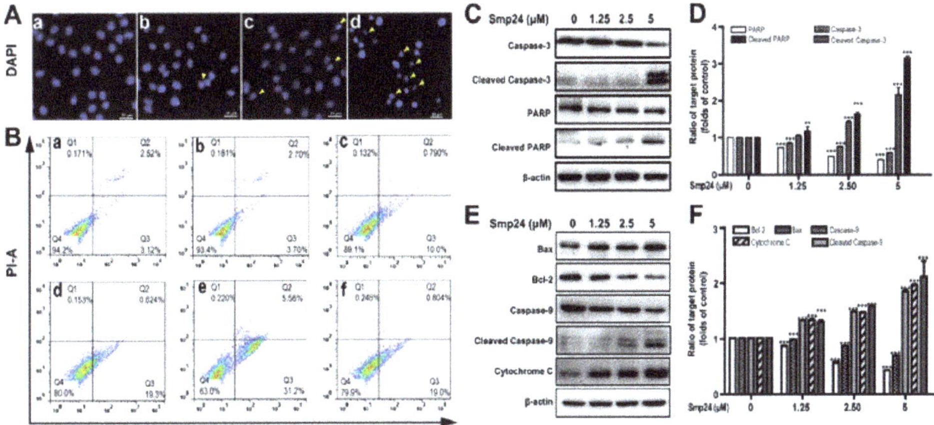

Figure 4. Apoptosis of A549 cells induced by Smp24. (**A**) Morphology of apoptotic A549 cells treated with Smp24 for 24 h. The nucleus was stained with DAPI and observed under fluorescence microscopy. Panel a: the control cells; Panels b–d: A549 cells in the presence of Smp24 (1.25, 2.5, or 5 μM), respectively. The apoptotic bodies are indicated by yellow triangles. Scale bar, 20 μm. (**B**) Representative cytometry analysis of apoptotic A549 cells after treatment with Smp24 or NAC for 24 h. Panel a: the control cells; Panels b–f: A549 cells in the presence of NAC, Smp24 (1.25, 2.5, and 5 μM), or NAC + 5 μM Smp24, respectively. (**C, E**) Representative western blots of caspase-3, cleaved caspase-3, PARP, cleaved PARP, Bax, Bcl-2, caspase-9, cleaved caspase-9 and cytochrome c. (**D, F**) Quantification of band densities in C, E. Bars represent the ratio of the target protein to the control group. Negative control: cells treated with equivalent solvent. Band densities were analyzed by Image J software and values were presented as mean ± SEM (*n* = 3). ** *p* < 0.01 and *** *p* < 0.001 are considered statistically significant when compared with the control group without Smp24.

Due to the important role in apoptosis and necrotic cell death, the mitochondria-mediated apoptotic pathway was investigated to define the mechanism of apoptotic signaling in A549 cells stimulated by Smp24 [16]. As shown in Figure 4C–F, cleaved caspase-3, cleaved caspase-9 and cleaved PARP expression were raised in concentration-dependent manner (Figure 4C–F). Bcl-2 and Bax, which belong to the Bcl-2 family, can regulate the release of mitochondrial proteins, which are closely related to apoptosis. In agreement, an increase in the expression of the pro-apoptotic protein Bax and a decrease in that of the anti-apoptotic protein Bcl-2 were observed in the presence of Smp24 (Figure 4E,F). It is well known that the rise in Bax/Bcl-2 ratio can accelerate cytochrome c release, consequently inducing apoptosis. In agreement, the expression of cytochrome c leaked from mitochondria was dose-dependently increased, following treatment with Smp24 (Figure 4E,F). Together, our data coincidentally demonstrated that Smp24 activates mitochondria apoptotic pathway.

2.5. Smp24 Arrests Cycle Distribution of A549 Cells in S Phase and G2/M Phase via Expression Regulation of Phase-related Proteins

Cell cycle distribution in Smp24-treated A549 cells was investigated to explore whether the proliferation inhibition induced by Smp24 was related to cell cycle arrest. As shown in Figure 5A, when compared with the control group, Smp24 (1.25, 2.5 and 5 µM) increased the number of A549 cells accumulating in S phase from 11.78% to 13.68%, 15.76% and 23.71% and in the G2/M phase from 8.08% to 8.23%, 10.2% and 27.04% after co-incubation for 24 h, respectively (Figure 5A). Consistently, in the G0/G1 phase, a decrease in the ratio of Smp24-treated cells was also observed from 76.00% to 68.2%, 56.5%, and 39.65%. These results indicated that treatment with Smp24 induces the S and G2/M phase arrest in A549 cells.

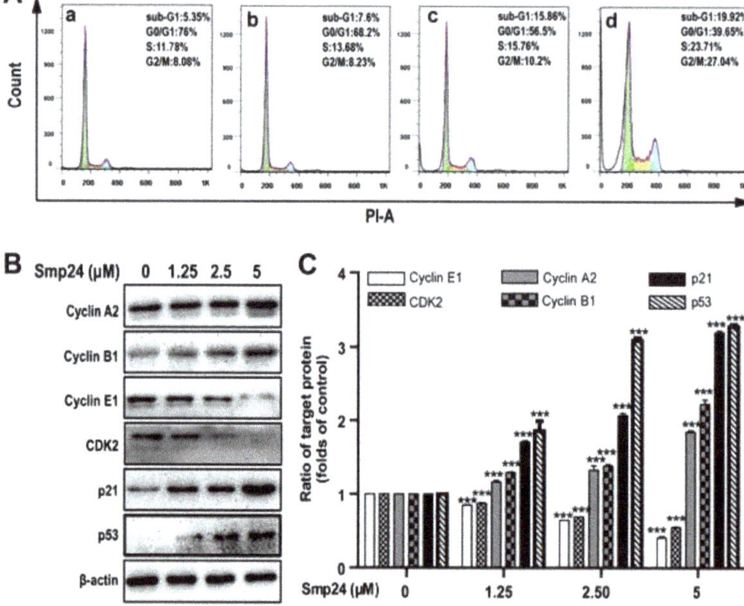

Figure 5. Effect of Smp24 on cell cycle distribution in A549 cells. (**A**) Flow cytometry analysis of cell cycle stages. A549 cells were exposed to Smp24 (1.25, 2.5, and 5 µM) for 24 h and followed by analysis with flow cytometry. Panel a: the control cells; Panels b–d: A549 cells in the presence of Smp24 (1.25, 2.5, or 5 µM), respectively. (**B**) Representative western blots of Cyclin A2, Cyclin B1, Cyclin E1, CDK2, p21[Waf1/Cip1], and p53. (**C**) Quantification of band densities in B. Bars represent the ratio of the target protein to the control group. Negative control: cells treated with equivalent solvent for corresponding time. Data are presented as mean ± SEM (n = 3). *** $p < 0.001$ are considered statistically significant while compared with the control group without Smp24.

To identify the underlying mechanism of cell cycle accumulation induced by Smp24, the expression alteration of crucial cell cycle regulation proteins was investigated. In line with its effects on cell cycle arrest, Smp24 upregulated the levels of Cyclin A2, Cyclin B1, p53, and the cyclin-dependent kinase inhibitor, p21[Waf1/Cip1] but downregulated the levels of Cyclin E1 and CDK2 in concentration-dependent manner, following 24 h treatment. In detail, compared with the control group, 5 µM Smp24 caused approximately a 0.85-, 1.22-, 2.20-, and 2.28-fold rises of Cyclin A2, Cyclin B1, p53, and p21$^{Waf1/Cip1}$ contents in A549 cells, while there were approximately 0.15-, 0.36-, and 0.59-fold and 0.13-, 0.31-, and 0.45-fold declines of CDK2 and Cyclin E1 contents in A549 cells after being exposed to Smp24 (1.25, 2.5 and 5 µM) for 24 h (Figure 5B,C).

2.6. Smp24 Induces Autophagy in A549 Cells via Inhibition of the PI3K/Akt/mTOR/FAK and p38/ERK/JNK Signaling Pathways

The formation of autophagosomes in Smp24-treated A549 cells, the hallmark of autophagy, was examined with TEM. As shown in Figure 6A, the control cells displayed normal cytoplasmic organelle and uncondensed chromatin, while Smp24-treated cells contained the increased number of vacuoles and autophagosomes with damaged cellular organelles. To elucidate the underlying molecular mechanisms of Smp24-induced autophagy in A549 cells, the expressions of autophagy-associated protein were measured. In comparison to the control cells, when A549 cells were treated with Smp24 (1.25, 2.5 and 5 µM) for 24 h, the expressions of phosphorylated Akt and mTOR as well as the total Akt were decreased in a concentration-dependent manner, which contrasted sharply with the expression trend of LC3A/B-I/II (Figure 6B,C), reflecting that Smp24 could induce autophagosome formation. Surprisingly, with the increasing concentration of Smp24, the contents of p62, a selective receptor of autophagy substrates, were increased first and then declined (Figure 6B,C).

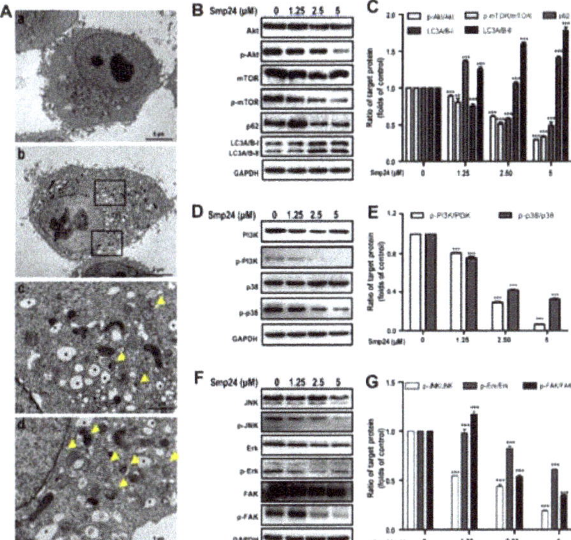

Figure 6. Regulation of autophagy and signaling pathways by Smp24. (**A**) TEM analysis of morphological structure of A549 cells with Smp24 treatment for 24 h. Panel a: the control cells, Panel b: A549 cells in the presence of 5 µM Smp24. Panels c and d: the magnified local areas in the corresponding upper and lower black squares of panel b. The vacuoles and autophagosomes were marked by black asterisk and yellow triangles, respectively. Magnification: 1200 × in panels a and b, 6000 × in panels c and d. (**B,D,F**) Representative western blots of proteins belonging to PI3K/Akt/mTOR/FAK and p38/ERK/JNK signaling pathways. (**C,E,G**) Quantification of band densities in B, D, and F. Bars represent the ratio of the target protein to the control group. Negative control: cells treated with equivalent solvent for corresponding time. Data are presented as mean ± SEM (*n* = 3). ** *p* < 0.01 and *** *p* < 0.001 are considered statistically significant as compared to the control group without Smp24.

PI3K and p38 are the upstream of the Akt/mTOR pathway and play vital roles in regulation of cell growth, cycle, apoptosis, migration, and survival [17]. As shown in Figure 6D,E, Smp24 dramatically decreased the phosphorylation of PI3K and p38 in a concentration-dependent manner, while having no effect on total p38 and PI3K expression in A549 cells. The ratios of p-p38/p38 and p-PI3K/PI3K respectively declined to approximately 32.40% and 5.66% following exposure to 5 μM Smp24 (Figure 6D,E).

The phosphorylation of ERK, JNK, and FAK are also important to regulate cancer cell adhesion, invasion, migration and proliferation [18,19]. In comparison to the control cells, Smp24 did not affect the expressions of the total FAK, JNK, and ERK, but it significantly decreased their phosphorylation in a concentration-dependent manner (Figure 6F,G). These findings indicated that Smp24 might induce autophagy in A549 cells via inhibiting the PI3K/Akt/mTOR/FAK and p38/ERK/JNK signaling pathways.

3. Discussion

Scorpion-derived peptides have become a rich source for new drug development against various diseases due to their multiple capabilities, especially in cancer treatment [20]. We previously reported the antitumor effect of Smp24, a venom-derived AMP of *Scorpio maurus palmatus*, against A549 human lung cancer cells via damaging the membrane and cytoskeleton. In agreement with the reported study, Smp24 significantly reduces the proliferation of A549 cells. Furthermore, Smp24 is internalized into A549 cells and subsequently interacts with mitochondria, leading to its dysfunction and ROS accumulation.

It is generally accepted that the existence of exclusive anionic components in the cancer cell membranes, such as phosphatidylserine, sialylated gangliosides, O-glycosylated mucins, and heparan sulfate, is associated with the cancer-selective toxicity of cationic AMPs [21]. The electrostatic binding of AMPs to cancer cells are enhanced by the interaction between cationic residues of peptides and anionic components in cell membranes. In line with this, Smp24 significantly increases the zeta potential of A549 cells in a concentration-dependent manner. Furthermore, the presence of heparan sulfate, ammonium chloride, and sodium azide markedly suppresses the cellular uptake of Smp24. These findings suggest the role of endocytosis in the internalization of Smp24 into A549 (Figure 2).

Mitochondria are vital for cancer development and their dysfunction can reduce cancer metabolism [15]. It is well known that the internalization of AMPs into cancer cells are followed by interaction with the mitochondrial membrane and forming of the transition pore, which accordingly induces swelling and rupture of mitochondria due to penetration of cytosolic ions and solutes into the inner membrane. Consequently, a series of events, such as the decline of mitochondrial membrane potential, elimination of ATP generation, accumulation of ROS, and damage of mitochondria happen [16]. Therefore, some AMPs can irreversibly lead to cell death via regulating the mitochondrial pathway [2,16]. For instance, in addition to the membrane pore formation mechanism [22], melittin might induce mitochondrial membrane depolarization, leading to the overproduction of pro-apoptotic factors, such as cytochrome c and ROS, and causing oxidative damage within the cell [5]. Similarly, Smp24 dramatically decreases mitochondria membrane potential (Figure 3B) and accumulation of ROS in a concentration-dependent manner in A549 cells (Figure 3A). The accumulated ROS can induce the expression of p53, which has substantial effects on the initiation of apoptosis via transactivating pro-apoptotic proteins (e.g., Bax) or interacting with anti-apoptotic mitochondrial proteins (e.g., Bcl-2) [23,24]. In agreement, Smp24 upregulates the p53, Bax, and cytochrome c while downregulating Bcl-2 in a concentration-dependent manner (Figure 4E,F and Figure 5B,C), and caspase inhibitors, including z-VAD-FMK and Z-DEVD-FMK, as well as necroptosis inhibitors, such as necrostatin-1, can reduce the suppressive effects of Smp24 on the viability of A549 cells (Figure 1B). Furthermore, after short-term treatment, NAC inhibits the accumulation of ROS in A549 cells (Figure 3A) and reverses Smp24-induced apoptosis (Figure 4B). Thus, Smp24 can induce the apoptosis of A549 cells involved in the mitochondrial pathway and ROS production. Excessive ROS can lead to cell death through necrosis [16], including major characteristics, such as swelling and dysfunction of the cytoplasm and the mitochondrial matrix, chromatin condensation, poration of the cellular membrane and effusion of the cytoplasmic contents into the extracellular space [25]. These findings coincide with non-specific mitochondrial membrane disruption and necrotic cell death in A549 cells caused by Smp24 reported earlier [10].

The crucial role of p53 in regulating the expression of various proteins responsible for both the G1/S and the G2/M transitions was reported earlier [26]. For example, p21$^{Waf1/Cip1}$ might serve as a negative modulator of the G1/S transition by inhibiting the kinase activity of CDK2 complexes [27]. Further, the accumulation of G2/M phase in cell cycle distribution is caused by the binding of p21$^{Waf1/Cip1}$ to the CDK1/cyclin B compound [28]. In addition, CDK2, cyclin A, cyclin B, and cyclin E participate regulation of G1 through early S phase and G2/M phase, respectively [29]. In our experiments, we have observed that the ratios of A549 cells in the S and G2/M phase are enhanced in a dose-dependent manner (Figure 5A). In addition, Smp24 significantly increases the expressions of Cyclin A2, Cyclin B1, p53, and p21$^{Waf1/Cip1}$, while inhibiting the expression of Cyclin E1 and CDK2 (Figure 5B,C), which are consistent with cell cycle arrest phenomenon.

It is well known that there is a complicated association between autophagy and apoptosis, which share common regulation proteins. For instance, PUMA can simultaneously induce apoptosis and autophagy via mitochondrial Bax pathway in return for mitochondrial perturbation [30]. Notably, various AMPs have been reported to exert antitumor activity via concurrently inducing cell apoptosis as well as autophagy processes, such as FK-16 [31], CTLEW [32], and brevenin-2R [33]. In the current study, necrostatin-1, a well-known inhibitor that suppresses both autophagy and apoptosis, significantly decreases the suppressive effects of Smp24 on A549 cells viability. Consistently, the downregulation in the expression of mTOR and Akt is observed in Smp24 treatment, while that of LC3A/B-II/I is significantly enhanced (Figure 6B,C). Interestingly, the protein expression of p62, a substrate of autophagy, is increased by treatment with 1.25 μM Smp24 but decreases in the higher concentration of Smp24 (Figure 6B,C). This phenomenon might be due to the boosting of autophagic flux of Smp24 at a low concentration, while high-concentration Smp24 has more potent effects on degrading p62 than promoting the autophagic flux. The MAPK signaling pathways also contribute a major role in both the autophagy and apoptosis process [34,35], and some compounds have been reported to exert antitumor effects against the A549 cell via inducing apoptosis and autophagy by suppressing both MAPK and PI3K/Akt/mTOR signaling pathways [36]. Consistently, the suppression of MAPK signaling pathways caused by Smp24 is observed in the current study. Thus, Smp24 exerts its cytotoxicity toward human lung cancer cells via inducing apoptosis and autophagy, too.

4. Conclusions

In summary, the current study reveals the mode and molecular mechanisms of Smp24 targeting mitochondria in A549 cells. Smp24 is internalized into A549 cells and interacts with mitochondria, leading to the loss of mitochondrial membrane potential, accumulating ROS, causing mitochondrial dysfunctions, which subsequently result in apoptosis, cell cycle arrest, and autophagy. Our findings further expose the antitumor mechanism of Smp24 against NSCLC A549, suggesting its application in lung carcinoma therapy.

5. Materials and Methods

5.1. Chemicals and Cell Culture

Phosphate-buffered saline (PBS), fetal bovine serum (FBS), RPMI-1640, and trypsin were obtained from Gibco (Grand Island, NY, USA). The A549 cell line was obtained from the American Type Culture Collection (Manassas, VA, USA) and was cultured in RPMI-1640 medium containing 10% FBS and 1% penicillin-streptomycin at 37 °C in an atmosphere of 5% CO_2. Cytochrome c, Bax, Bcl-2, PARP, cleaved PARP, caspase-3, cleaved caspase-3, caspase-9, cleaved caspase-9, p53, p21, Cyclin A2, Cyclin E1, Cyclin B1, CDK2, Erk, p-Erk, JNK, p-JNK, p38, p-p38, mTOR, p-mTOR, Akt, p-Akt, FAK, p-FAK, PI3K, p-PI3K, p62, LC3A/B- I/II, β-actin, GAPDH and all secondary antibodies were acquired from Cell Signaling Technology (Beverly, MA, USA). Necrostatin-1 (Nec-1, an inhibitor of RIP1K), Z-VAD-FMK (an inhibitor of pan caspase), Z-DEVD-FMK (an inhibitor of caspase-3), N-Acetyl-L-cysteine (NAC), mitochondrial membrane potential assay kit for JC-1, reactive oxygen species (ROS) assay kit, cell cycle and apoptosis analysis kit, and DAPI were obtained from Beyotime Institute of Biotechnology (Shanghai, China). Smp24 and FITC-labeled Smp24 were synthesized as reported in our previous study [11].

5.2. Cell Viability and Proliferation Assays

Cellular viability was analyzed via MTT method as reported in our previous study [37]. Briefly, A549 cells (1×10^4 cells/well) were cultured in 96-well plates and treated with a gradient concentration of Smp24 (1.25–20 µM) at different time intervals (12, 24, and 48 h). A549 cells were pre-incubated with different inhibitors including 40 µM Nec-1, 40 µM Z-DEVD-FMK, 40 µM Z-VAD-FMK, and 2 mM NAC for 30 min before 5 µM Smp24 was incubated with the cells for 12 h to investigate the responsibility of different signaling pathways in the cytotoxic effects of Smp24. After incubation, 10 µL MTT (5 mg/mL) was then added and incubated for 4 h at 37 °C in the dark. Subsequently, 200 µL DMSO was applied into each well after discarding the cell medium, and the cell viability was estimated by calculating the absorbance value at 490 nm via a microplate reader (Tecan Company, Männedorf, Switzerland). The experiments were performed in triplicate.

5.3. Peptide Internalization Analysis

After being seeded onto a 24-well plate overnight (1×10^5 cells/well), A549 cells were subsequently incubated with 1.25, 2.5, and 5 µM of FITC-labeled Smp24 at 37 °C for 1 h and 6 h. Flow cytometry (Becton Dickinson Company, Bedford, MA, USA) was used to measure the cell fluorescence intensity after washing with PBS. The location of Smp24 within cells was observed with fluorescence microscope at 400 × magnification. In detail, the cells were washed with PBS after treatment with 5 µM FITC-labeled Smp24 at 37 °C for 6, 12, and 24 h, respectively. Cells were then fixed with 4% paraformaldehyde (PFA) for 30 min, followed by staining with DAPI for 10 min. Approximately three single-plane pictures of each well were obtained.

To identify the effects of heparan sulfate on peptide internalization, 5 µM FITC-labeled Smp24 was pre-incubated with 5, 10, or 20 µg/mL heparan sulfate in RPMI-1640 medium for 30 min before the mixture was incubated with cells (1×10^5 cells/well). After 1 h and 6 h, flow cytometry (Becton Dickinson Company, Bedford, MA, USA) was used to measure the cell fluorescence intensity.

To identify whether the cellular energy state affects internalization, A549 cells were placed at 4 °C or 37 °C for 30 min before incubated with 5 µM FITC-labeled Smp24 for 1 h and 6 h. Thereafter, the cells were harvested, washed, and re-suspended in 200 µL of PBS for flow cytometry analysis. As a further confirmation of the role of energy in the translocation of Smp24, the FITC-labeled Smp24-treated A549 cells were pre-incubated with 40 µM NaN_3 or 50 mM NH_4Cl for 30 min and then subjected to flow cytometry. All experiments were detected in triplicate.

5.4. Zeta Potential Measurement

Zeta potential analysis was performed to evaluate the interaction between Smp24 and cancer cells. In short, after re-suspended in PBS, 1×10^5 A549 cells were mixed with Smp24 (0, 1.25, 2.5, and 5 µM). Then, the mixtures were loaded into Folded Capillary cell (DTS1070, Malvern Instruments Ltd., Worcestershire, UK) for measurement of zeta (ξ) potential with the Zetasizer system (Nano ZS; Malvern Instruments Ltd., Worcestershire, UK). The A549 cell suspension without Smp24 was considered as a control. All experiments had been detected in triplicate.

5.5. Transmission Electron Microscopy Analysis

After cultivation in a 6-well plate (2×10^5 cells/well) for 24 h, A549 cells were co-incubated with 5 µM Smp24 for another 24 h. Negative controls were defined as cells without Smp24 treatment. The cells were harvested and fixed with electron microscope fixative (2.5% glutaraldehyde at 0.1 M PBS) at room temperature for 2 h and then at 4 °C for 48 h. After removing the fixative, cells were incubated with 8% sucrose in PBS, followed by post-fixation with 1% osmium tetraoxide for 1 h at 4 °C. The cells were subsequently washed with PBS three times for 10 min. After being dehydrated with a series of gradient ethanol/water solutions, the cells were embedded in Poly/Bed 812 resin (Pelco, Redding, CA, USA). Lead citrate was used to stain the ultrathin sections, and samples were examined on a transmission electron microscope (TEM, Hitachi Company, Tokyo, Japan) at 80 kV.

5.6. Fluorescence Microscopy Analysis

For intracellular ROS content measurement, after being pretreated with Smp24 (0, 1.25, 2.5, and 5 µM) for 12 h, A549 cells were then washed with PBS before incubation with 10 µM of 2, 7-dichlorodihydro-fluoresceindiacetate (DCFH-DA) at 37 °C for 30 min in the dark. Inverted fluorescence microscopy (Axio Observer, Zeiss, Oberkochen, Germany) was used to observe cells at 200 × magnification. NAC was used as the positive control.

The changes on the mitochondrial membrane potential of A549 cells caused by Smp24 were identified using JC-1 staining (Beyotime Institute of Biotechnology, Shanghai, China). After being incubated with different concentrations of Smp24 for 12 h, A549 cells were then stained with JC-1 for 30 min at 37 °C according to the manufacturer's instructions. The changes were observed under fluorescence microscopy (Axio Observer, Zeiss, Oberkochen, Germany) at 400 × magnification. NAC was use as the positive control and approximately three random photos of each well were captured.

5.7. Cell Cycle and Apoptosis Analysis

Flow cytometry was used to identify the effects of Smp24 on cell cycle distributions and apoptosis in A549 cells. A549 cells with 2×10^5/well in density were cultured in 6-well plates overnight and treated with a gradient concentration of Smp24 or 2 mM NAC for 24 h. After being collected by centrifugation, cells were subsequently washed with cold PBS and followed by fixing with 70% ethanol overnight at 4 °C. The cells were stained with PI and RNase A at 37 °C for 30 min after being washed with cold PBS. For the apoptosis analysis, harvested cells were stained with Annexin V-FITC and PI for 15 min at room temperature and then detected by flow cytometry (Becton Dickinson Company, Bedford, MA, USA). All experiments were conducted in triplicate.

5.8. Western Blot Analysis

After culturing in 6-well plates, the 2×10^5 A549 cells were incubated with Smp24 (0, 1.25, 2.5, and 5 µM) for 24 h. After treatment, the harvested cells were extracted with RIPA lysis solution containing 1% phosphatase and protease inhibitors (FDbio Science Biotech Co., Ltd., Hangzhou, China) on ice for 15 min and then the supernatant was used for SDS-PAGE analysis and transferred to polyvinylidene difluoride (PVDF) membrane (Millipore, Billerica, MA, USA). After mixing with appropriate primary antibodies (1:1000) at 4 °C overnight and horseradish peroxidase-conjugated secondary antibodies (1: 2000) at 25 °C for 1 h, blot bands were observed by the hypersensitive ECL chemiluminescence agent and calculated using Image J software (64-bit, National Institutes of Health, MD, USA) in triplicate.

5.9. Statistical Analysis

All data were presented as mean $\pm$ SEM. Data were analyzed using one-way ANOVA followed by Bonferroni's test through GraphPad Prism 5.0 (GraphPad Software, Inc., La Jolla, CA, USA). Statistical significances were shown as * $p < 0.05$, ** $p < 0.01$, and *** $p < 0.001$.

Supplementary Materials: The following supporting information can be downloaded at: https://www.mdpi.com/article/10.3390/toxins14090590/s1, Figure S1: Internalization of Smp24 into A549 cells. (A) The internalization of FITC-labeled Smp24 into A549 cells after 6 h of treatment. (B–E) Effect of heparan sulfate, NaN$_3$, NH$_4$Cl, and temperature on the intracellular uptake of FITC-labeled Smp24 for 6 h, respectively. The values are presented as mean ± SEM (n = 3).

Author Contributions: X.X. and X.C. (Xin Chen): experiment design, study supervision, data evaluation, writing and revision of manuscript for publication. R.G. and X.C. (Xuewen Chen): data curation, formal analysis, investigation, validation, visualization, writing of original draft. T.N., J.C. and J.W.: data curation, formal analysis, investigation, validation, visualization. Y.G. and J.L.: data curation, formal analysis. M.A.A.-R.: resources, revision of original draft. All authors have read and agreed to the published version of the manuscript.

Funding: This study was funded by the Chinese National Natural Science Foundation (No. 31861143050, 31772476, 31911530077, 82070038) and in part by the Academy of Scientific Research and Technology (ASRT, Egypt; China-Egypt Scientific and Technological Cooperation Program).

Institutional Review Board Statement: Not applicable.

Informed Consent Statement: Not applicable.

Data Availability Statement: All data are available on request.

Conflicts of Interest: The authors declare that they have no known competing financial interests or personal relationships that could have appeared to influence the work reported in this paper.

References

1. Herbst, R.S.; Morgensztern, D.; Boshoff, C. The biology and management of non-small cell lung cancer. *Nature* **2018**, *553*, 446–454. [CrossRef] [PubMed]
2. Tornesello, A.L.; Borrelli, A.; Buonaguro, L.; Buonaguro, F.M.; Tornesello, M.L. Antimicrobial Peptides as Anticancer Agents: Functional Properties and Biological Activities. *Molecules* **2020**, *25*, 2850. [CrossRef] [PubMed]
3. Xu, X.; Lai, R. The chemistry and biological activities of peptides from amphibian skin secretions. *Chem. Rev.* **2015**, *115*, 1760–1846. [CrossRef] [PubMed]
4. Cho, J.; Hwang, I.S.; Choi, H.; Hwang, J.H.; Hwang, J.S.; Lee, D.G. The novel biological action of antimicrobial peptides via apoptosis induction. *J. Microbiol. Biotechnol.* **2012**, *22*, 1457–1466. [CrossRef] [PubMed]
5. Kong, G.M.; Tao, W.H.; Diao, Y.L.; Fang, P.H.; Wang, J.J.; Bo, P.; Qian, F. Melittin induces human gastric cancer cell apoptosis via activation of mitochondrial pathway. *World J. Gastroenterol.* **2016**, *22*, 3186–3195. [CrossRef]
6. Ju, X.; Fan, D.; Kong, L.; Yang, Q.; Zhu, Y.; Zhang, S.; Su, G.; Li, Y. Antimicrobial Peptide Brevinin-1RL1 from Frog Skin Secretion Induces Apoptosis and Necrosis of Tumor Cells. *Molecules* **2021**, *26*, 2059. [CrossRef]
7. Kuo, H.M.; Tseng, C.C.; Chen, N.F.; Tai, M.H.; Hung, H.C.; Feng, C.W.; Cheng, S.Y.; Huang, S.Y.; Jean, Y.H.; Wen, Z.H. MSP-4, an Antimicrobial Peptide, Induces Apoptosis via Activation of Extrinsic Fas/FasL- and Intrinsic Mitochondria-Mediated Pathways in One Osteosarcoma Cell Line. *Mar. Drugs* **2018**, *16*, 8. [CrossRef]
8. Chen, Y.P.; Shih, P.C.; Feng, C.W.; Wu, C.C.; Tsui, K.H.; Lin, Y.H.; Kuo, H.M.; Wen, Z.H. Pardaxin Activates Excessive Mitophagy and Mitochondria-Mediated Apoptosis in Human Ovarian Cancer by Inducing Reactive Oxygen Species. *Antioxidants* **2021**, *10*, 1883. [CrossRef]
9. Chai, J.; Yang, W.; Gao, Y.; Guo, R.; Peng, Q.; Abdel-Rahman, M.A.; Xu, X. Antitumor Effects of Scorpion Peptide Smp43 through Mitochondrial Dysfunction and Membrane Disruption on Hepatocellular Carcinoma. *J. Nat. Prod.* **2021**, *84*, 3147–3160. [CrossRef]
10. Guo, R.; Liu, J.; Chai, J.; Gao, Y.; Abdel-Rahman, M.A.; Xu, X. Scorpion Peptide Smp24 Exhibits a Potent Antitumor Effect on Human Lung Cancer Cells by Damaging the Membrane and Cytoskeleton In Vivo and In Vitro. *Toxins* **2022**, *14*, 438. [CrossRef]
11. Elrayess, R.A.; Mohallal, M.E.; Mobarak, Y.M.; Ebaid, H.M.; Haywood-Small, S.; Miller, K.; Strong, P.N.; Abdel-Rahman, M.A. Scorpion Venom Antimicrobial Peptides Induce Caspase-1 Dependant Pyroptotic Cell Death. *Front. Pharmacol.* **2021**, *12*, 788874. [CrossRef] [PubMed]
12. Harrison, P.L.; Heath, G.R.; Johnson, B.; Abdel-Rahman, M.A.; Strong, P.N.; Evans, S.D.; Miller, K. Phospholipid dependent mechanism of smp24, an α-helical antimicrobial peptide from scorpion venom. *Biochim. Biophys. Acta* **2016**, *1858*, 2737–2744. [CrossRef] [PubMed]
13. Freire, J.M.; Domingues, M.M.; Matos, J.; Melo, M.N.; Veiga, A.S.; Santos, N.C.; Castanho, M.A. Using zeta-potential measurements to quantify peptide partition to lipid membranes. *Eur. Biophys. J.* **2011**, *40*, 481–487. [CrossRef] [PubMed]
14. Yang, J.; Tsutsumi, H.; Furuta, T.; Sakurai, M.; Mihara, H. Interaction of amphiphilic α-helical cell-penetrating peptides with heparan sulfate. *Org. Biomol. Chem.* **2014**, *12*, 4673–4681. [CrossRef]
15. Drin, G.; Cottin, S.; Blanc, E.; Rees, A.R.; Temsamani, J. Studies on the internalization mechanism of cationic cell-penetrating peptides. *J. Biol. Chem.* **2003**, *278*, 31192–31201. [CrossRef]

16. Yang, Y.; Karakhanova, S.; Hartwig, W.; D'Haese, J.G.; Philippov, P.P.; Werner, J.; Bazhin, A.V. Mitochondria and Mitochondrial ROS in Cancer: Novel Targets for Anticancer Therapy. *J. Cell Physiol.* **2016**, *231*, 2570–2581. [CrossRef]
17. Arthur, J.S.; Ley, S.C. Mitogen-activated protein kinases in innate immunity. *Nat. Rev. Immunol.* **2013**, *13*, 679–692. [CrossRef]
18. Park, J.H.; Jeong, Y.J.; Park, K.K.; Cho, H.J.; Chung, I.K.; Min, K.S.; Kim, M.; Lee, K.G.; Yeo, J.H.; Park, K.K.; et al. Melittin suppresses PMA-induced tumor cell invasion by inhibiting NF-kappaB and AP-1-dependent MMP-9 expression. *Mol. Cells* **2010**, *29*, 209–215. [CrossRef]
19. Lim, H.N.; Baek, S.B.; Jung, H.J. Bee Venom and Its Peptide Component Melittin Suppress Growth and Migration of Melanoma Cells via Inhibition of PI3K/AKT/mTOR and MAPK Pathways. *Molecules* **2019**, *24*, 929. [CrossRef]
20. Hmed, B.; Serria, H.T.; Mounir, Z.K. Scorpion peptides: Potential use for new drug development. *J. Toxicol.* **2013**, 958797. [CrossRef]
21. Schweizer, F. Cationic amphiphilic peptides with cancer-selective toxicity. *Eur. J. Pharmacol.* **2009**, *625*, 190–194. [CrossRef] [PubMed]
22. Yang, L.; Harroun, T.A.; Weiss, T.M.; Ding, L.; Huang, H.W. Barrel-stave model or toroidal model? A case study on melittin pores. *Biophys. J.* **2001**, *81*, 1475–1485. [CrossRef] [PubMed]
23. Li, N.; Zhan, X. Mitochondrial Dysfunction Pathway Networks and Mitochondrial Dynamics in the Pathogenesis of Pituitary Adenomas. *Front. Endocrinol.* **2019**, *10*, 690. [CrossRef]
24. Dai, C.Q.; Luo, T.T.; Luo, S.C.; Wang, J.Q.; Wang, S.M.; Bai, Y.H.; Yang, Y.L.; Wang, Y.Y. p53 and mitochondrial dysfunction: Novel insight of neurodegenerative diseases. *J. Bioenerg. Biomembr.* **2016**, *48*, 337–347. [CrossRef] [PubMed]
25. Jain, M.V.; Paczulla, A.M.; Klonisch, T.; Dimgba, F.N.; Rao, S.B.; Roberg, K.; Schweizer, F.; Lengerke, C.; Davoodpour, P.; Palicharla, V.R.; et al. Interconnections between apoptotic, autophagic and necrotic pathways: Implications for cancer therapy development. *J. Cell. Mol. Med.* **2013**, *17*, 12–29. [CrossRef]
26. Anand, S.K.; Sharma, A.; Singh, N.; Kakkar, P. Entrenching role of cell cycle checkpoints and autophagy for maintenance of genomic integrity. *DNA Repair* **2020**, *86*, 102748. [CrossRef]
27. Warfel, N.A.; El-Deiry, W.S. p21WAF1 and tumourigenesis: 20 years after. *Curr. Opin. Oncol.* **2013**, *25*, 52–58. [CrossRef]
28. Wu, G.; Lin, N.; Xu, L.; Liu, B.; Feitelson, M.A. Feitelson, UCN-01 induces S and G2/M cell cycle arrest through the p53/p21(waf1) or CHK2/CDC25C pathways and can suppress invasion in human hepatoma cell lines. *BMC Cancer* **2013**, *13*, 167. [CrossRef]
29. Sánchez, I.; Dynlacht, B.D. New insights into cyclins, CDKs, and cell cycle control. *Semin. Cell Dev. Biol.* **2005**, *16*, 311–321. [CrossRef]
30. Yee, K.S.; Wilkinson, S.; James, J.; Ryan, K.M.; Vousden, K.H. PUMA- and Bax-induced autophagy contributes to apoptosis. *Cell Death Differ.* **2009**, *16*, 1135–1145. [CrossRef]
31. Ren, S.X.; Shen, J.; Cheng, A.S.; Lu, L.; Chan, R.L.; Li, Z.J.; Wang, X.J.; Wong, C.C.; Zhang, L.; Ng, S.S.; et al. FK-16 derived from the anticancer peptide LL-37 induces caspase-independent apoptosis and autophagic cell death in colon cancer cells. *PLoS ONE* **2013**, *8*, e63641. [CrossRef]
32. Ma, S.; Huang, D.; Zhai, M.; Yang, L.; Peng, S.; Chen, C.; Feng, X.; Weng, Q.; Zhang, B.; Xu, M. Isolation of a novel bio-peptide from walnut residual protein inducing apoptosis and autophagy on cancer cells. *BMC Complement. Altern. Med.* **2015**, *15*, 413. [CrossRef]
33. Ghavami, S.; Asoodeh, A.; Klonisch, T.; Halayko, A.J.; Kadkhoda, K.; Kroczak, T.J.; Gibson, S.B.; Booy, E.P.; Naderi-Manesh, H.; Los, M. Brevinin-2R(1) semi-selectively kills cancer cells by a distinct mechanism, which involves the lysosomal-mitochondrial death pathway. *J. Cell Mol. Med.* **2008**, *12*, 1005–1022. [CrossRef]
34. Chen, C.; Gao, H.; Su, X. Autophagy-related signaling pathways are involved in cancer (Review). *Exp. Ther. Med.* **2021**, *22*, 710. [CrossRef] [PubMed]
35. Santarpia, L.; Lippman, S.M.; El-Naggar, A.K. Targeting the MAPK-RAS-RAF signaling pathway in cancer therapy. *Expert Opin. Ther. Targets* **2012**, *16*, 103–119. [CrossRef] [PubMed]
36. Guo, M.; Liu, Z.; Si, J.; Zhang, J.; Zhao, J.; Guo, Z.; Xie, Y.; Zhang, H.; Gan, L. Cediranib Induces Apoptosis, G1 Phase Cell Cycle Arrest, and Autophagy in Non-Small-Cell Lung Cancer Cell A549 In Vitro. *Biomed. Res. Int.* **2021**, 5582648. [CrossRef]
37. Tian, M.; Liu, J.; Chai, J.; Wu, J.; Xu, X. Antimicrobial and Anti-inflammatory Effects of a Novel Peptide From the Skin of Frog Microhyla pulchra. *Front. Pharmacol.* **2021**, *12*, 783108. [CrossRef]

Article

Scorpion Peptide Smp24 Exhibits a Potent Antitumor Effect on Human Lung Cancer Cells by Damaging the Membrane and Cytoskeleton In Vivo and In Vitro

Ruiyin Guo [1,2,†], Junfang Liu [1,†], Jinwei Chai [2], Yahua Gao [2], Mohamed A. Abdel-Rahman [3] and Xueqing Xu [2,*]

1   Department of Pulmonary and Critical Care Medicine, Zhujiang Hospital, Southern Medical University, Guangzhou 510280, China; 21920154gry@i.smu.edu.cn (R.G.); ljf00157@163.com (J.L.)
2   Guangdong Provincial Key Laboratory of New Drug Screening, School of Pharmaceutical Sciences, Southern Medical University, Guangzhou 510515, China; vivian20@i.smu.edu.cn (J.C.); gaoyahua7721@163.com (Y.G.)
3   Zoology Department, Faculty of Science, Suez Canal University, Ismailia 41522, Egypt; mohamed_hassanain@science.suez.edu.eg
*   Correspondence: xu2003@smu.edu.cn
†   These authors contributed equally to this work.

Abstract: Smp24, a cationic antimicrobial peptide identified from the venom gland of the Egyptian scorpion *Scorpio maurus palmatus*, shows variable cytotoxicity on various tumor (KG1a, CCRF-CEM and HepG2) and non-tumor (CD34[+], HRECs, HACAT) cell lines. However, the effects of Smp24 and its mode of action on lung cancer cell lines remain unknown. Herein, the effect of Smp24 on the viability, membrane disruption, cytoskeleton, migration and invasion, and MMP-2/-9 and TIMP-1/-2 expression of human lung cancer cells have been evaluated. In addition, its in vivo antitumor role and acute toxicity were also assessed. In our study, Smp24 was found to suppress the growth of A549, H3122, PC-9, and H460 with IC_{50} values from about 4.06 to 7.07 μM and show low toxicity to normal cells (MRC-5) with 14.68 μM of IC_{50}. Furthermore, Smp24 could induce necrosis of A549 cells via destroying the integrity of the cell membrane and mitochondrial and nuclear membranes. Additionally, Smp24 suppressed cell motility by damaging the cytoskeleton and altering MMP-2/-9 and TIMP-1/-2 expression. Finally, Smp24 showed effective anticancer protection in a A549 xenograft mice model and low acute toxicity. Overall, these findings indicate that Smp24 significantly exerts an antitumor effect due to its induction of membrane defects and cytoskeleton disruption. Accordingly, our findings will open an avenue for developing scorpion venom peptides into chemotherapeutic agents targeting lung cancer cells.

Keywords: lung cancer; antimicrobial peptide; scorpion; venom; tumor

Key Contribution: Smp24 significantly exerts an antitumor effect due to its induction of membrane defects and cytoskeleton disruption in vivo and in vitro.

Citation: Guo, R.; Liu, J.; Chai, J.; Gao, Y.; Abdel-Rahman, M.A.; Xu, X. Scorpion Peptide Smp24 Exhibits a Potent Antitumor Effect on Human Lung Cancer Cells by Damaging the Membrane and Cytoskeleton In Vivo and In Vitro. *Toxins* **2022**, *14*, 438. https://doi.org/10.3390/toxins14070438

Received: 31 May 2022
Accepted: 23 June 2022
Published: 28 June 2022

Publisher's Note: MDPI stays neutral with regard to jurisdictional claims in published maps and institutional affiliations.

1. Introduction

Lung cancer, especially non-small cell lung cancer (NSCLC), has become the major cause of cancer death around the world, resulting in a mortality of more than 1.7 million annually [1]. Although chemotherapy, radiation, targeted therapy, and immune therapy have improved the survival rate of patients, lung adenocarcinoma is still one of the most aggressive and rapidly malignant types of cancer [1]. Therefore, developing new anti-cancer drugs with high selectivity, low toxicity, and adequate membrane permeability is of great importance.

Antimicrobial peptides (AMPs) are generally acknowledged as important components of the innate immune defense and are the most promising candidates to replace conventional antibiotics. Approximately 3000 native AMPs have been purified and identified from animals, plants, protozoa, bacteria, fungi and protists [2]. Of these peptides, 230 members among them are grouped into anticancer peptides (ACPs) that possess antitumor activity [2]. Most of ACPs target cell membranes rather than specific receptors, which make them less possible for cancer cells to develop drug resistance compared with traditional chemotherapy [3,4]. Therefore, the development of these peptides into potent anticancer agents, whether alone or in combination with other classical chemotherapy, has been considered as an innovative strategy for cancer treatment [3]. The action mechanism of most ACPs depends on the electrostatic interactions between cationic peptides and anionic lipids on cancer cell membranes [3,4]. As a result, they selectively disrupt the negatively charged plasma membranes of cancer cells without affecting zwitterionic cell membranes of non-cancer cells [3,4]. Furthermore, tumor cell migration is essential for invasion and dissemination from primary solid tumors [5]. Tumor migration suppression is also a therapy strategy and some AMPs, such as the short peptide RK1 from *Buthus occitanus*, can inhibit tumor cell migration and show therapeutic potential [6].

Scorpion venoms reveal potent suppressing effects on various types of cancers including lung, hepatoma, leukemia, breast, neuroblastoma, prostate, pancreas, glioma and lymphoma [7]. Our previous studies have described that Smp24, an amphipathic cationic AMP isolated from the Egyptian scorpion *Scorpio maurus palmatus*, can suppress microbes by membrane disruption [8,9]. The peptide shows stronger cytotoxicity against leukemic tumor cell lines (KG1-a and CCRF-CEM) than the normal cell lines (CD34$^+$, HRECs and HACAT) [10], which indicates that Smp24 may be a potential ACP. However, its effects and underlying mechanism on lung cancer cells remain to be clarified. In this study, we first elucidated the cytotoxicity of Smp24 on human lung cancer cells via membrane destruction. Furthermore, its effects on the migration ability of A549 cells via the regulation of filamentous actin (F-actin) and the alteration of MMP-2/9 and TIMP-1/2 protein expressions were assessed. Finally, Smp24 showed antitumor efficacy and low acute toxicity in the A549 xenograft mice model. To the best of our knowledge, this is the first detailed study describing the potent in vitro and in vivo efficacy of Smp24 against lung cancer cells.

2. Results

2.1. Smp24 Inhibits the Proliferation of Human Lung Cancer Cells

MTT was used to measure the cytotoxic activity of Smp24 against four human lung cancer cell lines. The IC$_{50}$ values and selectivity index of Smp24 for 24 h treatment against A549, H3122, PC-9 and H460 cells were about 4.06 ± 0.60, 4.64 ± 0.18, 6.34 ± 0.53, and 7.07 ± 0.81 μM, and 3.61, 3.16, 2.32, and 2.07, respectively (Figure 1A and Table S1). Nonetheless, Smp24 exhibited less inhibitory potency toward the proliferation of normal cells as evidenced from the higher IC$_{50}$ value of around 14.68 ± 0.79 μM for MRC-5 cells (Figure 1A), indicating that it is more selective toward cancer cells. Furthermore, the IC$_{50}$ value of Smp24 for 24 h treatment against A549 is lower than 31.98 ± 1.79 μM of cisplatin (Figures 1A and S1). Because the A549 cell line is most sensitive to Smp24 and NSCLC accounts for about 80–85% of human lung carcinoma [1], the NSCLC cell line A549 was selected for further investigation of the antitumor effect of Smp24. EdU incorporation assay revealed that Smp24 dramatically decreased the EdU fluorescence intensities of A549 cells compared with untreated cells, which further suggested that Smp24 inhibited cell proliferation in a dose-dependent manner (Figure 1B). The morphology of A549 cells incubated with Smp24 for 24 h is shown in Figure 1C. The control cells formed normal and fusiform shape with smooth surfaces. However, A549 cells treated with Smp24 were obviously smaller and rounder than the control cells. Moreover, their number decreased in a concentration-dependent manner with many floated cells and cellular debris. The above data suggest that Smp24 has both cytotoxic and cytostatic effects on lung cancer cells.

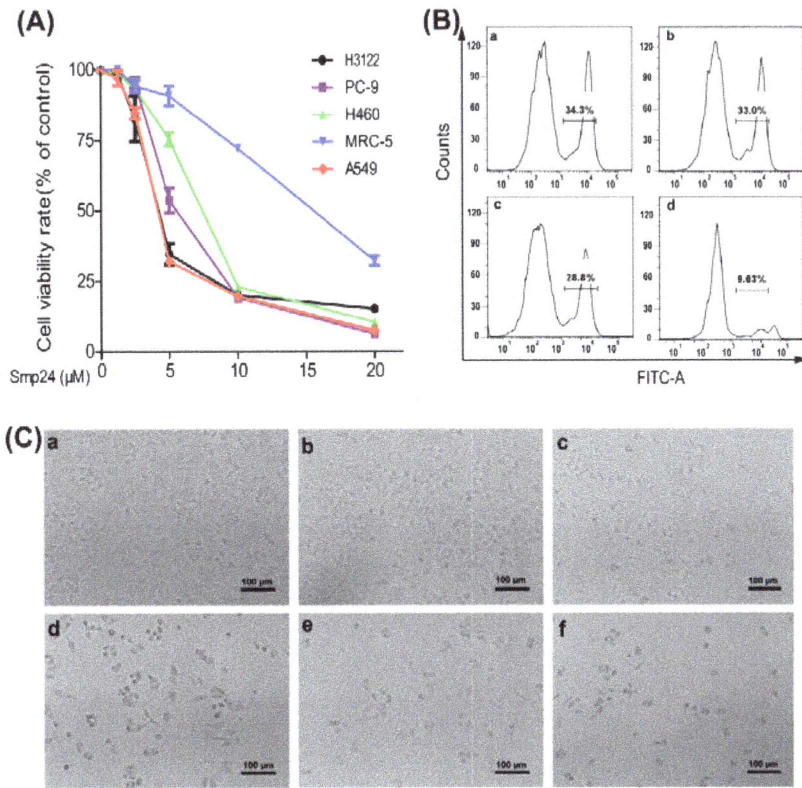

Figure 1. Viability and morphology changes of cancer cells induced by Smp24. (**A**) Viability of A549, H3122, PC-9, H460 and MRC-5 cells treated with Smp24 for 24 h. (**B**) Proliferation of A549 cells treated with the indicated concentrations of Smp24 for 24 h. Panels (**a–d**) are sequentially the cells treated with 0, 1.25, 2.5, and 5 µM of Smp24. (**C**) Cell morphology changes after treatment with Smp24 for 24 h. Panels (**a–f**) are sequentially the cells treated with 0, 1.25, 2.5, 5, 10 and 20 µM of Smp24. Scale bar, 100 µm. Data are normalized to control and presented as mean ± SEM (n = 3).

2.2. Smp24 Induces Necrosis of A549 Cells by Membrane Damage

To decipher the mechanism behind the antiproliferative effects of Smp24, its necrotizing effect was examined. The release of LDH was increased in Smp24-treated cells in both dose- and time-dependent manners (Figure 2A). The release of LDH evidenced cell membrane leakage throughout. Thus, SEM technology was used to observe morphological alteration of the cell membranes induced by Smp24. As shown in Figure 2B, the control cells displayed complete and smooth cell membranes, whereas Smp24-treated cells appeared with fragmented cell membranes and curly edges. The above findings indicated that the decrease in cell viability induced by Smp24 resulted from its cytolytic effect. To further confirm the membrane permeability changes following treatment with Smp24, calcein AM release assay was carried out. As shown in Figure 2C, Smp24 exposure concentration dependently accelerated the release of calcein AM from A549 cells, indicating the damage of membrane integrity in A549 cells. $CoCl_2$ is a quencher of cytosolic calcein fluorescence to selectively label mitochondria. Consistently, in the presence of $CoCl_2$, the cytosolic fluorescence was reduced from 90.70% to 74.70%, while for mitochondria, it increased from 1.5% to 8.17% when compared to calcein AM alone. Furthermore, this change in trend of the fluorescence intensity in the cytoplasm and mitochondria was further strengthened by

Smp24 in a concentration-dependent manner when compared with the control without Smp24 treatment (Figure 2C). Additionally, it is the necroptosis inhibitor necrostatin-1 rather than mitochondrial permeability transition pore inhibitor, Cyclosporine A, that can inhibit Smp24-induced proliferation and necrosis of A549 (Figure 2D,E). To further confirm the membranolytic effect of Smp24 on the nuclear membrane, DAPI staining assay was performed. As shown in Figure 2F, the untreated A549 cells were found to contain intact nuclei. However, the volume of A549 cells treated by Smp24 for 24 h was reduced, and the nuclear integrity was broken. Furthermore, in FITC channel observation, all cells showed a high green fluorescence due to the peptide in the cells and some green fluorescence co-localized blue fluorescence-derived DAPI staining, indicating that a portion of Smp24 entered into the nucleus, thereby damaging the cell nucleus (Figure 2F). Together, these results suggested that Smp24 induced non-specifically cellular and mitochondrial membrane damage in A549.

2.3. Smp24 Inhibits Motility by Damaging the Cytoskeleton and Altering MMP-2/-9 and TIMP-1/-2 Expression

Structural stability of cells and the integrity of their plasma membrane mostly rely on the dynamics and function of the cytoskeleton, which allow cells to maintain or adaptably modify their morphology to facilitate cell division, motility, and other biological activities [11,12]. Thus, the cytoskeleton was examined via staining F-actin with rhodamine–phalloidin. As displayed in Figure 3A, the cells without Smp24 treatment were smooth with obvious staining of flat microfilament bundles that were collateral to the edge of the cells. However, F-actin microfilaments became disorganized and polarized feathers of cytoskeleton variation following treatment with Smp24. Especially, after co-culture with 5 µM Smp24 for 24 h, the microfilament bundles of A549 cells were shortened and disordered with visibly red fluorescence. The above data indicate that Smp24 suppresses the cell motility via the regulation of F-actin type in A549 cells.

Previous reports have confirmed that the cytoskeleton is crucially involved with cell motility, which is importantly responsible for high cancer mortality, and its crucial steps include cell migration and invasion [13,14]. For this reason, the wound-healing and transwell assays were performed to measure the effects of Smp24 on migration and invasion of A549 cells. In comparison with the control cells, A549 cell migration was suppressed by Smp24 in a dose-dependent manner after treatment for 24 h, and the suppression rates of 0.3, 0.6 and 1.2 µM of Smp24 were about 62.23%, 78.63%, and 94.52%, respectively (Figure 3B,C). Consistently, Smp24 decreased the invasion of A549 cells at a concentration-dependent manner in the invasion assay, respectively (Figure 3D). Therefore, Smp24 could effectively inhibit cell invasion and migration of A549 cells. The regulation proteins of the extracellular matrix, such as MMPs and TIMPs, play important roles in cell motility [13]. Therefore, the influence of Smp24 on the mRNA expression of MMP-2/-9 and TIMP-1/-2 were measured by qPCR. As shown in Figure 3E, the expressions of TIMP-1/-2 mRNA were significantly upregulated while the expressions of MMP-2/-9 mRNA were significantly downregulated in a concentration-dependent manner following incubation with Smp24 for 12 h. In short, these results confirmed that Smp24 is in a position to suppress A549 cell invasion and migration.

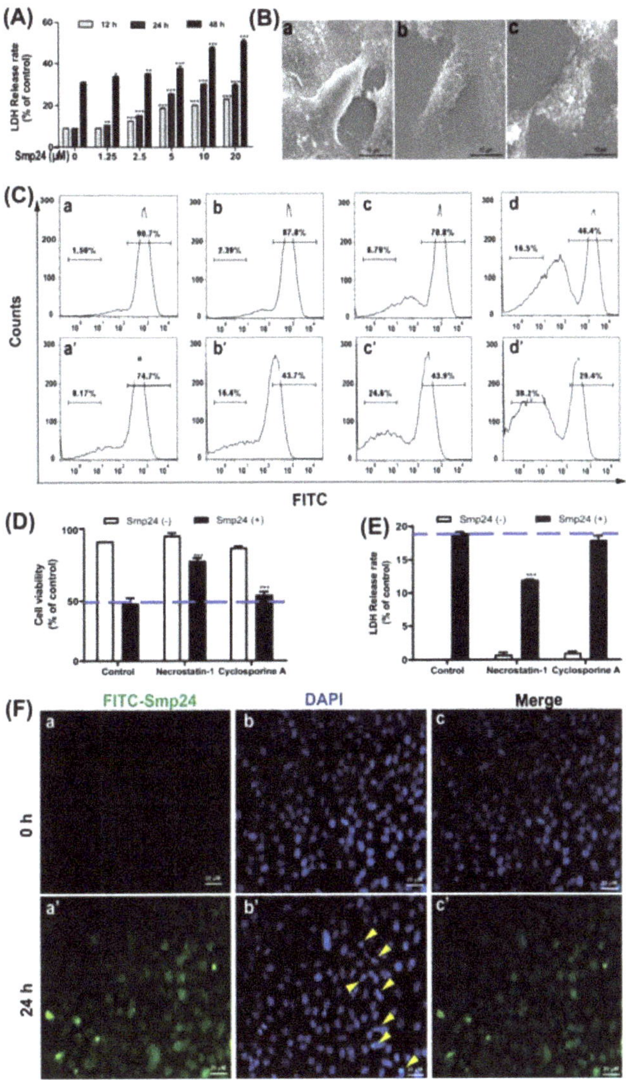

Figure 2. Necrosis induced by Smp24 in A549 cells. (**A**) LDH release of A549 cells induced by Smp24 (0–20 μM) for 12, 24 and 48 h. (**B**) SEM analysis of morphological structure of A549 cells. Panel (**a**): control A549 cells; panels (**b,c**): A549 cells treated by 2.5 and 5 μM of Smp24. Scale bar, 10 μm. (**C**) Representative flow cytometry analysis of calcein AM changes in A549 cells treated with Smp24 for 24 h. Panels (**a–d**): cells stained with calcein AM; panels (**a′–d′**): cells treated by calcein AM + $CoCl_2$; Panels (**a–a′,b–b′,c–c′,d–d′**): cells treated by 0, 1.25, 2.5 and 5 μM of Smp24, respectively. Results are presented as mean ± SEM (n = 3). * $p < 0.05$, ** $p < 0.01$, *** $p < 0.001$ are considered statistically significant as compared to the control. (**D,E**) Effects of inhibitors on the viability and the LDH release of Smp24-treated A549 cells. A549 cells were pre-incubated with the inhibitors (40 μM Necrostatin-1, 1 μM Cyclosporine A) for 30 min before being further incubated with 5 μM Smp24 for 12 h. (**F**) Fluorescence microscope observation of 5 μM FITC-labeled Smp24 internalized in A549 cells after co-incubation for 24 h. (**a,b**) Control (no treatment) group; (**a′,b′**) 24 h treatment group; panels (**c,c′**) are the merged figure. Yellow arrow: nuclear fragmentation or the apoptotic body. Scale bar, 20 μm.

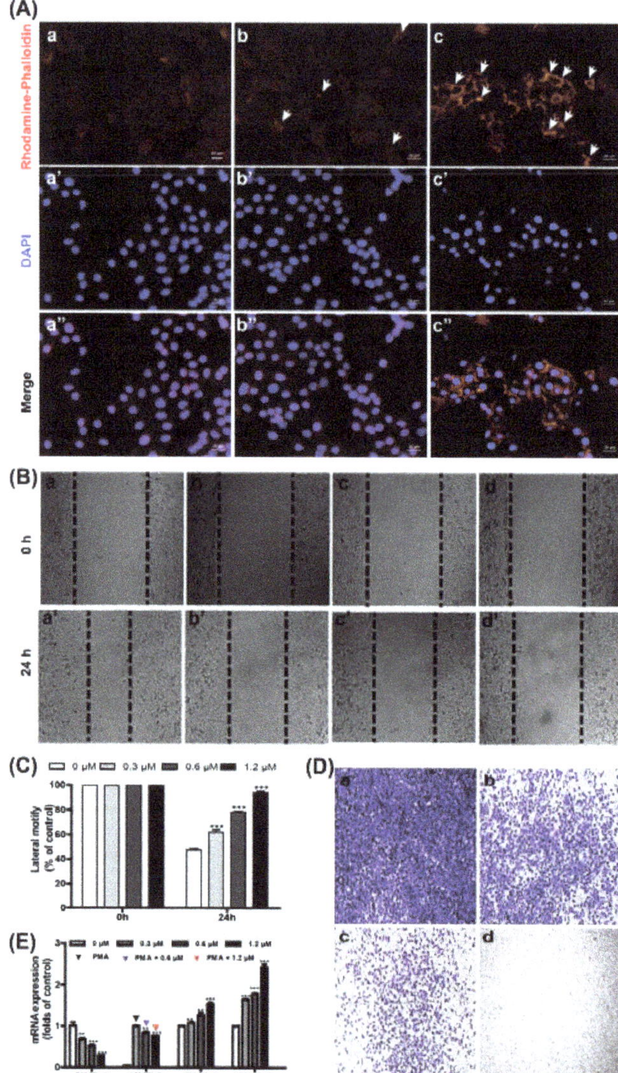

Figure 3. Effects of Smp24 on the motility and cytoskeleton reorganization of A549 cells. (**A**) Fluorescence staining images of F-actin. **Upper** panel: cells stained with rhodamine-phalloidin; **middle** panel: cells stained with DAPI; **lower** panel: the merging images of cells stained by rhodamine–phalloidin and DAPI. A549 cells were treated with Smp24 (0, 2.5, and 5 μM) for 24 h and successively stained by rhodamine-phalloidin as well as DAPI before fluorescence microscopy observation. White arrow: disordered microfilament bundles. Scale bar, 20 μm. (**B**) Representative pictures of A549 cells in the scratch migration assay at 0 and 24 h following incubation with Smp24 and PBS. Panels (**a–a′,b–b′,c–c′,d–d′**): the cells treated by 0, 0.3, 0.6 and 1.2 μM of Smp24, respectively. (**C**) Statistical analysis for the scratch migration assay. (**D**) Typical images of A549 cells in the transwell invasion assay following culture with Smp24 and PBS for 24 h. Panels (**a–d**): the cells treated by 0, 0.3, 0.6 and 1.2 μM of Smp24, respectively. (**E**) Statistical analysis of relative mRNA contents of MMP-2, MMP-9, TIMP-1 and TIMP-2. Results are mean ± SEM (n = 3). ** $p < 0.01$ and *** $p < 0.001$ are considered statistically significant as compared to the control group without Smp24.

2.4. Smp24 Inhibits Tumor Growth in Mice

The in vivo pharmacodynamic effects of Smp24 were assessed using A549 xenograft mice model. As indicated in Figure 4, in comparison to the control group, the tumor volume and weight in the Smp24-treated group were reduced by approximately 65.4% and 64.3%, respectively (Figure 4B–D). However, the treatment with Smp24 did not affect the weight changes of body and important organs such as heart, lung, liver, spleen and kidney (Figure 4E,F). As shown in Figure 4G, HE staining displayed that tumor cells in the control group generally were enlarged with fairly complete structure and regular shapes, while massive cell shrinkage and the inflammatory cell infiltration appeared in the Smp24-treated group. Furthermore, immuno-histochemical analysis demonstrated that the expression of cleaved caspase-3 (brown area) in the Smp24-treated group was dramatically enhanced when compared to the control group (Figure 4H). The above results suggested that Smp24 can exert potent antitumor effects in vivo.

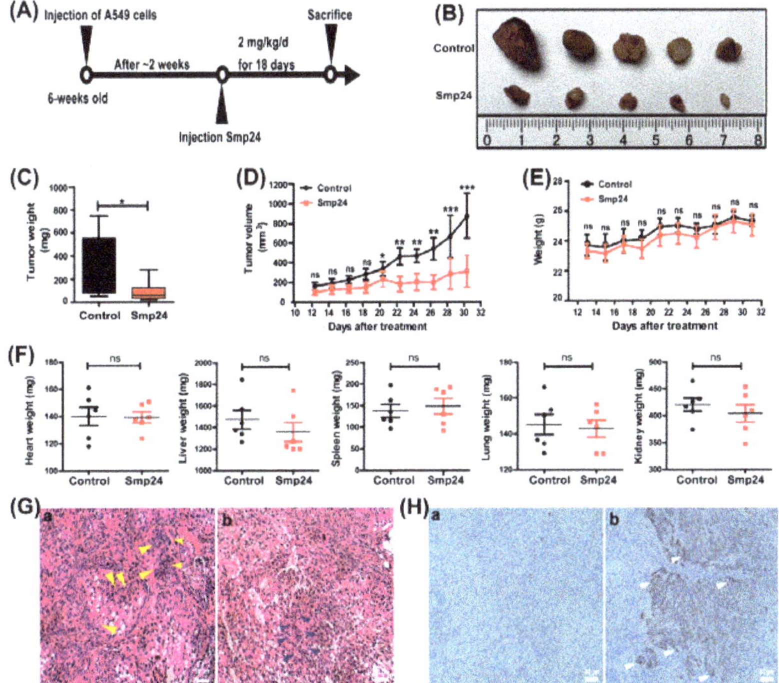

Figure 4. In vivo antitumor effects of Smp24. (**A**) Experimental schedule for xenograft mice. (**B**) Images of tumors from sacrificed nude mice. (**C**) Tumor weight. (**D**) Tumor volume. (**E**) Body weight change. (**F**) The weight of heart, liver, spleen, lung and kidney. (**G**) HE staining analysis of tumor tissue. Scale bar, 50 μm. (**H**) Immunohistochemical analysis of cleaved caspase-3 in tumor tissue from lung carcinoma xenografts. Panel (**a**): control group; panel (**b**): Smp24-treated group. Scale bar, 50 μm. Data presented are mean ± SEM (n = 5). ns: no significance, * $p < 0.05$, ** $p < 0.01$ and *** $p < 0.001$ are considered statistically significant as compared to the control group without Smp24 treatment. Yellow arrow: enlarged tumor cells. Blue arrow: shrinking cells. White arrow: positive apoptotic staining (brown areas).

2.5. Smp24 Has No Hepatotoxicity and Nephrotoxicity in Acute Toxicity Doses

In the acute toxicity test of mice, physiological saline was selected as the normal control group. Doses of 5 and 10 mg/kg were chosen for the acute toxicity test. After the first day's injection, there was no animal death in any group. Plasma AST, ALT, BUN and

Cre are key biomarkers reflecting liver and kidney damage in the clinic [15]. All of them did not change significantly in the mice of the 5 and 10 mg/kg Smp24 groups after injection 48 h (Figure 5A–D). Compared with the important organs weights of mice pre-injection, the heart, lung, liver, spleen and kidney weights of mice in the 5 and 10 mg/kg Smp24 groups did not change significantly (Figure 5E–I). Therefore, the results indicated that Smp24 might not have hepatotoxicity and nephrotoxicity in the acute toxicity doses.

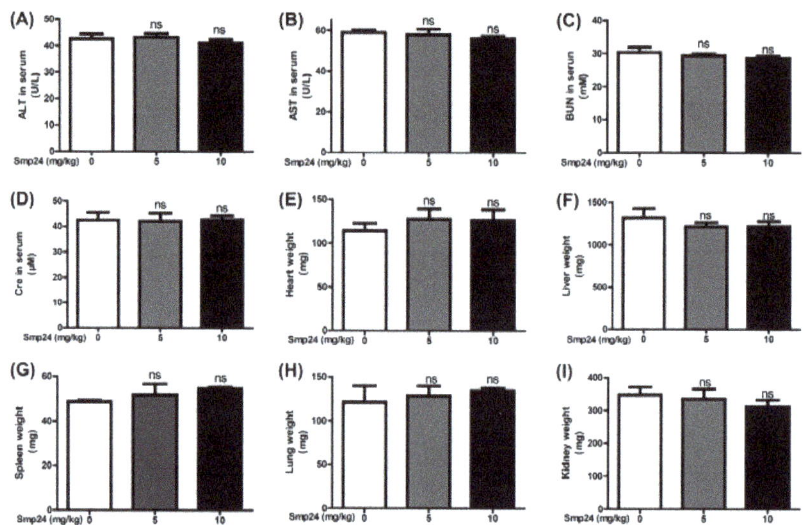

Figure 5. Acute toxicity analysis of Smp24. (**A–D**) The level of ALT, AST, BUN, Cre in serum of normal mice after 48 h treatment with 0, 5 and 10 mg/kg Smp24. (**E–I**) The weight of heart, liver, spleen, lung and kidney of mice after 48 h treatment with 0, 5 and 10 mg/kg Smp24. ns: no significance.

3. Discussion

Despite the great improvement that has been achieved in its treatment in recent decades, lung cancer remains the most lethal type of cancer and still is one of the biggest threats to the health care system globally [1]. AMPs derived from natural or designed peptides possess selective cytotoxicity against various tumor cells [16]. Smp24 is an amphipathic and cationic AMP from Egyptian scorpion *Scorpio maurus palmatus*, which exhibits the cytotoxicity to leukemic tumor cell lines [10]. In line with the previous study, we have found that Smp24 preferentially inhibits the proliferation of A549, H3122, PC-9 and H460 cells in vitro (Figure 1A). Moreover, its antitumor capability is further proven in the xenograft mice (Figure 4).

A previous study has also proven that Smp24 can induce pore formation of membrane [9]. In line with them, our compelling evidence from LDH release, calcein AM staining and SEM observation (Figure 2) demonstrate that cationic Smp24 as a pore-forming peptide can interact with tumor cell membranes, elevate their permeability, and induce pore formation in the plasma membrane and cell death [3,17,18]. Recent findings suggest that some cationic AMP such as Brevivin-1RL1 [19], KL15 [20], D-K6L9 [21] and P18 [22] can exhibit their induction of tumor necrosis via membrane destruction. Furthermore, after exposure to A549 cells for 24 h, Smp24 causes mitochondrial damage (Figure 2C) and cytoskeleton disruption (Figure 3A). In addition, Smp24 can distribute into the cell nucleus, thereby damaging the nucleus (Figure 2F), and consequently resulting in cell death. Finally, the signal pathway inhibitors such as Necrostatin-1 can inhibit Smp24-induced proliferation and necrosis of A549 (Figure 2D,E). Cyclosporine A, the permeability transition inhibitor, cannot inhibit the cytotoxicity of Smp24 to A549 cells (Figure 2D,E), suggesting that the non-specific mitochondrial membrane disruption of Smp24 should be responsible for its induction of the mitochondrial dysfunction. These data further suggest that the cytotoxicity of Smp24 to tumor cells might result from its damage of the plasma membranes and other intracellular cascades.

The different charge distributions of the cell membrane surface between cancer and non-cancer cells may partially contribute to the selectivity of AMPs. For example, the rich anionic components such as heparan sulfate, PS, and sialylated gangliosides provide a net negative charge on the surface of cancer cells. For this reason, AMPs can electrostatically bind cancer cell membranes and especially kill tumor cells [3]. Consistently, a series of short amphiphilic triblock AMPs shows that anti-tumor effects and $K_4F_6K_4$ with high charge most strongly affects A549 cells among them. Differently, although Smp24 has significantly lower overall charge (+3) than $K_4F_6K_4$, the IC_{50} value (32.21 µM) of $K_4F_6K_4$ against A549 cells is higher than 4.06 µM of Smp24. It is reported that the net charge need not be as high as possible to improve the anticancer activity, and some mutant peptides with low charge showed strong activity among a series of peptides [23]. Thus, net charge and anticancer activity are not always positively correlated. Instead, some α-helical peptides with cationic and amphipathic character derived from animal venoms can penetrate the membrane of tumor cells despite their difference in spatial structure and the content of disulfide bonds. Thus, other physical and chemical properties such as hydrophobicity and helicity may also contribute to the antitumor capability of AMPs [24,25].

Generally, the cytoskeleton cannot counteract the external pressure from the extracellular environment, the plasma membrane may be irreversibly deformed, leading to plasma membrane leakiness [12]. Specifically, actin fragmentation eliminates the cytoskeletal support of the plasma membrane, which then causes the appearance of pores in the plasma membrane [11]. Plasma membrane pores unbalance the controlled trafficking of ions (e.g., calcium ion), metabolites and waste products in and out of the cell and, thereby, disrupt the intracellular homeostatic conditions. Clearly, calcium concentrations that are above critical are cytotoxic, as they lead to mitochondrial failure [26]. In this study, both the cell membrane and mitochondrial membrane were damaged after Smp24 treatment (Figure 2). Hence, persistent perturbations to the cytoskeleton may lead to breakdown of cell function, permeabilization of the plasma membrane, loss of cell homeostasis, and eventual cell death [12].

Cell migration is a highly dynamic process controlled by the cytoskeleton, which mainly comprises the actin microfilaments, microtubules and intermediate filaments. During migration, cells polarize and form protrusions at the front where new adhesions are formed, which transmit the traction forces required for movement. All of these steps are coupled to major cytoskeletal rearrangements and are controlled by a wide array of signaling cascades [27]. In addition, F-actin is a functional indicator of cell migration and can mostly affect the motility of cancer cells [28,29]. In our experiment, after treatment with Smp24 for 24 h, both F-actin reorganization and cell motility are inhibited (Figure 3A–D). Recent studies have linked the mitochondrial ROS increase to suppression of lung cancer cell motility [30–32]. Moreover, mitochondria dysfunction, ROS accumulation and perturbation of F-actin fibers in the liver cancer cells treated by antitumor molecules coincidently occurred [33,34]. Considering that Smp24 can increase the contents of ROS in A549 cells (data not shown), Smp24 damages the mitochondrial membrane and increases ROS content, which disrupts the cytoskeleton. Focal adhesions (FAs) are dynamic signaling structures that anchor the cytoskeleton to the extracellular matrix via numerous effector proteins such as integrins, cadherins and other adhesion molecules [35]. Accordingly, microtubules can also adjust the integrin activation, the recruitment of specific adhesion complex components, and the assembly of FAs [36,37]. Furtherly, the assembly and disassembly of Fas are chiefly regulated by focal adhesion kinase (FAK), which can be inhibited by cytoskeleton disorganization. In addition, FAs disassembly has been reported to require dephosphorylation of P-FAK [38], and the stiffness-dependent activations of the FAK–p130Cas–Rac and PI3K-Akt pathway lead to the changed expression and distribution of N-cadherin and integrins in cells [39,40]. Considering that Smp24 can inhibit the phosphorylation of FAK and PI3K (data not shown) and damage the cytoskeleton, we can deduce the reduced adhesive potential of these cells and inhibit the mobility of cells, which is consistent with the effects of polypeptides reported by Xia et al. [41]. The migration and invasion of NSCLC, which are greatly affected by the MMPs protein family, especially MMP-2 and MMP-9, are important for its metastasis to other body regions such as bone, brain and spleen. Thus, their suppression is a matter of importance for the treatment of NSCLC [1,13]. Smp24 is demonstrated to own the obvious suppression effects on the migration and invasion of A549 cells (Figure 3B–D). Meanwhile, Smp24 can dramatically decrease the mRNA levels of MMP-2/-9 while increasing those of TIMP-1/-2 in A549 cells (Figure 3E). These data confirm that Smp24 can markedly restrain the metastasis of A549 cells. Therefore, the motility suppression of lung cancer cells by Smp24 may be associated with its changes of mRNA levels of MMP-2/-9 and TIMP-1/-2 plus F-actin reorganization.

For most peptides, despite the cogent evidence about their antitumor effects in vitro, their clinical applications have met with major obstacles owing to several issues, including their non-specific cellular cytotoxicity and rapid degradation in the blood [42]. In order to further verify the antitumor effects of Smp24 in vivo, a tumor-burdened experiment was carried out. In our study, 18 d after treatment of 2 mg/kg Smp24, the tumor weight and volume were reduced about 64.3% and 65.4%, respectively, which demonstrate its stability and antitumor potency in vivo (Figure 4B–D). Furthermore, it does not affect the weight changes of body and important organs such as heart, liver, spleen, lung and kidney (Figure 4E,F), confirming its low cytotoxicity to normal mammalian cells and biological safety in vivo. Melittin is a cationic AMP with 26 amino acids and presents antitumor effects toward various tumor cells via occurrence of necrosis and motility [43]. However, subcutaneous injection of melittin at doses of 0.5 and 10 mg/kg significantly suppresses NSCLC tumor growth by 27% and 61%, respectively [44]. Considering that Smp24 contains 24 amino acids in its primary sequence and has stronger antitumor capacity against NSCLC than melittin in vivo, and the high dose of Smp24 (10 mg/kg) lacks damaging effects on the liver and kidney (Figure 5), Smp24 has lower synthesis cost and is more efficient than melittin. Together, Smp24 has the anti-A549 cell effects in vivo without remarkable toxicity, which makes it more in reference to clinical trials. However, most natural peptides are difficult to directly apply in clinics without modification to improve their therapeutic efficacy and stability in plasma. The effects of Smp24 with cyclization, hybridization and nanodrug modification on tumor cells should be explored in the future.

4. Conclusions

In conclusion, the present findings indicate that Smp24 exerts a cytotoxic effect in vitro and an antitumor activity in in vivo models of lung carcinoma cells by suppression of cell motility and induction of necrosis. Mechanistically, Smp24 can destroy the integrity of the A549 cell membrane, mitochondrial and nuclear membrane, inhibit cytoskeleton organization and change the expression of MMP-2/-9 and TIMP-1/-2 in A549 cells. This discovery will extend the antitumor mechanism of AMPs and open an avenue for developing scorpion venom peptides into chemo-therapeutic agents targeting lung cancer cells.

5. Materials and Methods

5.1. Animals and Ethics Statement

A total of 10 BALB/c nude mice of both sexes (18–20 g, 6 weeks) were acquired from the Laboratory Animal Center of Southern Medical University and were raised in a specific pathogen-free mini-barrier system of the central animal facility of Southern Medical University under controlled conditions (60% humidity, 21 ± 2 °C room temperature, and 12 h light-dark cycle). The experimental protocols (ethical approval number: L2019226 on 11 November 2019) involving animals were approved by the Animal Ethics Committee of Southern Medical University and were implemented in the light of the international regulations for animal research. The animals were placed in an induction chamber, and anesthesia was induced with 5% isoflurane before sacrificed by cervical dislocation.

5.2. Chemicals and Cell Culture

RPMI-1640, fetal bovine serum (FBS), phosphate-buffered saline (PBS) and trypsin were all obtained from Gibco (New York, NY, USA). A549, H3122, PC-9, H460 and MRC-5 cells were purchased from the American Type Culture Collection (Manassas, WV, USA). Cells were grown in RPMI-1640 medium containing 1% penicillin–streptomycin and 10% FBS under the condition of 37 °C and 5% CO_2. Lactate dehydrogenase (LDH) release assay kit, rhodamine–phalloidin and DAPI were obtained from Beyotime Institute of Biotechnology, Shanghai, China. Smp24 (IWSFLIKAATKLLPSLFGGGKKDS) was synthesized by GL Biochem Ltd. (Shanghai, China) and purified with an Inertsil ODS-SP (C-18) RP-HPLC column (Shimazu, Japan) to over 95% purity. The high-purity peptide was collected, lyophilized, and further identified by MALDI-TOF mass spectrometry (Figures S2 and S3). The theoretical mass of 2578.09 Da of the peptide coincides with the experimentally determined one, 2578.30 Da.

5.3. Cell Viability and Proliferation Assays

MTT assay was performed to analyze cellular viability as previously reported by us [45]. Briefly, A549, H3122, PC-9, H460 and MRC-5 cells (1 × 10^4 cells/well) were grown in 96-well plates and exposed to the different concentrations of Smp24 (1.25–20 μM) with 24 h. To determine which signal pathway is primarily responsible for the cytotoxic effects of Smp24 on A549 cells, various inhibitors including 40 μM Necrostatin-1, 40 μM Z-DEVD-FMK, and 1 μM cyclosporine A were pre-incubated with A549 cells for 30 min before 5 μM Smp24 was incubated with the cells for 12 h. After incubation, 10 μL MTT (5 mg/mL) was added before incubation for 4 h at 37 °C in the dark. Subsequently, the cell medium was discarded, and 200 μL DMSO was applied into each well before the absorbance value at 490 nm was determined via the microplate reader (Tecan Company, Männedorf, Switzerland). Cell proliferation was further detected by the BeyoClick™ EdU cell proliferation kit with Alexa Fluor 488 (Beyotime Institute of Biotechnology, Shanghai, China) according to the protocol of manufacturer. In brief, A549 cells (2 × 10^5 cells/well) were grown in 6-well plates and exposed to different concentrations of Smp24 (0, 1.25, 2.5 and 5 μM) for 24 h. Then, the cells were incubated with EdU for 2 h and stained with Alexa Fluor 488 for 30 min in the dark before cell fluorescence intensity was measured by flow cytometry (Becton Dickinson Company, Franklin Lakes, NJ, USA). The experiments had been performed in triplicate.

5.4. Membrane Integrity Assay

To examine the membrane integrity, A549 cells were stained with calcein AM following the manufacturer's manual (Beyotime Institute of Biotechnology, Shanghai, China). Briefly, 1 × 10^5 A549 cells were cultured in a 12-well plate overnight. Then, the cells were exposed to Smp24 (0, 1.25, 2.5, and 5 μM) for 24 h and detached by trypsinization, washed with PBS, followed by incubated with 1 μM calcein AM at 37 °C for 30 min. After calcein, AM in the medium was cleared and cells were washed with PBS, the levels of calcein AM were detected by flow cytometry (Becton Dickinson Company, Franklin Lakes, NJ, USA). To measure the integrity of the mitochondrial membrane, after cells were incubated with calcein AM for 30 min and washed with PBS for three times, 1 μM $CoCl_2$ was added into the cells and incubated at 37 °C for another 15 min, followed by flow cytometry analysis. All experiments had been detected in triplicate.

5.5. Cell Morphology Observation

First, 2 × 10^5 A549 cells were seeded onto a 6-well plate overnight. Subsequently, the cells were exposed to Smp24 at the different concentrations (0–10 μM) for 24 h before the cellular morphology was examined under the inverted phase contrast microscope (100× magnification). Approximately 3 single pictures of each well were captured.

5.6. LDH Release Assay

The LDH release assay was carried out with a commercial kit (Beyotime Institute of Biotechnology, Shanghai, China) following the manufacturer's manual. In short, A549 cells were incubated with Smp24 (1.25, 2.5, 5, 10 and 20 µM) for 12, 24, and 48 h, respectively. Then, 10 µL of LDH release solution was applied per well and incubated for 1 h. The supernatant was transferred to a new plate and mixed with 60 µL of substrate solution per well by gentle shaking for 30 min in the dark. In some experiments, A549 cells were pre-incubated with the inhibitors (40 µM Necrostatin-1, 40 µM Z-DEVD-FMK, and 1 µM cyclosporine A) for 30 min before the cells were incubated with 5 µM of Smp24 for 12 h. Thereafter, a microplate reader (Tecan Company, Männedorf, Switzerland) was used to measure the absorbance value at 490 nm. All experiments were conducted in triplicate.

5.7. Scanning Electron Microscopy Analysis

A549 cells (1.2×10^5 cells/well) were cultured on the glass coverslips inserted in a 12-well plate for 24 h. Next, a range of concentrations of Smp24 (0, 2.5, and 5 µM) was added to the 12-well plate and co-cultured for 24 h. The cells without Smp24 were considered as the negative control. The cells were sequentially fixed with 4% glutaric dialdehyde at room temperature for 2 h and 2.5% glutaric dialdehyde at 4 °C for 8 h, followed by dehydrating with a series of gradient ethanol/water solutions. Subsequently, the samples were dried and coated with gold before observation with Phenom ProX instrument at 15 kV. The experiments were conducted in triplicate.

5.8. Fluorescence Microscopy Analysis

For F-actin reorganization analysis, 7×10^4 A549 cells were seeded in a 24-well plate overnight and incubated with Smp24 (0, 2.5, and 5 µM) for 24 h. After being fixed with 4% PFA, the cell F-actin was stained with rhodamine–phalloidin for 30 min, washed with PBS, and then stained with DAPI for another 10 min. The F-actin was observed under fluorescence microscope at 400× magnification after washing.

To further locate Smp24 within cells, after treatment with 5 µM FITC-labeled Smp24 at 37 °C for 24 h, the cells were washed with PBS, fixed with 4% paraformaldehyde (PFA) for 30 min, stained with DAPI for 10 min and followed by observation with fluorescence microscope at 400× magnification. About 3 single-plane pictures for each well were obtained.

5.9. Cell Motility Assay

Cellular motility was determined using both the scratch migration and transwell invasion assays as previously described by us [45]. For the scratch migration assay, 2×10^5 A549 cells were plated onto 6-well plates overnight. After scratching the cell monolayer with a P10 pipette tip on the second day, the cell fragments were removed by washing with PBS. A549 cells were incubated in serum-free 1640 medium with Smp24 (0, 0.3, 0.6 and 1.2 µM) for 24 h. Three single-plane photos were used to detect scratch widths at 0 h (W_0) and 24 h (W_t). Per point, the migration index (M_I) was calculated as $M_I = W_t/W_0$. The cell transwell invasion assay was conducted with a transwell plate (8 µm pore size; Corning, Inc., New York, NY, USA). The upper compartment was seeded with 5×10^4 cells in 100 µL 1640 serum-free medium with Smp24 (0, 0.3, 0.6, and 1.2 µM). The lower compartment was added by 500 µL of 1640 containing 10% FBS as an attractant for 24 h. Subsequently, the noninvasive cells (the upper side) were removed with cotton swabs, and the invasive cells on the lower side were fixed with 4% formaldehyde, followed by crystal violet staining before being photographed under a light microscope (200× magnification). Next, the crystal violet was dissolved by 100 µL of 10% (*v/v*) acetic acid, and the absorbance value at 570 nm was measured to quantify cell invasion via a microplate reader (Tecan Company, Männedorf, Switzerland). All experiments were conducted in triplicate.

5.10. Quantitative Real-Time Polymerase Chain Reaction (qRT-PCR) Assay

First, 2×10^5 A549 cells were grown in 6-well plates overnight and mixed with Smp24 (0, 0.3, 0.6, and 1.2 μM) for 12 h. Next, the cells were centrifuged and the precipitate was harvested for qRT-PCR to measure the mRNA levels of MMP-2, MMP-9, TIMP-1 and TIMP-2 with the qPCR instrument (Light Cycler480II, ROCHE Ltd., Basel, Switzerland) as reported previously by us (Table 1) [45]. To upregulate the expression of MMP-9, the cells were pretreated with PMA (50 nM) for 1 h before harvest. The reaction cycles for all genes were: 95 °C for 3 min, 40 cycles at 94 °C for 15 s and 60 °C for 45 s. *GAPDH* gene was quantified as a control to verify equal initial quantities of RNA and as an internal standard to quantify PCR products. Results were calculated with the $2^{-\Delta\Delta CT}$ method, and target gene expression was standardized to *GAPDH*. All experiments were conducted in triplicate.

Table 1. The primers sequences of qRT-PCR.

Genes	Forward	Reverse
GAPDH	5′-CGGAGTCAACGGATTTGGTCGTAT-3′	5′-AGCCTTCTCCATGGTGGTGAAGAC-3′
MMP-2	5′-AGCGAGTGGATGCCGCCTTTAA-3′	5′-CATTCCAGGCATCTGCGATGAG-3′
MMP-9	5′-GCCACTACTGTGCCTTTGAGTC-3′	5′-CCCTCAGAGAATCGCCAGTACT-3′
TIMP-1	5′-CTGTTGTTGCTGTGGCTGATAG-3′	5′-CGCTGGTATAAGGTGGTCTGG-3′
TIMP-2	5′-ACCCTCTGTGACTTCATCGTGC-3′	5′-GGAGATGTAGCACGGGATCATG-3′

5.11. In Vivo Antitumor Experiments

First, 5×10^6 A549 cells with viability over 90% were inoculated subcutaneously into the right flank per mouse. Tumor growth was examined daily by palpation and the diameter was measured with calipers in two planes. Tumor volumes were counted with the following formula: volume (mm^3) = (smallest diameter)^2 × (largest diameter)/2. Then, the mice were randomized into two groups (n = 5 mice/group) for physiological saline and Smp24 treatment. When the tumor grew to about 50 mm^3, Smp24 (2 mg/kg) or physiological saline was administrated near the tumor site every three days for 18 days. The body weight of mice and tumor volumes were recorded daily. On the 18th day of administration, the tumors and organs (heart, liver, spleen, lung, kidney) were removed, weighed and photographed. After embedding in paraffin, the tumors were stained with HE staining reagent to analyze the morphological alterations with lung cancer cells. To further detect the tumor tissue apoptosis levels, the apoptosis-related cleaved caspase-3 was also examined by immunohistochemical staining followed by observation under inverted-phase contrast microscope (200× magnification). Optical density analysis was performed on the immunohistochemical images using Image J software. The results were acquired from five mice for each group.

5.12. Acute Toxicity Analysis

To detect the acute toxicity of Smp24 in vivo, the mice were weighed and randomly divided into three groups with 3 animals each, and Smp24 (5 mg/kg and 10 mg/kg) and saline were intraperitoneally administered into mice of different groups. At 48 h after injection, serum and organ tissues from mice were collected for blood biochemical detection and organ weight measurement, respectively.

5.13. Statistical Analysis

Statistical analysis was performed by one-way ANOVA with Bonferroni's multiple comparison with GraphPad Prism 5.0 (GraphPad Software, Inc., La Jolla, CA, USA). Data were recorded as mean ± SEM. Statistical significance was shown as * $p < 0.05$, ** $p < 0.01$ and *** $p < 0.001$.

6. Patents

The study has been patented, grant number 202110599012.0.

Supplementary Materials: The following supporting information can be downloaded at: https://www.mdpi.com/article/10.3390/toxins14070438/s1. Table S1: IC_{50} values and selectivity index of cancer cell lines treated with Smp24 for 24 h. Figure S1: RP-HPLC purification of Smp24. Figure S2: MALDI-TOF-MS identification of Smp24. Figure S3: Viability of A549 cells treated with cisplatin for 24 h.

Author Contributions: Conceptualization, X.X.; methodology, X.X., R.G. and J.C.; software, R.G. and J.C.; validation, R.G., J.C. and Y.G.; formal analysis, R.G. and J.L.; investigation, X.X.; resources, R.G.; data curation, R.G. and J.L.; writing—original draft preparation, X.X., R.G. and J.L.; writing—review and editing, X.X. and M.A.A.-R.; visualization, R.G. and J.L.; supervision, X.X.; project administration, X.X.; funding acquisition, X.X. All authors have read and agreed to the published version of the manuscript.

Funding: This research was funded by the Chinese National Natural Science Foundation, grant number 31861143050, 31772476, 31911530077, 82070038 and in part by the Academy of Scientific Research and Technology (ASRT, Egypt; China–Egypt Scientific and Technological Cooperation Program).

Institutional Review Board Statement: The animal study protocol was approved by the Animal Ethics Committee of Southern Medical University and were implemented in light of the international regulations for animal research (ethical approval number: L2019226), on 11 November 2019.

Informed Consent Statement: Not applicable.

Data Availability Statement: All data are available on request.

Conflicts of Interest: The authors declare no conflict of interest.

References

1. Herbst, R.S.; Morgensztern, D.; Boshoff, C. The biology and management of non-small cell lung cancer. *Nature* **2018**, *553*, 446–454. [CrossRef]
2. Wang, G.; Li, X.; Wang, Z. APD3: The antimicrobial peptide database as a tool for research and education. *Nucleic Acids Res.* **2016**, *44*, D1087–D1093. [CrossRef]
3. Tornesello, A.L.; Borrelli, A.; Buonaguro, L.; Buonaguro, F.M.; Tornesello, M.L. Antimicrobial peptides as anticancer agents: Functional properties and biological activities. *Molecules* **2020**, *25*, 2850. [CrossRef]
4. Xu, X.; Lai, R. The chemistry and biological activities of peptides from amphibian skin secretions. *Chem. Rev.* **2015**, *115*, 1760–1846. [CrossRef]
5. Polacheck, W.J.; Zervantonakis, I.K.; Kamm, R.D. Tumor cell migration in complex microenvironments. *Cell Mol. Life Sci.* **2013**, *70*, 1335–1356. [CrossRef]
6. Khamessi, O.; Ben, M.H.; Elfessi-Magouri, R.; Kharrat, R. RK1, the first very short peptide from Buthus occitanus tunetanus inhibits tumor cell migration, proliferation and angiogenesis. *Biochem. Biophys. Res. Commun.* **2018**, *499*, 1–7. [CrossRef]
7. Raposo, C. Scorpion and spider venoms in cancer treatment: State of the art, challenges, and perspectives. *J. Clini. Transl. Res.* **2017**, *3*, 233–249. [CrossRef]
8. Harrison, P.L.; Abdel-Rahman, M.A.; Strong, P.N.; Tawfik, M.M.; Miller, K. Characterisation of three alpha-helical antimicrobial peptides from the venom of Scorpio maurus palmatus. *Toxicon* **2016**, *117*, 30–36. [CrossRef]
9. Harrison, P.L.; Heath, G.R.; Johnson, B.; Abdel-Rahman, M.A.; Strong, P.N.; Evans, S.D.; Miller, K. Phospholipid dependent mechanism of smp24, an alpha-helical antimicrobial peptide from scorpion venom. *Biochim. Biophys. Acta* **2016**, *1858*, 2737–2744. [CrossRef]
10. Elrayess, R.A.; Mohallal, M.E.; El-Shahat, Y.M.; Ebaid, H.M.; Miller, K.; Strong, P.N.; Abdel-Rahman, M.A. Cytotoxic effects of smp24 and smp43 scorpion venom antimicrobial peptides on tumour and non-tumour cell lines. *Int. J. Pept. Res. Ther.* **2020**, *26*, 1409–1415. [CrossRef]
11. Gefen, A.; Weihs, D. Mechanical cytoprotection: A review of cytoskeleton-protection approaches for cells. *J. Biomech.* **2016**, *49*, 1321–1329. [CrossRef]
12. Gefen, A.; Weihs, D. Cytoskeleton and plasma-membrane damage resulting from exposure to sustained deformations: A review of the mechanobiology of chronic wounds. *Med. Eng. Phys.* **2016**, *38*, 828–833. [CrossRef]
13. Friedl, P.; Wolf, K. Tumour-cell invasion and migration: Diversity and escape mechanisms. *Nat. Rev. Cancer* **2003**, *3*, 362–374. [CrossRef]
14. Svitkina, T. The actin cytoskeleton and Actin-Based motility. *Cold Spring Harb. Perspect. Biol.* **2018**, *10*, a18267. [CrossRef]

15. Zhang, X.; Wu, J.Z.; Lin, Z.X.; Yuan, Q.J.; Li, Y.C.; Liang, J.L.; Zhan, J.Y.; Xie, Y.L.; Su, Z.R.; Liu, Y.H. Ameliorative effect of supercritical fluid extract of Chrysanthemum indicum Linnen against D-galactose induced brain and liver injury in senescent mice via suppression of oxidative stress, inflammation and apoptosis. *J. Ethnopharmacol.* **2019**, *234*, 44–56. [CrossRef]

16. Maraming, P.; Klaynongsruang, S.; Boonsiri, P.; Peng, S.F.; Daduang, S.; Leelayuwat, C.; Pientong, C.; Chung, J.G.; Daduang, J. The cationic cell-penetrating KT2 peptide promotes cell membrane defects and apoptosis with autophagy inhibition in human HCT 116 colon cancer cells. *J. Cell Physiol.* **2019**, *234*, 22116–22129. [CrossRef]

17. Leite, N.B.; Aufderhorst-Roberts, A.; Palma, M.S.; Connell, S.D.; Ruggiero, N.J.; Beales, P.A. PE and PS lipids synergistically enhance membrane poration by a peptide with anticancer properties. *Biophys. J.* **2015**, *109*, 936–947. [CrossRef]

18. Paredes-Gamero, E.J.; Martins, M.N.; Cappabianco, F.A.; Ide, J.S.; Miranda, A. Characterization of dual effects induced by antimicrobial peptides: Regulated cell death or membrane disruption. *Biochim. Biophys. Acta* **2012**, *1820*, 1062–1072. [CrossRef]

19. Ju, X.; Fan, D.; Kong, L.; Yang, Q.; Zhu, Y.; Zhang, S.; Su, G.F.; Li, Y. Antimicrobial peptide Brevinin-1RL1 from frog skin secretion induces apoptosis and necrosis of tumor cells. *Molecules* **2021**, *26*, 2059. [CrossRef]

20. Chen, Y.C.; Tsai, T.L.; Ye, X.H.; Lin, T.H. Anti-proliferative effect on a colon adenocarcinoma cell line exerted by a membrane disrupting antimicrobial peptide KL15. *Cancer Biol. Ther.* **2015**, *16*, 1172–1183. [CrossRef]

21. Cichon, T.; Smolarczyk, R.; Matuszczak, S.; Barczyk, M.; Jarosz, M.; Szala, S. D-K6L 9 peptide combination with IL-12 inhibits the recurrence of tumors in mice. *Arch. Immunol. Ther. Exp.* **2014**, *62*, 341–351. [CrossRef]

22. Huang, W.; Lu, L.; Shao, X.; Tang, C.; Zhao, X. Anti-melanoma activity of hybrid peptide P18 and its mechanism of action. *Biotechnol. Lett.* **2010**, *32*, 463–469. [CrossRef]

23. Yang, D.; Zhu, L.; Lin, X.; Zhu, J.; Qian, Y.; Liu, W.; Chen, J.; Zhou, C.; He, J. Therapeutic effects of synthetic triblock amphiphilic short antimicrobial peptides on human lung adenocarcinoma. *Pharmaceutics* **2022**, *14*, 929. [CrossRef]

24. Radis-Baptista, G. Cell-penetrating peptides derived from animal venoms and toxins. *Toxins* **2021**, *13*, 147. [CrossRef]

25. Ma, R.; Wong, S.W.; Ge, L.; Shaw, C.; Siu, S.; Kwok, H.F. In vitro and MD simulation study to explore physicochemical parameters for antibacterial peptide to become potent anticancer peptide. *Mol. Ther. Oncolytics* **2020**, *16*, 7–19. [CrossRef]

26. Zoeteweij, J.P.; van de Water, B.; de Bont, H.J.; Mulder, G.J.; Nagelkerke, J.F. Calcium-induced cytotoxicity in hepatocytes after exposure to extracellular ATP is dependent on inorganic phosphate effects on mitochondrial calcium. *J. Biol. Chem.* **1993**, *268*, 3384–3388. [CrossRef]

27. Seetharaman, S.; Etienne-Manneville, S. Cytoskeletal crosstalk in cell migration. *Trends Cell Biol.* **2020**, *30*, 720–735. [CrossRef]

28. Gu, S.; Kounenidakis, M.; Schmidt, E.M.; Deshpande, D.; Alkahtani, S.; Alarifi, S.; Föller, M.; Alevizopoulos, K.; Lang, F.; Stournaras, C. Rapid activation of FAK/mTOR/p70S6K/PAK1-signaling controls the early testosterone-induced actin reorganization in colon cancer cells. *Cell Signal.* **2013**, *25*, 66–73. [CrossRef]

29. Jeong, Y.J.; Choi, Y.; Shin, J.M.; Cho, H.J.; Kang, J.H.; Park, K.K.; Choe, J.Y.; Bae, Y.S.; Han, S.M.; Kim, C.H.; et al. Melittin suppresses EGF-induced cell motility and invasion by inhibiting PI3K/Akt/mTOR signaling pathway in breast cancer cells. *Food Chem. Toxicol.* **2014**, *68*, 218–225. [CrossRef]

30. Zhang, T.; Li, S.M.; Li, Y.N.; Cao, J.L.; Xue, H.; Wang, C.; Jin, C.H. Atractylodin induces apoptosis and inhibits the migration of a549 lung cancer cells by regulating ROS-Mediated signaling pathways. *Molecules* **2022**, *27*, 2946. [CrossRef]

31. Hua, R.; Pei, Y.; Gu, H.; Sun, Y.; He, Y. Antitumor effects of flavokawain-B flavonoid in gemcitabine-resistant lung cancer cells are mediated via mitochondrial-mediated apoptosis, ROS production, cell migration and cell invasion inhibition and blocking of PI3K/AKT Signaling pathway. *J. BUON* **2020**, *25*, 262–267.

32. Sun, M.; Hong, S.; Li, W.; Wang, P.; You, J.; Zhang, X.; Tang, F.; Wang, P.; Zhang, C. MiR-99a regulates ROS-mediated invasion and migration of lung adenocarcinoma cells by targeting NOX4. *Oncol. Rep.* **2016**, *35*, 2755–2766. [CrossRef]

33. Yu, L.; Chen, W.; Tang, Q.; Ji, K.Y. Micheliolide inhibits liver cancer cell growth via inducing apoptosis and perturbing actin cytoskeleton. *Cancer Manag. Res.* **2019**, *11*, 9203–9212. [CrossRef]

34. Ko, C.Y.; Shih, P.C.; Huang, P.W.; Lee, Y.H.; Chen, Y.F.; Tai, M.H.; Liu, C.H.; Wen, Z.H.; Kuo, H.M. Sinularin, an anti-cancer agent causing mitochondria-modulated apoptosis and cytoskeleton disruption in human hepatocellular carcinoma. *Int. J. Mol. Sci.* **2021**, *22*, 3946. [CrossRef]

35. Fife, C.M.; Mccarroll, J.A.; Kavallaris, M. Movers and shakers: Cell cytoskeleton in cancer metastasis. *Br. J. Pharmacol.* **2014**, *171*, 5507–5523. [CrossRef]

36. Zhou, X.; Li, J.; Kucik, D.F. The microtubule cytoskeleton participates in control of beta2 integrin avidity. *J. Biol. Chem.* **2001**, *276*, 44762–44769. [CrossRef]

37. Ng, D.H.; Humphries, J.D.; Byron, A.; Millon-Fremillon, A.; Humphries, M.J. Microtubule-dependent modulation of adhesion complex composition. *PLoS ONE* **2014**, *9*, e115213. [CrossRef]

38. Ezratty, E.J.; Partridge, M.A.; Gundersen, G.G. Microtubule-induced focal adhesion disassembly is mediated by dynamin and focal adhesion kinase. *Nat. Cell Biol.* **2005**, *7*, 581–590. [CrossRef]

39. Mui, K.L.; Bae, Y.H.; Gao, L.; Liu, S.L.; Xu, T.; Radice, G.L.; Chen, C.S.; Assoian, R.K. N-Cadherin induction by ECM stiffness and FAK overrides the spreading requirement for proliferation of vascular smooth muscle cells. *Cell Rep.* **2015**, *10*, 1477–1486. [CrossRef]

40. Maes, H.; Van Eygen, S.; Krysko, D.V.; Vandenabeele, P.; Nys, K.; Rillaerts, K.; Garg, A.D.; Verfaillie, T.; Agostinis, P. BNIP3 supports melanoma cell migration and vasculogenic mimicry by orchestrating the actin cytoskeleton. *Cell Death Dis.* **2014**, *5*, e1127. [CrossRef]

41. Xia, L.; Wu, Y.; Kang, S.; Ma, J.; Yang, J.; Zhang, F. CecropinXJ, a silkworm antimicrobial peptide, induces cytoskeleton disruption in esophageal carcinoma cells. *Acta Biochim. Biophys. Sin.* **2014**, *46*, 867–876. [CrossRef] [PubMed]
42. Liscano, Y.; Onate-Garzon, J.; Delgado, J.P. Peptides with dual antimicrobial-anticancer activity: Strategies to overcome peptide limitations and rational design of anticancer peptides. *Molecules* **2020**, *25*, 4245. [CrossRef] [PubMed]
43. Liu, C.C.; Hao, D.J.; Zhang, Q.; An, J.; Zhao, J.J.; Chen, B.; Zhang, L.L.; Yang, H. Application of bee venom and its main constituent melittin for cancer treatment. *Cancer Chemother. Pharm.* **2016**, *78*, 1113–1130. [CrossRef]
44. Zhang, S.F.; Chen, Z. Melittin exerts an antitumor effect on nonsmall cell lung cancer cells. *Mol. Med. Rep.* **2017**, *16*, 3581–3586. [CrossRef]
45. Deng, Z.; Chai, J.; Zeng, Q.; Zhang, B.; Ye, T.; Chen, X.; Xu, X. The anticancer properties and mechanism of action of tablysin-15, the RGD-containing disintegrin, in breast cancer cells. *Int. J. Biol. Macromol.* **2019**, *129*, 1155–1167. [CrossRef]

Article

A Pore Forming Toxin-like Protein Derived from Chinese Red Belly Toad *Bombina maxima* Triggers the Pyroptosis of Hippomal Neural Cells and Impairs the Cognitive Ability of Mice

Qingqing Ye [1], Qiquan Wang [2], Wenhui Lee [3], Yang Xiang [2], Jixue Yuan [1], Yun Zhang [3,4,*] and Xiaolong Guo [1,*]

[1] School of Physical Education, Yunnan Normal University, Kunming 650500, China
[2] Human Aging Research Institute, School of Life Science, Nanchang University, Jiangxi Key Laboratory of Human Aging, Nanchang 330031, China
[3] Key Laboratory of Animal Models and Human Disease Mechanisms of the Chinese Academy of Sciences, Key Laboratory of Bioactive Peptides of Yunnan Province, Kunming Institute of Zoology, Chinese Academy of Sciences, Kunming 650201, China
[4] Center for Excellence in Animal Evolution and Genetics, Chinese Academy of Sciences, Kunming 650223, China
* Correspondence: zhangy@mail.kiz.ac.cn (Y.Z.); 210091@ynnu.edu.cn (X.G.)

Abstract: Toxin-like proteins and peptides of skin secretions from amphibians play important physiological and pathological roles in amphibians. βγ-CAT is a Chinese red-belly toad-derived pore-forming toxin-like protein complex that consists of aerolysin domain, crystalline domain, and trefoil factor domain and induces various toxic effects via its membrane perforation process, including membrane binding, oligomerization, and endocytosis. Here, we observed the death of mouse hippocampal neuronal cells induced by βγ-CAT at a concentration of 5 nM. Subsequent studies showed that the death of hippocampal neuronal cells was accompanied by the activation of Gasdermin E and caspase-1, suggesting that βγ-CAT induces the pyroptosis of hippocampal neuronal cells. Further molecular mechanism studies revealed that the pyroptosis induced by βγ-CAT is dependent on the oligomerization and endocytosis of βγ-CAT. It is well known that the damage of hippocampal neuronal cells leads to the cognitive attenuation of animals. The impaired cognitive ability of mice was observed after intraperitoneal injection with 10 μg/kg βγ-CAT in a water maze assay. Taken together, these findings reveal a previously unknown toxicological function of a vertebrate-derived pore-forming toxin-like protein in the nerve system, which triggers the pyroptosis of hippocampal neuronal cells, ultimately leading to hippocampal cognitive attenuation.

Keywords: pore-forming toxin; red-belly toad; *Bombina maxima*; hippocampal neuronal cell; pyroptosis; cognitive function

Key Contribution: The paper provides first-hand evidence for the role of vertebrate-derived aerolysin-like pore-forming toxins in triggering pyroptosis of hippocampal neuronal cell and inducing hippocampal cognitive functional impairment.

Citation: Ye, Q.; Wang, Q.; Lee, W.; Xiang, Y.; Yuan, J.; Zhang, Y.; Guo, X. A Pore Forming Toxin-like Protein Derived from Chinese Red Belly Toad *Bombina maxima* Triggers the Pyroptosis of Hippomal Neural Cells and Impairs the Cognitive Ability of Mice. *Toxins* **2023**, *15*, 191. https://doi.org/10.3390/toxins15030191

Received: 15 January 2023
Revised: 22 February 2023
Accepted: 28 February 2023
Published: 3 March 2023

1. Introduction

Pore-forming toxins (PFTs) are important exotoxins which are usually secreted by pathogenic bacteria and have a toxic function, mainly by perforating the target cell membrane and causing cell death [1]. Aerolysin is a β-barrel-type PFT primarily produced by *Aeromonas* species. Interestingly, an increasing number of genomic data and bioinformatic analyses have shown that aerolysin-like proteins (ALPs) widely exist in all living things, including vertebrates [2,3]. It is well known that pathogenic bacteria-derived ALPs exert various toxic effects (such as hemolysis, cell death, and others) via channel formation in a target cell's membrane [4]. Regrettably, the detailed pathophysiologic functions of vertebrate-derived ALPs are mostly unknown; one of the most important reasons for this is that naturally purified ALPs are difficult to obtain [5]. Recently, emerging evidence has suggested that recombinant vertebrate-derived ALPs play pivotal toxicological and pathophysiological roles in living organisms. The gene encoding natterin-like ALPs from the *Danio rerio* fish has been found to play important physiological roles in the embryonic development of zebrafish [6,7]. Moreover, the recombinant ALP of *Danio rerio*, named Aep1, has also been found to play a crucial role in the antimicrobial immunity of zebrafish [8]. It has been reported that an ALP from lamprey, named LIP, can selectively kill tumor cells [9]. All of these findings indicate that ALPs from vertebrates are involved in many pathophysiologic processes, while the detailed biological functions and molecular mechanisms of natural ALPs remain unclear.

The hippocampal region of the brain is responsible for learning and memory, and the integrity and plasticity of hippocampal neurons are crucial to the maintenance of the learning and cognitive function of the hippocampus [10,11]. Damage to hippocampal neurons will obviously affect the learning and cognitive abilities of animals. A number of exotoxins secreted by bacteria have been reported to produce significant toxicity to neurons; one of the most representative types is neurotoxins [12,13], including botulinum neurotoxins, tetanus neurotoxins, cholera toxins, etc. Both botulinum neurotoxins and tetanus neurotoxins cause severe neuroparalytic syndromes dependent on their metalloprotease activities [14]. As the main member of the bacterial exotoxin family, PFTs are also toxic to the nervous system. The epsilon toxin (ETX), an ALP produced by *Clostridium perfringens*, has been identified to induce perivascular oedema of the brain and lead to the firing of the neural network by binding to certain neurons or oligodendrocytes [15]. However, the roles of vertebrate-derived ALPs in the nervous system are largely unclear, and to date, the effect of ALPs from vertebrates on the hippocampus has not been reported.

In our previous studies, an ALP complex composed of one ALP subunit and two trefoil factor (TFF) subunits, named βγ-CAT, was identified from the skin secretions of *Bombina maxima*, a red-belly toad distributed specifically in southwest China [16]. The results of previous studies revealed that βγ-CAT has diverse toxic effects via pore formation in the membrane of target cells, especially mammalian cells. βγ-CAT not only strongly induces the hemolysis of red blood cells but also triggers the Ca^{2+}-dependent apoptosis of platelets [17]. Not surprisingly, an increasing amount of evidence has shown that βγ-CAT also displays diverse pharmacological activities in addition to typical toxic effects. Further studies showed that βγ-CAT can selectively kill tumor cells, promote wound healing and tissue repair, and counteract enveloped virus invasion by directly killing enveloped virus [18,19]. The study of detailed molecular mechanisms revealed that acidic glycosphingolipids in target cell membranes are the receptors of βγ-CAT and mediate the membrane binding, oligomerization, and endocytosis of βγ-CAT [20,21]. Acidic glycosphingolipids are mainly distributed in the nervous system, especially in the myelin sheath of neurons [22,23]. Unfortunately, the toxic roles of βγ-CAT in the nervous system remain elusive.

Here, the detailed role of βγ-CAT in the brain was explored, and our findings showed that βγ-CAT can cross the blood–brain barrier and induce the cell death of mouse hippocampal neuronal cells. Subsequent studies indicated that βγ-CAT-induced cell death was gasdermin-E- and caspase-1-dependent, so the βγ-CAT-induced cell death of mouse hippocampal neuronal cells is pyroptosis. Further studies revealed that the pyroptosis of mouse hippocampal neuronal cells induced by βγ-CAT relied on the membrane binding, oligomerization, and endocytosis of βγ-CAT. Finally, the cognitive ability of an animal was shown to be impaired after being treated with βγ-CAT. These findings provide previously unknown data which enable better understanding of the roles of vertebrate ALPs in the nervous system.

2. Results

2.1. βγ-CAT Displays Strong Cytotoxicity and Induces Death of Mouse Hippocapmal Neuronal Cells

It is well known that acidic glycosphingolipids (mainly gangliosides and sulfatides) are mostly distributed in neurons or glial cells [24]. Acidic glycosphingolipids in cell membrane are the receptors of βγ-CAT and are involved in the endocytosis of βγ-CAT, as described previously [20]. Thus, the cytotoxic effect of βγ-CAT on neurons was studied first. To test the effect of the cytotoxicity of βγ-CAT on neurons, the lactic dehydrogenase (LDH) release assay was performed, and the HT-22 mouse hippocampal neuronal cell line was used. The findings showed that the LDH release of HT-22 cells induced by βγ-CAT was largely increased at a concentration no lower than 5 nM, and the LDH release of HT-22 cells induced by βγ-CAT was shown to occur in a dose-dependent manner (Figure 1A). In addition, an MTT assay was also performed to determine the viability of HT-22 cells after being treated with βγ-CAT, and the findings showed that cell viability was significantly reduced after treatment with βγ-CAT at a starting concentration of 5 nM (Figure 1B). These findings suggest that βγ-CAT is highly cytotoxic to hippocampal neuronal cells and impairs cell survival. To further study HT-22 cell death induced by βγ-CAT, flow cytometry was performed. HT-22 cells underwent progressive cell death after being treated by 5 nM βγ-CAT at different treatment times, and this was shown in a time-dependent manner (Figure 2A). The further quantitative analysis of the dead cells revealed that significant cell death occurred after HT-22 cells were treated with 5 nM βγ-CAT for 3 min or more (Figure 2B). Taken together, these results show that βγ-CAT presented strong cytotoxicity to mouse hippocampal neuronal cells and further led to cell death. However, the type of cell death of hippocampal neuronal cells that βγ-CAT induces still requires further study.

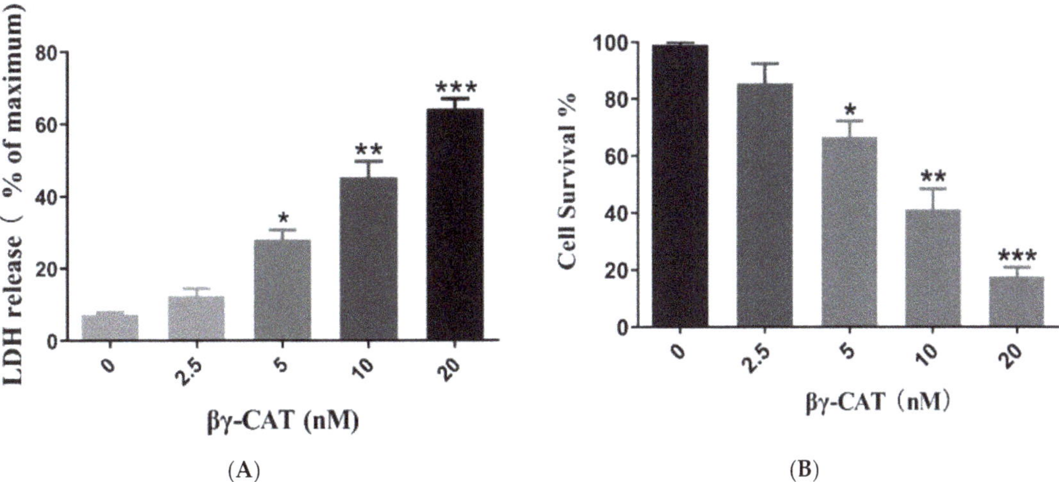

Figure 1. βγ-CAT has strong cytotoxicity to HT-22 mouse hippocampal neuronal cells. (**A**) HT-22 cells were treated with a gradient of concentrations of βγ-CAT (from 2.5 to 20 nM) at 37 °C for 10 min; then, the LDH release of HT-22 cells and the cytotoxicity of βγ-CAT was determined using LDH release assay. (**B**) HT-22 cells were treated with a gradient of concentrations of βγ-CAT (from 2.5 to 20 nM) at 37 °C for 10 min; then, the survival rate of HT-22 cells was measured using the MTS assay. * $p < 0.05$, ** $p < 0.01$, and *** $p < 0.001$ versus the respective control. An unpaired two-tailed Student's *t*-test was used in LDH release assay and MTT assay.

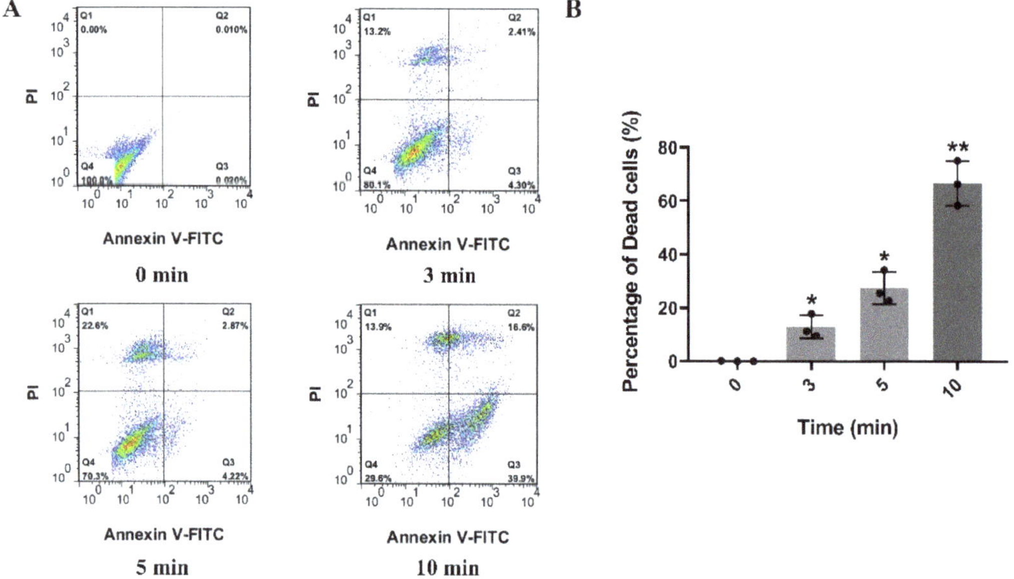

Figure 2. Flow cytometry analysis revealed that βγ-CAT can trigger cell death of HT-22 cells. (**A**) HT-22 cells were treated with 5 nM βγ-CAT at 37 °C with a gradient of times (from 3 to 10 min); then, the HT-22 cells were collected and washed three times and incubated with FITC-labeled Annexin V monoclonal antibody and PI. Finally, the cell death was determined via flow cytometry. (**B**) HT-22 cells were treated with 5 nM βγ-CAT at 37 °C with a gradient of times (from 3 to 10 min), and the percentage of dead cells was calculated and quantified. * $p < 0.05$ and ** $p < 0.01$ versus the control. An unpaired two-tailed Student's *t*-test was used in flow cytometry assay. The data of flow cytometry are representative of three independent experiments.

2.2. βγ-CAT Triggers Gasdermin-E-Dependent Pyroptosis of HT-22 Cells

It is well known that, to date, many types of cell death have been identified, such as apoptosis, programmed cell necrosis, ferroptosis, etc. [25,26]. Pyroptosis is a type of inflammatory cell death triggered by excessive inflammation, which is generally considered to be accompanied by the activation of inflammasomes and the mature cleavage of gasdermins. Thus, pyroptosis is also defined as gasdermin-mediated programmed necrotic cell death [27]. To study the type of cell death induced by βγ-CAT, the transcriptomic analysis of HT-22 cells after treatment with βγ-CAT revealed that the expression of gasdermin E is upregulated after being treated with βγ-CAT, suggesting that the βγ-CAT-triggered cell death of HT-22 cells is pyroptosis. However, more evidence is still needed to support our hypothesis. In order to determine the type of gasdermin expressed on HT-22 cells, a polymerase chain reaction (PCR) and real-time quantification PCR (RT-qPCR) were performed, and the findings showed that gasdermin E (GSDME) was the predominantly expressed gasdermin on HT-22 cells (Figure 3A,B). A subsequent Western blotting assay revealed that both GSDMD and GSDME are expressed in Raw 264.7 cells, while only GSDME is expressed in HT-22 cells (Figure 3C). Furthermore, the expression of a cleaved GSDME N-terminal in HT-22 cells was increased significantly after treatment with βγ-CAT at a concentration of 3 nM for different lengths of time, and the expression of a cleaved GS-DME N-terminal was presented in a time-dependent manner. In addition to GSDME, the expression levels of cleaved caspase 1, IL-1β, and IL-18 were also significantly increased in HT-22 cells after βγ-CAT treatment (Figure 3D). As mentioned above, the occurrence of pyroptosis is usually accompanied by the cleavage of gasdermin and the maturation and release of IL-1β or IL-18. A subsequent enzyme-linked immunosorbent assay (ELISA) showed that both IL-1β and IL-18 were largely increased in the cell supernatant after being treated with 3 nM βγ-CAT (Figure 3E,F). These findings showed that the cell death of HT-22 cells induced by βγ-CAT was GSDME- and caspase-1-dependent, suggesting that βγ-CAT has the ability to trigger the GSDME- and caspase-1-dependent pyroptosis of HT-22 cells.

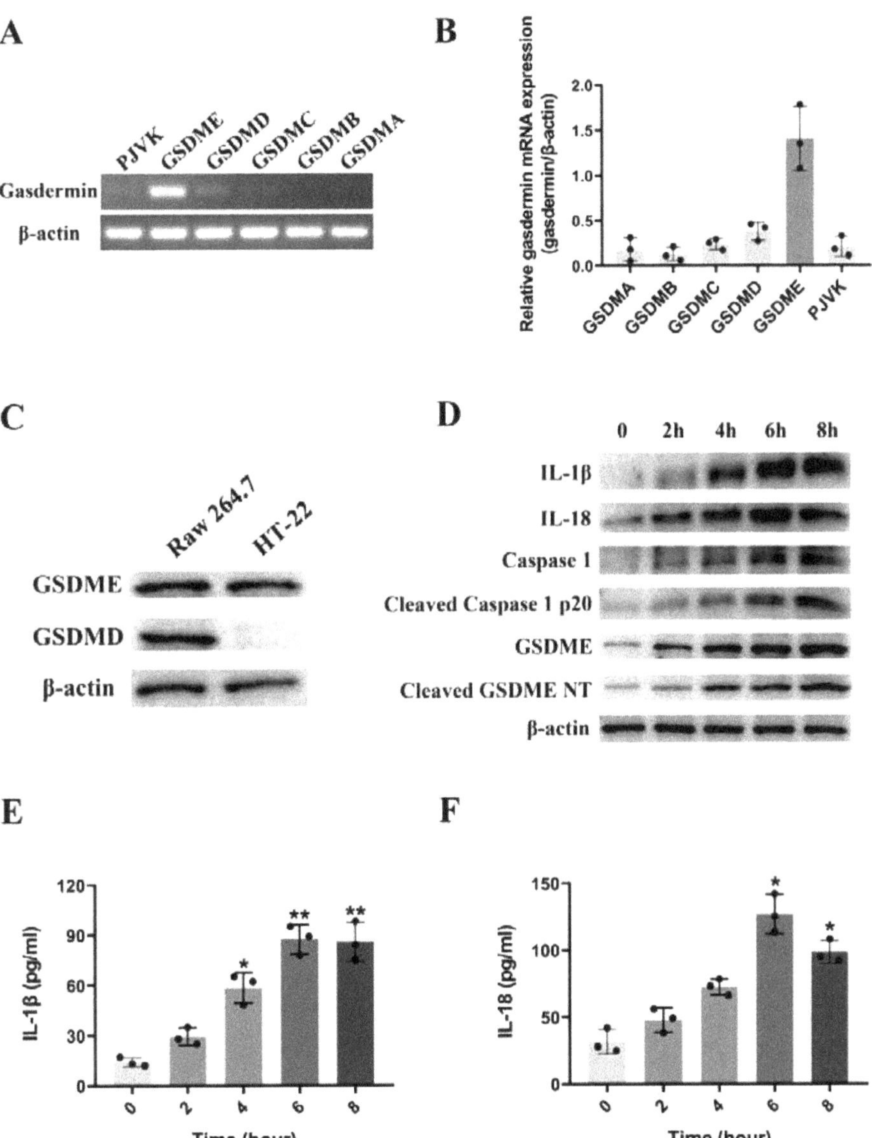

Figure 3. βγ-CAT triggers GSDME-dependent pyroptosis of HT-22 mouse hippocampal neuronal cells. (**A**) The expression of gasdermins in HT-22 cells was determined via PCR assay using specific primers. (**B**) The expression of gasdermins in HT-22 cells was determined via RT-qPCR assay using specific primers. (**C**) The expression of gasdermins in HT-22 cells was determined via Western blotting assay using specific antibodies of GSDMD and GSDME. (**D**) HT-22 cells were incubated with 3 nM βγ-CAT at 37 °C for different lengths of time (0 h, 2 h, 4 h, 6 h, and 8 h), and the full-length GSDME,

cleaved GSDME N-terminal, caspase 1, cleaved caspase 1 p20, IL-1β, and IL-18 were detected via Western blotting using specific antibodies. (**E**) HT-22 cells were incubated with 3 nM βγ-CAT at 37 °C for different lengths of time (0 h, 2 h, 4 h, 6 h, and 8 h); then, the IL-1β in cell supernatant was measured via ELISA. * $p < 0.05$ and ** $p < 0.01$ versus the control. An unpaired two-tailed Student's *t*-test was used in ELISA assay. (**F**) HT-22 cells were incubated with 3 nM βγ-CAT at 37 °C for different times (0 h, 2 h, 4 h, 6 h, and 8 h); then, the IL-18 in cell supernatant was measured via ELISA. * $p < 0.05$ versus the control. An unpaired two-tailed Student's *t*-test was used in ELISA assay. All immunoblots in (**C,D**) are representative of three independent experiments. The 3 dots on each column in (**B,E,F**) represent the data of 3 independent repeated experiments.

2.3. The Pyroptosis of HT-22 Cells Triggered by βγ-CAT Depend on Its Membrane Binding, Oligomerization and Endocytosis

A previous study showed that βγ-CAT functions through a general mode of action, namely membrane binding, oligomerization, and endocytosis [20]. Here, the pyroptosis of HT-22 cells triggered by βγ-CAT was found like in the studies mentioned above, but the detailed molecular mechanisms remain unclear. First, flow cytometry analysis revealed that βγ-CAT binds strongly with the plasma membranes of HT-22 cells at the concentration of 2.5 or 5 nM (Figure 4A). In general, oligomerization and pore formation will occur once ALPs bind to the cell membrane. Therefore, Western blotting was used to detect the oligomerization of βγ-CAT. A sodium dodecyl sulfate (SDS)-stable oligomer at approximately 180 kDa could be observed after HT-22 cells were treated with βγ-CAT at the concentrations of 5 or 10 nM (Figure 4B), suggesting that βγ-CAT not only undergoes membrane binding but also undergoes allosteric and oligomerization processes. A previous study revealed that the endocytosis of βγ-CAT is pivotal for the functions of βγ-CAT, so βγ-CAT was determined to be a secretory endolysosome channel [28]. Thus, the endocytosis of βγ-CAT was further studied using confocal microscopy. The co-localization of βγ-CAT and lamp-1 could be observed after HT-22 cells were treated with 5 nM βγ-CAT (Figure 4C). The further quantitative analysis of the co-localization region revealed that the Pearson's coefficient of the βγ-CAT treatment group was much higher than that of the control group (Figure 4D), suggesting that after endocytosis, βγ-CAT functions by regulating the endocytolysosomal pathway. Taken together, these findings suggest that the pyroptosis induced by βγ-CAT relies on the membrane binding, oligomerization, and endocytosis of βγ-CAT.

2.4. βγ-CAT Crosses the Blood-Brain Barrier and Reduces Cognitive Function of Mice

As mentioned above, βγ-CAT triggers the GSDME-dependent pyroptosis of mouse hippocampal neuronal cells through its membrane binding, oligomerization, and endocytosis. It is well known that the excessive death of hippocampal neuronal cells impairs cognitive function in the hippocampus [29,30]. Therefore, it is necessary to explore the role of βγ-CAT on the cognitive function of mice. It has been reported that many toxin-like proteins cannot cross the blood–brain barrier (BBB) because of their high molecular weight. βγ-CAT is a 72 kDa protein complex, and theoretically, it crosses the BBB with difficulty. Here, the ability of βγ-CAT to cross BBB was determined using an in vitro BBB model built with a co-culture of the human cerebral microvascular endothelial cell line (hCMEC/D3) and human normal glial cell line (HEB). Our findings revealed that the transendothelial electrical resistance (TEER) reduction was increased significantly after treatment with βγ-CAT at the concentrations of 2.5 or 5 nM, and the peak of TEER reduction was present 1 h after βγ-CAT treatment (Figure 5A). Similarly, the penetration rate of Na–F was also increased after βγ-CAT treatment (Figure 5B). To further test the penetrability of βγ-CAT on the BBB, an in vivo BBB penetration assay was performed using the fluorescent tracer method. The whole brain of a mouse was removed two hours after the mice were intraperitoneally injected with 100 μL of 40 μg/mL βγ-CAT. The immunofluorescence analysis revealed that βγ-CAT crosses the BBB and is enriched in the hippocampus (Figure 5C). These findings suggest that βγ-CAT treatment augments the permeability of the BBB and then crosses the BBB. To further study the impaired hippocampal synaptic functions and spatial cognitive functions induced by βγ-CAT, a water maze assay was adopted. The findings from the water maze assay revealed that the escape latency of mice during the spatial learning stage increased significantly after intraperitoneal injection with 10 μg/kg βγ-CAT (Figure 6A). In addition to the escape latency, the dwell time during the memory test also increased after intraperitoneal injection with 10 μg/kg βγ-CAT (Figure 6B). In addition to time, the distance in the water maze was also measured, and the findings showed that the moving distance in the center also increased after intraperitoneal injection with 10 μg/kg βγ-CAT (Figure 6C). Curiously, the percentage time in the center did not significantly increase after intraperitoneal injection with 10 μg/kg βγ-CAT (Figure 6D). These findings reveal that βγ-CAT has the ability to pass through the BBB and then trigger the pyroptosis of hippocampal neuronal cells; finally, it impairs the cognitive functions of the hippocampus.

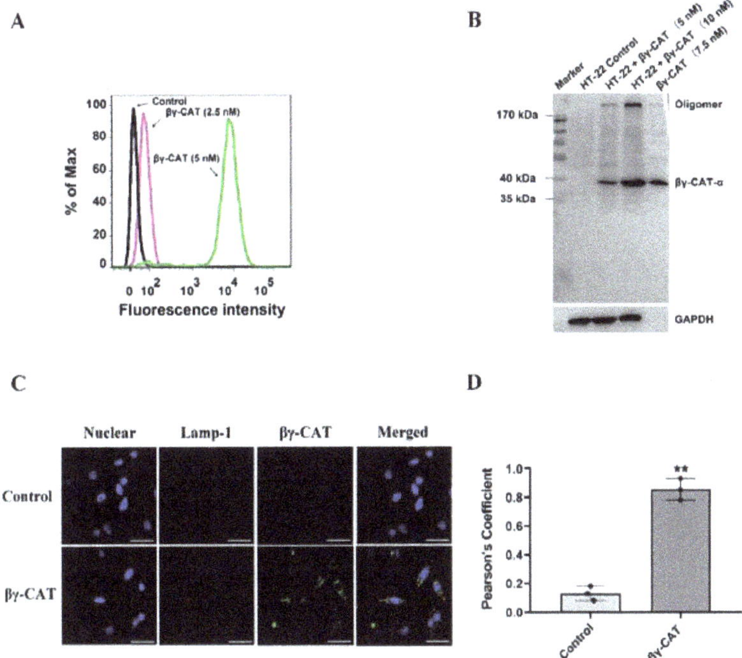

Figure 4. The pyroptosis triggered by βγ-CAT depended on the membrane, oligomerization, and endocytosis of βγ-CAT. (**A**) The HT-22 cells were treated with βγ-CAT at the concentrations of 2.5 and 5 nM for 15 min and then incubated with anti-βγ-CAT antibody; finally, the membrane binding of βγ-CAT was determined via flow cytometry. The untreated HT-22 cells were used as a negative control. (**B**) The HT-22 cells were treated with βγ-CAT at the concentrations of 5 and 10 nM for 30 min and then incubated with rabbit anti-βγ-CAT polyclonal antibody and goat anti-rabbit secondary antibody; finally, the oligomerization of βγ-CAT was determined via Western blotting. (**C**) The HT-22 cells were treated via βγ-CAT at the concentrations of 5 nM for 30 min and then incubated with rabbit anti-βγ-CAT polyclonal antibody and mouse anti-lamp-1 antibody. After being washed three times, the HT-22 cells were then incubated with FITC-labeled goat anti-rabbit secondary antibody and Cy3-labeled goat anti-mouse secondary antibody; finally, the endocytosis and lysosomal colocalization of βγ-CAT was determined via confocal microscopy. Scale bar = 25 μm. (**D**) Pearson's correlation coefficients for co-localization of Lamp-1 and βγ-CAT. Data are expressed as mean ± SD of 4–6 cells. ** $p < 0.01$ versus the control. All immunoblots in (**B**) are representative of three independent experiments.

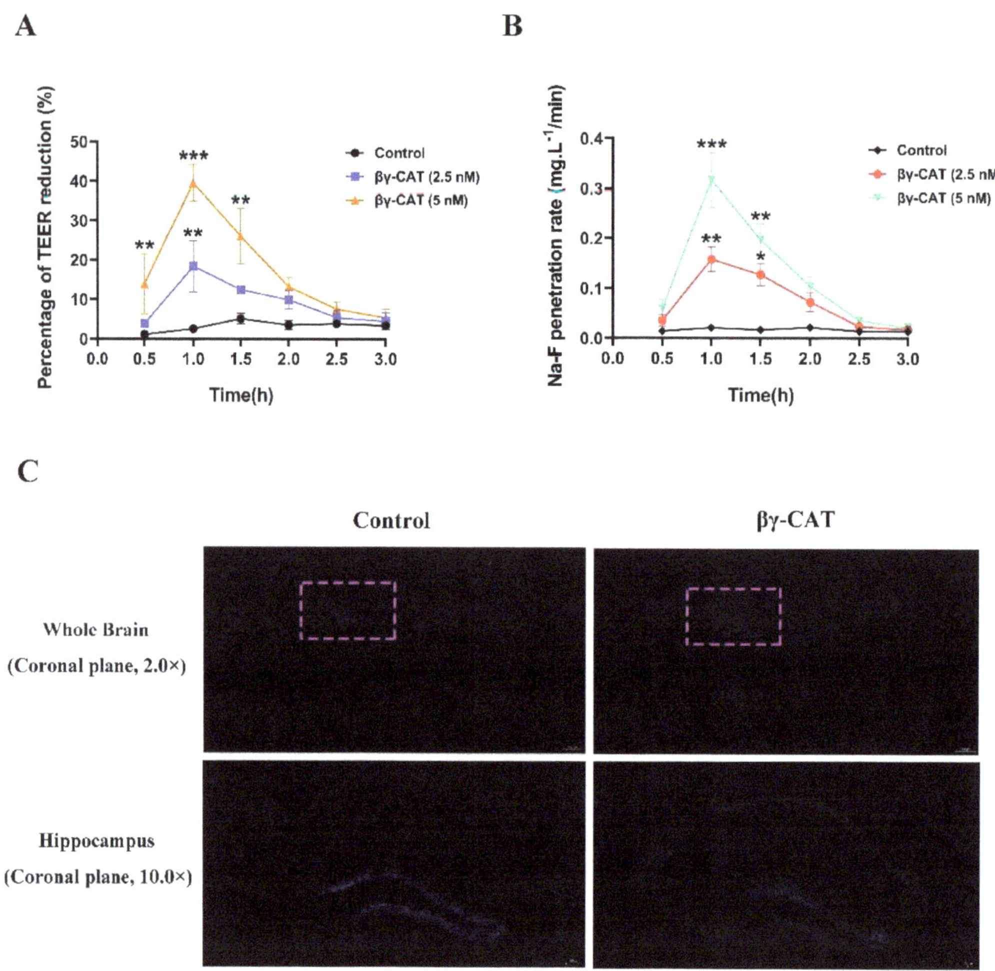

Figure 5. βγ-CAT crosses the BBB and is enriched in the hippocampus. It impairs the cognitive functions of mice. (**A**) The in vitro BBB cell models were treated with βγ-CAT at the concentrations of 2.5 and 5 nM for different lengths of time (0.5 h, 1.0 h, 1.5 h, 2.0 h, 2.5 h, and 3.0 h); then, the TEER values were determined with a Millicell ERS-2 instrument. The untreated BBB model was used as a negative control. * $p < 0.05$, ** $p < 0.01$ and *** $p < 0.001$ versus the control. (**B**) The in vitro BBB cell models were treated with βγ-CAT at the concentrations of 2.5 and 5 nM for different lengths of time (0.5 h, 1.0 h, 1.5 h, 2.0 h, 2.5 h, and 3.0 h); then, the Na–F penetration rate was determined using a fluorescence multiwall plate reader. The untreated BBB model was used as a negative control. * $p < 0.05$, ** $p < 0.01$ and *** $p < 0.001$ versus the control. (**C**) The mouse was intraperitoneally injected with 100 μL of 40 μg/mL βγ-CAT; then, the brain was extracted 2 h after intraperitoneal injection. The brain slices were fixed and incubated via rabbit anti-βγ-CAT antibody, then stained with 488-labeled goat anti-rabbit fluorescent secondary antibody, and finally, the fluorescence in the brain tissue was observed with a fluorescence microscope. The area marked by the magenta dotted frame is the hippocampus. Scale bar = 500 μm (upper) and 100 μm (lower).

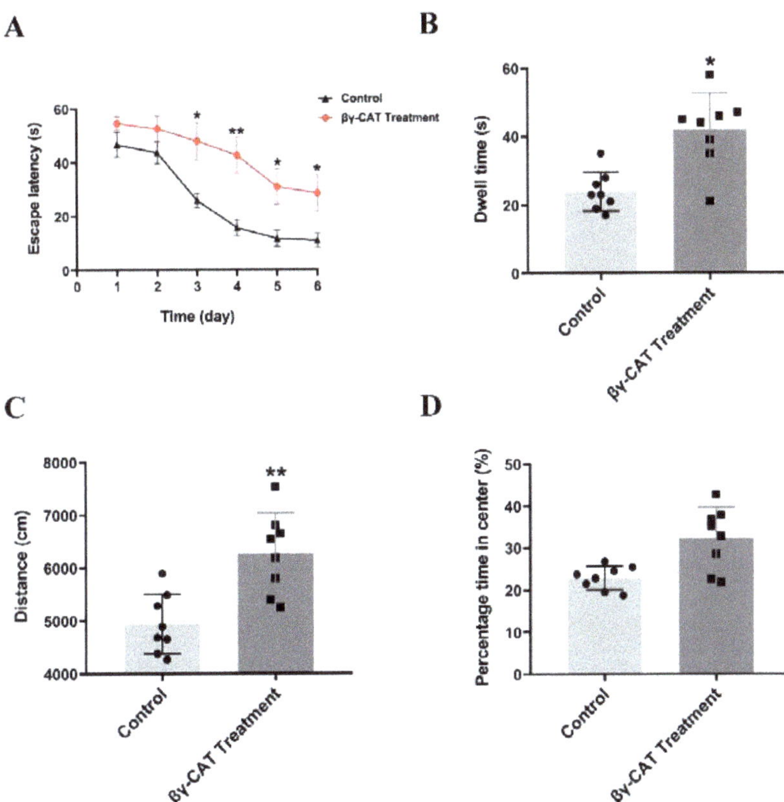

Figure 6. βγ-CAT impairs the hippocampal cognitive functions of mice. (**A**) The mice were intraperitoneally injected with 10 μg/kg βγ-CAT and then underwent a spatial learning test; the escape latency to reach the hidden platform was measured. The untreated mice were used as the negative control. * $p < 0.05$ and ** $p < 0.01$ versus control. (n = 8). (**B**) The memory test was conducted after the spatial learning test; the dwell time was recorded in a 60 s memory test. The untreated mice were used as the negative control. * $p < 0.05$ versus control. (n = 8). (**C**) The open field test was conducted after the mice were intraperitoneally injected with 10 μg/kg βγ-CAT; the total travel distance was measured. The untreated mice were used as the negative control. ** $p < 0.01$ versus control. (n = 8). (**D**) The open field test was conducted after the mice were intraperitoneally injected with 10 μg/kg βγ-CAT; the percentage of time spent in the center of the apparatus was measured. The untreated mice were used as the negative control (n = 8). The 8 dots on each column in (**B–D**) represent the data of 8 individual animals in each group in a representative experiment.

3. Discussion

Toxins are not only important weapons/arsenals to enable poisonous animals to hunt prey/defend themselves, but they are also efficient and active pharmacological molecules that can be used by humans [5]. In addition, the regulation of toxins on the physiological processes of organisms is also an important research field. βγ-CAT is a pore-forming toxin-like protein complex which was first identified in *Bombina maxima*, a Chinese red-belly toad. Previous studies revealed that βγ-CAT has a variety of physiological regulatory functions (such as immune regulation, wound healing, tissue repair, etc.). A further molecular mechanism study indicated that βγ-CAT functions predominantly by regulating the endolysosome pathway and the general action mode of βγ-CAT functions, including membrane binding, oligomerization, and endocytosis. Based on this action mode of βγ-

CAT, the roles of βγ-CAT in the nervous system were further studied, and the findings showed that βγ-CAT can trigger the gasdermin-E-dependent pyroptosis of hippocampal neuronal cells and impairs the cognitive functions of the hippocampus.

The effects of biotoxins on the nervous system have been reported previously; for example, α-bungarotoxin, a neurotoxin derived from the venom of the *Bungarus mul-ticinctus* snake, has been found to block neuromuscular transmission by binding the nicotinic acetylcholine receptor α-subunit in skeletal muscles [31]. The cholera toxins produced by *Vibrio cholerae* affect the normal neural functions of neurons and glial cells by interacting with ganglioside GM1 in neurons in the cerebral cortex [32]. As the main family of pore-forming toxins, bacterial-secreted ALPs have various pathophysiological roles in the nervous system [33]. One of the representative ALPs is the epsilon toxin (ETX), which is mainly produced by *Clostridium perfringens* and has been found to be a cause of central nervous system white matter disease in ruminant animals [34]. However, the detailed biological functions of vertebrate-derived ALPs remain elusive. Particularly, the roles of vertebrate-derived ALPs in the nervous system remain unclear to date. There are two primary reasons for this: (1) natural and active ALPs are difficult to obtain from vertebrates and (2) the huge molecular weights of most ALPs mean they are less likely to be able to cross the BBB. Indeed, the existence of the BBB is crucial to the protection of the function of the brain, and many macromolecular exogenous substances are isolated from the BBB because of their antigenicity and heterogeneity [35]. However, there is no absolute relationship between molecular weight and BBB penetration [36]. A recent study revealed that the epsilon toxin (ETX) can cross the BBB by binding to the lymphocyte protein (MAL) on the luminal surface of endothelial cells [37]. Thus, macromolecules with high molecular weights are also likely to cross the BBB in a unique way. Coincidentally, βγ-CAT is a toxin-like protein with a high molecular weight of 72 kDa, so theoretically, βγ-CAT penetrates the BBB with difficulty. Surprisingly, in the present study, first-hand evidence was provided which supported the concept that βγ-CAT can cross the BBB (Figure 5). This can be explained from the perspective of the receptor of βγ-CAT. Acidic glycosphingolipids in cell membranes are the receptors of βγ-CAT and are involved in the membrane binding, oligomerization, and endocytosis of βγ-CAT, as described previously [20]. In fact, acidic glycosphingolipids (mainly gangliosides and sulfatides in vertebrates) are predominantly distributed in the nervous system, especially in the brain [38]. In addition, gangliosides and sulfatides are also highly expressed by endothelial cells and astrocytes, which make up the BBB [39,40], which enables βγ-CAT to easily bind to the BBB and then cross it. Why is βγ-CAT specifically localized in the hippocampus after crossing the BBB and entering the brain? One possible reason for this is that besides acidic glycosphingolipids, an unknown protein receptor of βγ-CAT may specifically exist in the hippocampus. However, more detailed molecular mechanisms need to be further studied.

The central nervous system currently consists of three major cell types: neurons, astrocytes, and microglia [41,42]. Currently, microglia are reported to be mainly involved in immune function; the excessive release of pro-inflammatory factors and gasdermin activation in microglia will lead to the pyroptosis of the whole nervous system [43,44]. In the present study, evidence showed that only GSDME is expressed in hippocampal neuronal cells, not GSDMD. There have been controversial reports regarding the expression patterns of gasdermin families in HT-22 cells. Some scholars believe that HT-22 cells are expressed in GSDMD [45], while others believe that HT-22 cells are exclusively expressed in GSDME [46], while other gasdermin family members are not. In our study, both RT-qPCR and Western blotting showed that HT-22 cells express GSDME, rather than GSDMD (Figure 3A–C). Meanwhile, we further confirmed that βγ-CAT induced the pyroptosis of HT-22 cells in a GSDME-dependent manner (Figure 3D–F). Our findings provided convincing first-hand evidence of the involvement of hippocampal neurons in inflammatory process.

Some physiological pore-forming toxin-like proteins or peptides in the human body are also involved in the development of various neurodegenerative diseases under pathological conditions [47,48]. Amyloid β peptide (Aβ), a peptide with membrane-perforating activity which has been identified in the human brain, was found to be involved in the pathogenesis of Alzheimer's disease [49,50]. Detailed studies revealed that cholesterol and gangliosides in cell membranes are potential binding sites of Aβ [51–53]. The pore-forming characteristics and mechanisms of Aβ are obviously similar to those of βγ-CAT reported in this study. Unfortunately, the detailed pathogenic mechanisms of Aβ in Alzheimer's disease or Parkinson's disease remain elusive [54]. Therefore, βγ-CAT is expected to be an appropriate candidate for the exploration of the detailed pathogenic mechanism of Aβ in Alzheimer's disease or other neurodegenerative diseases.

The primary function of the toxins secreted by the skin of amphibians is defensive, rather than predatory [55,56]. However, the detailed roles of βγ-CAT in defense against predators remain elusive. Here, our findings showed that βγ-CAT impaired the cognitive memory functions of animals by triggering the GSDME-dependent pyroptosis of hippocampal neurons (Figure 6). Based on these findings, we hypothesize that after a predator preys on the red-belly toad for the first time, a symptom of poisoning, mainly entailing "decreased learning and memory function", will occur in the predator, meaning the predator will not be able to remember its prey clearly. Eventually, this will prevent the toad from being eaten by its predators. Perhaps this is the main reason for the large amount of βγ-CAT which exists in the frog's skin secretions.

In summary, in our study, we systematically explored the detailed toxic functions and mechanisms of βγ-CAT in hippocampal neuronal cells and the hippocampus of the brain, further enriching our knowledge of the toxic effects and physiological functions of this toxin. Importantly, this work will lay a theoretical foundation for the subsequent study of the biological functions of ALPs in vertebrates.

4. Conclusions

The toxic effects and biological functions of vertebrate-derived ALPs remain unclear. βγ-CAT was the first natural ALP and TFF complex to be identified in vertebrates. Previous studies revealed that βγ-CAT exerts various pathophysiological roles via its regulation of the endolysosome pathway. A further study showed that acidic glycosphingolipids are the receptors of βγ-CAT and mediate the membrane binding, oligomerization, and endocytosis of βγ-CAT. As acidic glycosphingolipids are mainly distributed in the nervous system, the functions of βγ-CAT in the nervous system were explored in the present study. The main conclusions from the study are listed below:

(1)   βγ-CAT shows strong cytotoxicity to HT-22 hippocampal neuronal cells and triggers the GSDME-dependent pyroptosis of HT-22 cells.
(2)   The pyroptosis of HT-22 cells induced by βγ-CAT relies on the membrane binding, oligomerization, and endocytosis of βγ-CAT.
(3)   βγ-CAT can cross the BBB and affects the cognitive function of mice.

5. Materials and Methods

5.1. Animals

The ICR mice weighing 18 to 22 g were used in the study and purchased from Hunan SJA Laboratory Animal Co., Ltd. (Changsha, China). The mice were housed in IVC cages under constant temperature (23 ± 1 °C) and humidity with a 12 h light/dark cycle. All procedures and the care and handling of animals were approved by the Ethics Committee of Yunnan Normal University (No.YSLL20220316).

5.2. Cell Lines, Antibodies and Reagents

The mouse hippocampal neuronal cell line HT-22 was purchased from the American Type Culture Collection (Manassas, VA, USA) and maintained in Dulbecco's modified Eagle's medium (DMEM) supplemented with 10% fetal calf serum (FCS). The mouse macrophage Raw 264.7 cells were cultured in DMEM medium supplemented with 10% FCS. Natural βγ-CAT was purified from the skin secretions of *Bombina maxima* as described previously [16]. The LDH cytotoxicity assay kit was purchased from Beyotime Biotechnology Inc. (Shanghai, China). The MTT assay kit (ab211091) was purchased from Abcam Inc. (Cambridge, CB2 0AX, UK). The Annexin V-FITC apoptosis detection kit (APOAF-50TST) was purchased from Sigma-Aldrich (St. Louis, MO, USA). The anti-GSDME (ab215191), anti-Lamp1 antibody (ab25630), anti-GSDMD (ab219800), and anti-IL-18 (ab191860) antibodies were purchased from Abcam Inc. (Cambridge, CB2 0AX, UK). The IL-1β (3A6) mouse monoclonal antibody (12242S) was purchased from Cell Signaling Technology, Inc. (Danvers, MA, USA). The anti β-actin antibody (81115-1-RR) was purchased from Proteintech Group, Inc. (Wuhan, Hubei, China). Mouse IL-18 ELISA kit (SEKM-0019) and IL-1β ELISA kit (SEKM-0002) were purchased from Solarbio Life Sciences (Beijing, China). Anti-βγ-CAT antibody was prepared by our lab. The caspase 1 p20 antibody (sc-398715) and GAPDH antibody (sc-365062) were purchased from Santa Cruz Biotechnology (Santa Cruz, CA, USA). The Cy3-conjugated goat anti-mouse IgG/IgM (H+L) secondary antibody (M30010) and the FITC-conjugated goat anti-rabbit IgG/IgM (H+L) secondary antibody (F-2765) were purchased from Thermo Fisher Scientific Inc. (Waltham, MA, USA). All other reagents were purchased from Sigma-Aldrich (St. Louis, MO, USA).

5.3. LDH Release Detection

The LDH release measurement of HT-22 cells after treated by βγ-CAT was used to assess the cytotoxicity of βγ-CAT to mouse hippocampal neuronal cells. The measurement procedure of LDH release in this study was performed as previously described [20]. First, the HT-22 cells were cultured in DMEM medium supplemented with 10% FCS and the cells were harvested and seeded into a 96-well plate when the cells grew to 80% confluence. After the cells grew to 80–90% confluence, the medium with 10% FCS was replaced by serum-free medium and then cultured for 2 h. Next, the serum-free medium of each well was removed and washed three times with cold phosphate buffer saline (PBS), then 100 μL βγ-CAT with different concentrations was added to each well and incubated at 37 °C for 10 min. At last, the cell supernatant of each well was collected and the LDH content in cell supernatant was detected according to the manufacturer's instructions. The HT-22 cells treated with 0.1% Triton X-100 was known as 100% LDH release, and the cells treated with PBS served as a negative control.

5.4. MTT Assay

To detect the cell viability of HT-22 cells after treated by βγ-CAT, the MTT assay was performed. In brief, the HT-22 cells were harvested and planted into a 96-well plate. After the cells grew to 80–90% confluence, the medium with 10% FCS was replaced by serum-free medium and then cultured for 2 h. Next, the cells were incubated with gradient concentrations of βγ-CAT (from 2.5 to 20 nM) at 37 °C in water bath for 10 min. After incubation, the βγ-CAT of each well was removed and washed three times with cold PBS. Then, 120 μL MTS reagents were added to each well and incubated at 37 °C for 2 h. Finally, the absortance at 490 nm of each well was recorded by using a 96-well plate reader. Cells only treated with βγ-CAT served as a positive control and the cells treated with FBS-free medium served as a negative control. The HT-22 cells treated with PBS were known as 100% survival and cells treated by 0.1% Triton X-100 were known as 0% survival.

5.5. Flow Cytometry Detection of HT-22 Cell Death

The method of flow cytometry was used to detect the cell death of HT-22 cells after being treated by βγ-CAT as previously described [57]. Briefly, the HT-22 cells were harvested when the cells grew to 80–90% confluence, and then the cells were washed three times with cold PBS. The washed cells were next treated with 5 nM βγ-CAT at 37 °C for gradient times (from 3 to 10 min), and then the HT-22 cells were collected and washed three times with cold PBS. The washed cells were then incubated with FITC labeled Annexin V and propidium iodide (PI). The mixture of cells and FITC Annexin V with PI was vortexed gently and incubated at 25 °C for 15 min in the dark. Finally, the cell supernatant was collected and analyzed by flow cytometry. The unstained cells served as a blank control.

The cells only stained with FITC Annexin V or only stained with PI also served as the control. The HT-22 cells untreated with βγ-CAT were served as a negative control.

5.6. PCR and RT-qPCR Assay

The methods of PCR and RT-qPCR were employed to detect the expression of gasdermins in HT-22 cells as previously described [58]. Briefly, the cultured HT-22 cells were collected by digesting with trypsin. Then, the total RNA was extracted by using the TRIzol reagent according to the manufacturer's instructions. Next, the cDNA first-strand was synthesized by using a reverse transcription system according to the manufacturer's instructions. To detect the expression profile of gasdermins in HT-22 cells, the specific gasdermin primers for PCR and RT-qPCR were designed using Oligo 7 software and listed in Table 1. It is worth mentioning that the GSDMB primer designed here was based on human GSDMB isoform since no remarkable GSDMB sequence can be found in GenBank. In PCR assay, 35 cycles were performed and amplified by using taq DNA polymerase. The RT-qPCR assay was performed by using the SYBR Premix Ex Taq II two-step qRT-PCR kit according to the manufacturer's instructions on a LightCycler 480 real-time PCR system (Roche LightCycler 480, Roche, Mannheim, Germany). The relative expression of gasdermins was determined by CT value analysis of reference genes and target genes by Pfaffl method.

Table 1. Sequences of primers used for detection of gasdermins of HT-22 cell.

Primer	Sequence	Product Length
PCR		
GSDMA-Forward	TCCCTCCTGGAGAAAAGCCA	338 bp
GSDMA-Reverse	ACTTAGCACTGTCAGAGCCC	
GSDMB-Forward	TGGATGCCGGCACTACACAAC	248 bp
GSDMB-Reverse	GGTAGTTCCCTCTTCAGCTTCC	
GSDMC-Forward	GATCTGAGGCCTGTCAAATGC	524 bp
GSDMC-Reverse	TCTGTTTGCCACTGTCCACT	
GSDMD-Forward	CCGGAGTGTTTTGGCTCCTT	260 bp
GSDMD-Reverse	ACCACAAACAGGTCATCCCC	
GSDME-Forward	GTCAGCGCACTAGCAGAAATG	173 bp
GSDME-Reverse	ATGCCAAACCTCTCTGTGTC	
PJVK-Forward	GCTGACAAGTACCAACCCCT	595 bp
PJVK-Reverse	CACAAATGTCGAAGGCACCG	
β-actin-Forward	CCACCATGTACCCAGGCATT	253 bp
β-actin-Reverse	AGGGTGTAAAACGCAGCTCA	
RT-qPCR		
GSDMA-Forward	TCCCTCCTGGAGAAAAGCCA	261 bp
GSDMA-Reverse	GTGCTTCCAGGGTCACTTCG	
GSDMB-Forward	CCGTTAGAAGCCTTGTTGATGC	180 bp
GSDMB-Reverse	CCGTTGAGTCTACATTATCCAG	
GSDMC-Forward	CAGATGCAACCAAATTCTGCC	207 bp
GSDMC-Reverse	TGGTTTCGACATCCAGGTCA	
GSDMD-Forward	GATCAAGGAGGTAAGCGGCA	195 bp
GSDMD-Reverse	CACTCCGGTTCTGGTTCTGG	
GSDME-Forward	AGTTTTCCTGGGGACTTGCT	170 bp
GSDME-Reverse	CAATGTCAGCAGAGGCAAACAA	
PJVK-Forward	TCAGCGAAGCTGACAAGTACC	300 bp
PJVK-Reverse	CCACCTCATGTTTGGTCACG	
β-actin-Forward	CCACCATGTACCCAGGCATT	253 bp
β-actin-Reverse	AGGGTGTAAAACGCAGCTCA	

5.7. Western Blotting Assay

Firstly, the Western blotting assay was adopted to detect the expression of GSDMD or GSDME in HT-22 cells. Briefly, the cultured HT-22 cells were collected by digesting with trypsin and lysed with cell lysis buffer containing protease inhibitor cocktail. The total proteins of HT-22 cells were extracted and quantified by BCA method. Then, the total protein sample was separated by 12% sodium dodecyl sulfate-polyacrylamide gel electrophoresis (SDS-PAGE) and then electrotransferred onto polyvinylidene difluoride (PVDF) membranes. The PVDF membrane was subsequently blocked with 5% skimmed milk containing 0.1% Tween-20 at room temperature for 2 h. Then, the membrane was incubated with rabbit anti-GSDMD and rabbit anti-GSDME antibodies (1:1000 diluted) at 4 °C overnight. Next, the primary antibodies were removed and then incubated with HRP-conjugated goat anti-rabbit secondary antibodies (1:5000 diluted). At last, the protein bands were visualized with the SuperSignal WestPico chemiluminescence substrate via Gel Imager System.

To detect the expression of GSDME, cleaved GSDME N-terminal (NT), caspase 1, cleaved caspase 1 p20, IL-1β, and IL-18 of HT-22 cells, the HT-22 cells were treated by 5 nM βγ-CAT in a time gradient from 2 h to 8 h, and then washed three times with cold PBS to remove the βγ-CAT. The collected HT-22 cells were then lysed to obtain the total proteins. Then, the total protein sample was separated by 12% sodium dodecyl sulfate-polyacrylamide gel electrophoresis (SDS-PAGE) and electrotransferred onto polyvinylidene difluoride (PVDF) membranes. The PVDF membrane was subsequently blocked with 5% skimmed milk containing 0.1% Tween-20 at room temperature for 2 h. Then, the membrane was incubated with rabbit anti-GSDMD antibody (1:1000 diluted), mouse anti-caspase-1 p20 antibody (1:500 diluted), mouse anti-IL-1β monoclonal antibody (1:1000 diluted), rabbit anti-IL-18 antibody (1:1000 diluted), and rabbit anti-β actin antibody (1:5000 diluted) at 4 °C overnight. Next, the primary antibodies were removed and then incubated with HRP-conjugated goat anti-rabbit secondary antibody and HRP-conjugated goat anti-mouse secondary antibody (1:5000 diluted). At last, the protein bands were visualized with the SuperSignal WestPico chemiluminescence substrate via Gel Imager System.

To detect the oligomerization of βγ-CAT on the membrane of HT-22 cells, the HT-22 cells were treated by 5 nM and 10 nM βγ-CAT at 37 °C for 30 min, in which the treated cells were collected and lysed. After SDS-PAGE and transfer to PVDF, the PVDF membrane was subsequently blocked with 5% skimmed milk containing 0.1% Tween-20 at room temperature for 2 h. Then, the membrane was incubated with rabbit anti-βγ-CAT antibody (1:1000 diluted) and HRP-conjugated goat anti-rabbit secondary antibody (1:5000 diluted). Finally, the protein bands were visualized with the SuperSignal WestPico chemiluminescence substrate via Gel Imager System.

5.8. Enzyme-Linked Immuno Sorbent Assay (ELISA)

To determine the IL-1β and IL-18 levels in the cell supernatant of HT-22 cells after being treated with βγ-CAT, the ELISA was performed. Briefly, the HT-22 cells were treated by 5 nM βγ-CAT at a time gradient from 2 h to 8 h, and then the cell supernatant was collected and the IL-1β and IL-18 of the supernatant were measured by the commercial ELISA kit according to the manufacturer's instructions.

5.9. Flow Cytometry Detection of the Membrane Binding of βγ-CAT

The method of flow cytometry was used to detect the membrane binding of βγ-CAT with HT-22 cells. Briefly, the HT-22 cells were treated by βγ-CAT at the concentrations of 2.5 and 5 nM for 15 min, and then βγ-CAT was removed and the cells were washed three times with cold PBS. The cells were incubated with rabbit anti-βγ-CAT antibody (1:500 diluted) and FITC-conjugated goat anti-rabbit secondary antibody (1:500 diluted). Finally, the cells were resuspended in 300 μL of PBS and analyzed on a flow cytometer (FACSVantage SE; Becton Dickinson, Franklin Lakes, NJ, USA). Data were analyzed using FlowJo software 7.6.1 (Tree Star Inc., Ashland, OH, USA)

5.10. Confocal Microscopy Assay of the Endocytosis and Co-Localization of βγ-CAT

To detect the endocytosis of βγ-CAT and the co-localization between βγ-CAT and endolysosome, the confocal microscopy assay was performed as previously described [20]. In brief, the HT-22 cells were treated by βγ-CAT at the concentrations of 5 nM for 30 min, and then incubated with rabbit anti-βγ-CAT polyclonal antibody (1:500 diluted) and mouse anti-lamp-1 antibody (1:200 diluted). After being washed three times, the HT-22 cells were then incubated with FITC-conjugated goat anti-rabbit secondary antibody (1:500 diluted) and Cy3-conjugated goat anti-mouse secondary antibody (1:500 diluted). Finally, the nuclei were stained with DAPI. After being washed three times, the slides were observed under a confocal microscope (Olympus FV1000, Olympus Corporation, Tokyo, Japan). For the co-localization analyses of βγ-CAT with Lamp-1, the region of interests in HT-22 cells were analyzed using the "Just Another Co-localization Plugin (JACoP)" of Image J. The offset of each image was set automatically to avoid an arbitrary judgement, and then the Pearson's correlation coefficient was calculated as previously described [59].

5.11. In Vitro BBB Model Establishment and βγ-CAT Penetration Detection

To detect the ability of βγ-CAT to penetrate the BBB, an in vitro BBB model was built with a co-culture of the human cerebral microvascular endothelial cell line (hCMEC/D3) and the human normal glial cell line (HEB), as previously described [60,61]. Briefly, the cultured HEB cells at cell numbers of 5×10^4 cells/cm^2 were first seeded on the bottom of the 24-well Transwell inserts for 24 h until they had adhered. Subsequently, the hCMEC/D3 cells at cell numbers of 5×10^5 cells/cm^2 were seeded on the top side of the Transwell chamber. After being co-cultured for 4–6 days, the integrity of the BBB model was determined via TEER measurement using Millicell ERS-2 (Millipore, MA, USA). The criterion for a successful in vitro BBB model was that the TEER value was higher than 200 $\Omega \cdot$cm^2.

Based on the in vitro BBB model, the ability of βγ-CAT to penetrate the BBB was determined. Briefly, the in vitro BBB cell models (in the Transwell chamber) were treated with βγ-CAT at the concentrations of 2.5 and 5 nM for different lengths of time (0.5 h, 1.0 h, 1.5 h, 2.0 h, 2.5 h, and 3.0 h); then, the TEER values were recorded using Millicell ERS-2 (Millipore, MA, USA). In addition, sodium fluorescein (Na-F) was also used to test the paracellular permeability of the BBB. After βγ-CAT treatment, a 10 mg/L Na–F solution was then added to the in vitro BBB model. Finally, the concentration of Na–F was measured using a fluorescence multiwall plate reader. The untreated BBB model was used as a negative control.

5.12. In Vivo BBB Penetration Assay

To further study the BBB penetration activity of βγ-CAT, the in vivo assay was performed according to previously described methods [62]. Briefly, C57 mice were divided into 2 groups. Each mouse in the control group was intraperitoneally injected with 100 µL of normal saline, and each mouse in the experimental group was intraperitoneally injected with 100 µL of 40 µg/mL βγ-CAT. Two hours after the intraperitoneal injection, the mice were euthanized via CO_2 inhalation, and the brain tissues were quickly extracted. The brain tissue was fixed with 4% paraformaldehyde and then embedded in paraffin and sectioned. The slices were then blocked with 5% goat serum, incubated with the rabbit anti-βγ-CAT primary antibody, incubated with 488-labeled goat anti-rabbit fluorescent secondary antibody, and finally, the fluorescence in the brain tissue was observed with a fluorescence microscope.

5.13. Water Maze Assay for Assessment of Spatial Learning and Memory of Mice

In our study, a water maze assay was performed to detect the cognitive functions (especially spatial learning and memory functions) of mice after intraperitoneal injection with 10 µg/kg βγ-CAT. The detailed procedure for the water maze assay was previously described [63,64]. A circular tank with a 150 cm diameter and 60 cm height served as the site of the water maze; then, the tank was divided into 4 equal quadrants (I, II, III, and IV). A platform with a 10 cm diameter was located at the center of quadrant III and 2 cm below the water surface. The water temperature in the tank was maintained at 24 ± 2 °C. The movement of the mice in the tank was monitored using a computerized video tracking system. It is well known that spatial learning is the first step in a water maze assay. In the period of spatial learning, the untreated and βγ-CAT-treated mice were required to undergo trials on each training day for 6 consecutive days. The mice were allowed to swim freely to find the platform in the tank. The time taken to find the platform (escape latency) was recorded. One day after spatial learning, memory training was performed. In this training procedure, the platform was taken out from the tank, and the mice were placed at the starting point of quadrant I. The quadrant dwell time was recorded during the 60 s memory test. Furthermore, an open field assay was also performed by using a square plastic arena. In this arena, mice were permitted to freely explore the arena for 20 min; then, the distance traveled and time spent in the central region were recorded.

5.14. Statistical Analysis

The data in this paper are presented as mean ± SD. The experiments were repeated independently at least two times. All data in the paper are representative of three independent experiments and were analyzed using GraphPad Prism 8.0 software. The significance of differences in LDH release assay and MTT assay was determined by unpaired two-tailed Student's *t*-test. The comparative analysis of escape latency and dwell time in water maze assay were performed by using unpaired two-tailed Student's *t*-test.

Author Contributions: X.G. and Y.Z. designed the research framework. Q.Y., Q.W., J.Y. and Y.X. performed most of the experiments, including LDH release assay, MTT assay, flow cytometry assay, Western blotting assay, and confocal microscopy assay. X.G. and Q.Y. performed the in vitro BBB model establishment assay and the subsequent data processing and analysis. W.L. and Q.W. designed the water maze assay and analyzed the data from the water maze. Q.Y. and X.G. wrote the paper. Y.Z. and Y.X. edited and revised the manuscript. All authors have read and agreed to the published version of the manuscript.

Funding: This work was funded by grants from the Project of Basic Research of Yunnan Province (grant number 202001AU070119 and 202101AT070297 to Q.Y., grant number 202101AT070292 to X.G.).

Institutional Review Board Statement: Ethics approval was obtained from the Ethics Committee of Yunnan Normal University with the approved number No.YSLL20220316 (approved date: 17 March 2022).

Informed Consent Statement: Not applicable.

Data Availability Statement: All data supporting the findings of this study are available within the published article. Additional materials related to this paper may be requested from the authors.

Conflicts of Interest: The authors declare no conflict of interest.

References

1. Dal Peraro, M.; van der Goot, F.G. Pore-forming toxins: Ancient, but never really out of fashion. *Nat. Rev. Microbiol.* **2016**, *14*, 77–92. [CrossRef] [PubMed]
2. Cirauqui, N.; Abriata, L.A.-O.; van der Goot, F.A.-O.X.; Dal Peraro, M. Structural, physicochemical and dynamic features conserved within the aerolysin pore-forming toxin family. *Sci. Rep.* **2017**, *7*, 13932. [CrossRef] [PubMed]
3. Szczesny, P.; Iacovache, I.; Muszewska, A.; Ginalski, K.; van der Goot, F.G.; Grynberg, M. Extending the aerolysin family: From bacteria to vertebrates. *PLoS ONE* **2011**, *6*, e20349. [CrossRef] [PubMed]
4. Cressiot, B.A.-O.; Ouldali, H.; Pastoriza-Gallego, M.; Bacri, L.A.-O.; Van der Goot, F.G.; Pelta, J. Aerolysin, a Powerful Protein Sensor for Fundamental Studies and Development of Upcoming Applications. *ACS Sens.* **2019**, *4*, 530–548. [CrossRef]
5. Zhang, Y. Why do we study animal toxins? *Zool. Res.* **2015**, *36*, 183–222.
6. Seni-Silva, A.A.-O.; Maleski, A.A.-O.; Souza, M.A.-O.; Falcao, M.A.-O.; Disner, G.A.-O.; Lopes-Ferreira, M.A.-O.; Lima, C.A.-O.X. Natterin-like depletion by CRISPR/Cas9 impairs zebrafish (Danio rerio) embryonic development. *BMC Genom.* **2022**, *23*, 123. [CrossRef]
7. Lima, C.A.-O.X.; Disner, G.A.-O.; Falcão, M.A.P.; Seni-Silva, A.C.; Maleski, A.A.-O.; Souza, M.M.; Reis Tonello, M.C.; Lopes-Ferreira, M. The Natterin Proteins Diversity: A Review on Phylogeny, Structure, and Immune Function. *Toxins* **2021**, *13*, 538. [CrossRef]
8. Chen, L.L.; Xie, J.; Cao, D.D.; Jia, N.; Li, Y.J.; Sun, H.; Li, W.F.; Hu, B.; Chen, Y.; Zhou, C.Z. The pore-forming protein Aep1 is an innate immune molecule that prevents zebrafish from bacterial infection. *Dev. Comp. Immunol.* **2018**, *82*, 49–54. [CrossRef]

9. Pang, Y.; Gou, M.; Yang, K.; Lu, J.; Han, Y.; Teng, H.; Li, C.; Wang, H.; Liu, C.; Zhang, K.; et al. Crystal structure of a cytocidal protein from lamprey and its mechanism of action in the selective killing of cancer cells. *Cell Commun. Signal.* **2019**, *17*, 54. [CrossRef]

10. Lisman, J.; Buzsáki, G.; Eichenbaum, H.; Nadel, L.; Ranganath, C.; Redish, A.D. Viewpoints: How the hippocampus contributes to memory, navigation and cognition. *Nat. Neurosci.* **2017**, *20*, 1434–1447. [CrossRef]

11. Ekstrom, A.D.; Ranganath, C.A.-O. Space, time, and episodic memory: The hippocampus is all over the cognitive map. *Hippocampus* **2018**, *28*, 680–687. [CrossRef] [PubMed]

12. Yang, N.J.; Chiu, I.M. Bacterial Signaling to the Nervous System through Toxins and Metabolites. *J. Mol. Biol.* **2017**, *429*, 587–605. [CrossRef]

13. Xu, X.; Lai, R. The chemistry and biological activities of peptides from amphibian skin secretions. *Chem. Rev.* **2015**, *115*, 1760–1846. [CrossRef] [PubMed]

14. Pirazzini, M.; Montecucco, C.A.-O.; Rossetto, O. Toxicology and pharmacology of botulinum and tetanus neurotoxins: An update. *Arch. Toxicol.* **2022**, *96*, 1521–1539. [CrossRef]

15. Wioland, L.; Dupont, J.; Bossu, J.-L.; Popoff, M.R.; Poulain, B. Attack of the nervous system by Clostridium perfringens Epsilon toxin: From disease to mode of action on neural cells. *Toxicon* **2013**, *75*, 122–135. [CrossRef]

16. Liu, S.B.; He, Y.Y.; Zhang, Y.; Lee, W.H.; Qian, J.Q.; Lai, R.; Jin, Y. A novel non-lens betagamma-crystallin and trefoil factor complex from amphibian skin and its functional implications. *PLoS ONE* **2008**, *3*, e1770. [CrossRef]

17. Gao, Q.; Xiang, Y.; Zeng, L.; Ma, X.; Lee, W.; Zhang, Y. Characterization of the bg-crystallin domains of bg-CAT, a non-lens bg-crystallin and trefoil factor complex, from the skin of the toad *Bombina maxima*. *Biochimie* **2011**, *93*, 1865–1872. [CrossRef] [PubMed]

18. Xiang, Y.; Yan, C.; Guo, X.L.; Zhou, K.F.; Li, S.A.; Gao, Q.; Wang, X.; Zhao, F.; Liu, J.; Lee, W.H. Host-derived, pore-forming toxin–like protein and trefoil factor complex protects the host against microbial infection. *Proc. Natl. Acad. Sci. USA* **2014**, *111*, 6702–6707. [CrossRef]

19. Liu, L.; Deng, C.J.; Duan, Y.L.; Ye, C.J.; Gong, D.H.; Guo, X.L.; Lee, W.; Zhou, J.; Li, S.A.; Zhang, Y. An Aerolysin-like Pore-Forming Protein Complex Targets Viral Envelope to Inactivate Herpes Simplex Virus Type 1. *J. Immunol.* **2021**, *207*, 888–901. [CrossRef]

20. Guo, X.L.; Liu, L.Z.; Wang, Q.Q.; Liang, J.Y.; Lee, W.H.; Xiang, Y.; Li, S.A.; Zhang, Y. Endogenous pore-forming protein complex targets acidic glycosphingolipids in lipid rafts to initiate endolysosome regulation. *Commun. Biol.* **2019**, *2*, 59. [CrossRef]

21. Wang, Q.; Bian, X.; Zeng, L.; Pan, F.; Liu, L.; Liang, J.; Wang, L.; Zhou, K.; Lee, W.; Xiang, Y.; et al. A cellular endolysosome-modulating pore-forming protein from a toad is negatively regulated by its paralog under oxidizing conditions. *J. Biol. Chem.* **2020**, *295*, 10293–10306. [CrossRef] [PubMed]

22. Schnaar, R.L. Gangliosides of the Vertebrate Nervous System. *J. Mol. Biol.* **2016**, *428*, 3325–3336. [CrossRef] [PubMed]

23. Schmitt, S.; Cantuti Castelvetri, L.; Simons, M. Metabolism and functions of lipids in myelin. *Biochim. Biophys. Acta (BBA) Mol. Cell Biol. Lipids* **2015**, *1851*, 999–1005. [CrossRef]

24. Bonetto, G.A.-O.; Di Scala, C.A.-O. Importance of Lipids for Nervous System Integrity: Cooperation between Gangliosides and Sulfatides in Myelin Stability. *J. Neurosci.* **2019**, *39*, 6218–6220. [CrossRef] [PubMed]

25. Bertheloot, D.; Latz, E.A.-O.; Franklin, B.A.-O. Necroptosis, pyroptosis and apoptosis: An intricate game of cell death. *Cell Mol. Immunol.* **2021**, *18*, 1106–1121. [CrossRef]

26. Li, J.; Cao, F.; Yin, H.-l.; Huang, Z.; Lin, Z.; Mao, N.; Sun, B.; Wang, G. Ferroptosis: Past, present and future. *Cell Death Dis.* **2020**, *11*, 88. [CrossRef]

27. Shi, J.; Gao, W.; Shao, F. Pyroptosis: Gasdermin-Mediated Programmed Necrotic Cell Death. *Trends Biochem. Sci.* **2017**, *42*, 245–254. [CrossRef]

28. Zhang, Y.; Wang, Q.Q.; Zhao, Z.; Deng, C.J. Animal secretory endolysosome channel discovery. *Zool. Res.* **2021**, *42*, 141–152. [CrossRef]

29. Opitz, B. Memory Function and the Hippocampus. *Front. Neurol. Neurosci.* **2014**, *34*, 51–59.

30. Sweatt, J.D. Hippocampal function in cognition. *Psychopharmacology* **2004**, *174*, 99–110. [CrossRef]

31. Brun, O.; Zoukimian, C.; Oliveira-Mendes, B.; Montnach, J.; Lauzier, B.; Ronjat, M.; Béroud, R.; Lesage, F.; Boturyn, D.; De Waard, M. Chemical Synthesis of a Functional Fluorescent-Tagged α-Bungarotoxin. *Toxins* **2022**, *14*, 79. [CrossRef] [PubMed]

32. Haigh, J.L.; Williamson, D.J.; Poole, E.; Guo, Y.; Zhou, D.J.; Webb, M.E.; Deuchars, S.A.; Deuchars, J.; Turnbull, W.B. A versatile cholera toxin conjugate for neuronal targeting and tracing. *Chem. Commun.* **2020**, *56*, 6098–6101. [CrossRef] [PubMed]

33. Alves, G.G.; de Ávila, R.A.M.; Chávez-Olórtegui, C.D.; Lobato, F.C.F. Clostridium perfringens epsilon toxin: The third most potent bacterial toxin known. *Anaerobe* **2014**, *30*, 102–107. [CrossRef] [PubMed]

34. Linden, J.R.; Ma, Y.; Zhao, B.; Harris, J.M.; Rumah, K.R.; Schaeren-Wiemers, N.; Vartanian, T. Clostridium perfringens Epsilon Toxin Causes Selective Death of Mature Oligodendrocytes and Central Nervous System Demyelination. *mBio* **2015**, *6*, e02513. [CrossRef]

35. Obermeier, B.A.-O.X.; Daneman, R.; Ransohoff, R.M. Development, maintenance and disruption of the blood-brain barrier. *Nat. Med.* **2013**, *19*, 1584–1596. [CrossRef]

36. Zhao, Y.; Gan, L.; Ren, L.; Lin, Y.; Ma, C.; Lin, X. Factors influencing the blood-brain barrier permeability. *Brain Res.* **2022**, *1788*, 147937. [CrossRef]

37. Adler, D.; Linden, J.R.; Shetty, S.V.; Ma, Y.; Bokori-Brown, M.; Titball, R.W.; Vartanian, T. Clostridium perfringens Epsilon Toxin Compromises the Blood-Brain Barrier in a Humanized Zebrafish Model. *iScience* **2019**, *15*, 39–54. [CrossRef]

38. Schnaar, R.L. Brain gangliosides in axon–myelin stability and axon regeneration. *FEBS Lett.* **2010**, *584*, 1741–1747. [CrossRef]

39. Profaci, C.P.; Munji, R.N.; Pulido, R.S.; Daneman, R. The blood-brain barrier in health and disease: Important unanswered questions. *J. Exp. Med.* **2020**, *217*, e20190062. [CrossRef]

40. Mayo, L.; Trauger, S.A.; Blain, M.; Nadeau, M.; Patel, B.; Alvarez, J.I.; Mascanfroni, I.D.; Yeste, A.; Kivisäkk, P.; Kallas, K.; et al. Regulation of astrocyte activation by glycolipids drives chronic CNS inflammation. *Nat. Med.* **2014**, *20*, 1147–1156. [CrossRef]

41. Goshi, N.; Morgan, R.K.; Lein, P.J.; Seker, E. A primary neural cell culture model to study neuron, astrocyte, and microglia interactions in neuroinflammation. *J. Neuroinflamm.* **2020**, *17*, 155. [CrossRef] [PubMed]

42. Anderson, M.A. Targeting Central Nervous System Regeneration with Cell Type Specificity. *Neurosurg. Clin. N. Am.* **2021**, *32*, 397–405. [CrossRef] [PubMed]

43. Colonna, M.; Butovsky, O. Microglia Function in the Central Nervous System During Health and Neurodegeneration. *Annu. Rev. Immunol.* **2017**, *35*, 441–468. [CrossRef]

44. McNamara, N.B.; Munro, D.A.D.; Bestard-Cuche, N.; Uyeda, A.; Bogie, J.F.J.; Hoffmann, A.; Holloway, R.K.; Molina-Gonzalez, I.; Askew, K.E.; Mitchell, S.; et al. Microglia regulate central nervous system myelin growth and integrity. *Nature* **2023**, *613*, 120–129. [CrossRef]

45. Liu, D.; Dong, Z.; Xiang, F.; Liu, H.; Wang, Y.; Wang, Q.; Rao, J. Dendrobium Alkaloids Promote Neural Function After Cerebral Ischemia-Reperfusion Injury Through Inhibiting Pyroptosis Induced Neuronal Death in both In Vivo and In Vitro Models. *Neurochem. Res.* **2020**, *45*, 437–454. [CrossRef]

46. Liu, Y.; Wen, D.; Gao, J.; Xie, B.; Yu, H.; Shen, Q.; Zhang, J.; Jing, W.; Cong, B.; Ma, C. Methamphetamine induces GSDME-dependent cell death in hippocampal neuronal cells through the endoplasmic reticulum stress pathway. *Brain Res. Bull.* **2020**, *162*, 73–83. [CrossRef]

47. Aguayo, S.; Schuh, C.M.A.P.; Vicente, B.; Aguayo, L.G. Association between Alzheimer's Disease and Oral and Gut Microbiota: Are Pore Forming Proteins the Missing Link? *J. Alzheimer's Dis.* **2018**, *65*, 29–46. [CrossRef] [PubMed]

48. Di Scala, C.; Armstrong, N.; Chahinian, H.; Chabrière, E.; Fantini, J.; Yahi, N. AmyP53, a Therapeutic Peptide Candidate for the Treatment of Alzheimer's and Parkinson's Disease: Safety, Stability, Pharmacokinetics Parameters and Nose-to Brain Delivery. *Int. J. Mol. Sci.* **2022**, *23*, 13383. [CrossRef]

49. Amodeo, G.F.; Pavlov, E.V. Amyloid β, α-synuclein and the c subunit of the ATP synthase: Can these peptides reveal an amyloidogenic pathway of the permeability transition pore? *Biochim. Biophys. Acta (BBA) Biomembr.* **2021**, *1863*, 183531. [CrossRef]

50. Amodeo, G.F.; Lee, B.Y.; Krilyuk, N.; Filice, C.T.; Valyuk, D.; Otzen, D.E.; Noskov, S.; Leonenko, Z.; Pavlov, E.V. C subunit of the ATP synthase is an amyloidogenic calcium dependent channel-forming peptide with possible implications in mitochondrial permeability transition. *Sci. Rep.* **2021**, *11*, 8744. [CrossRef]

51. Venko, K.; Novič, M.; Stoka, V.; Žerovnik, E. Prediction of Transmembrane Regions, Cholesterol, and Ganglioside Binding Sites in Amyloid-Forming Proteins Indicate Potential for Amyloid Pore Formation. *Front. Mol. Neurosci.* **2021**, *14*, 619496. [CrossRef]

52. Sepehri, A.; Lazaridis, T. Putative Structures of Membrane-Embedded Amyloid β Oligomers. *ACS Chem. Neurosci.* **2023**, *14*, 99–110. [CrossRef] [PubMed]

53. Pastore, A.; Raimondi, F.; Rajendran, L.; Temussi, P.A. Why does the Aβ peptide of Alzheimer share structural similarity with antimicrobial peptides? *Commun. Biol.* **2020**, *3*, 135. [CrossRef] [PubMed]

54. Tatulian, S.A. Challenges and hopes for Alzheimer's disease. *Drug Discov. Today* **2022**, *27*, 1027–1043. [CrossRef] [PubMed]

55. Daly, J.W. The chemistry of poisons in amphibian skin. *Proc. Natl. Acad. Sci. USA* **1995**, *92*, 9–13. [CrossRef]

56. O'Connell, L.A.; Course, L.S.; O'Connell, J.D.; Paulo, J.A.; Trauger, S.A.; Gygi, S.P.; Murray, A.W. Rapid toxin sequestration modifies poison frog physiology. *J. Exp. Biol.* **2021**, *224*, jeb230342. [CrossRef]

57. Crowley, L.C.; Marfell, B.J.; Scott, A.P.; Waterhouse, N.J. Quantitation of Apoptosis and Necrosis by Annexin V Binding, Propidium Iodide Uptake, and Flow Cytometry. *Cold Spring Harb. Protoc.* **2016**, *11*, pdb.prot087288. [CrossRef]

58. Ye, Q.Q.; Tian, H.F.; Chen, B.; Shao, J.R.; Qin, Y.; Wen, J.F. Giardia's primitive GPL biosynthesis pathways with parasitic adaptation 'patches': Implications for Giardia's evolutionary history and for finding targets against Giardiasis. *Sci. Rep.* **2017**, *7*, 9507. [CrossRef]

59. Takeuchi, A.; Matsuoka, S. Spatial and Functional Crosstalk between the Mitochondrial Na+-Ca2+ Exchanger NCLX and the Sarcoplasmic Reticulum Ca^{2+} Pump SERCA in Cardiomyocytes. *Int. J. Mol. Sci.* **2022**, *23*, 7948. [CrossRef]

60. Liang, W.; Xu, W.; Zhu, J.; Zhu, Y.; Gu, Q.; Li, Y.; Guo, C.; Huang, Y.; Yu, J.; Wang, W.; et al. Ginkgo biloba extract improves brain uptake of ginsenosides by increasing blood-brain barrier permeability via activating A1 adenosine receptor signaling pathway. *J. Ethnopharmacol.* **2020**, *246*, 112243. [CrossRef]

61. Czupalla, C.J.; Liebner, S.; Devraj, K. In vitro models of the blood-brain barrier. *Methods Mol. Biol* **2014**, *1135*, 415–437. [PubMed]

62. Devraj, K.; Guérit, S.; Macas, J.; Reiss, Y. An In Vivo Blood-brain Barrier Permeability Assay in Mice Using Fluorescently Labeled Tracers. *J. Vis. Exp.* **2018**, *132*, e57038.

63. Sun, W.; Yang, Y.; Wu, Z.; Chen, X.; Li, W.; An, L. Chronic Cyanuric Acid Exposure Depresses Hippocampal LTP but Does Not Disrupt Spatial Learning or Memory in the Morris Water Maze. *Neurotox. Res.* **2021**, *39*, 1148–1159. [CrossRef]
64. Dinel, A.L.; Lucas, C.; Guillemet, D.; Laye, S.; Pallet, V.; Joffre, C. Chronic Supplementation with a Mix of Salvia officinalis and Salvia lavandulaefolia Improves Morris Water Maze Learning in Normal Adult C57Bl/6J Mice. *Nutrients* **2020**, *12*, 1777. [CrossRef] [PubMed]

Article

The Bi-Functional Paxilline Enriched in Skin Secretion of Tree Frogs (*Hyla japonica*) Targets the KCNK18 and BK$_{Ca}$ Channels

Chuanling Yin [†], Fanpeng Zeng [†], Puyi Huang *[,†], Zhengqi Shi, Qianyi Yang, Zhenduo Pei, Xin Wang, Longhui Chai, Shipei Zhang, Shilong Yang, Wenqi Dong, Xiancui Lu and Yunfei Wang *

College of Wildlife and Protected Area, Northeast Forestry University, Harbin 150040, China
* Correspondence: puyihuang@nefu.edu.cn (P.H.); wangyunfei@nefu.edu.cn (Y.W.)
† These authors contributed equally to this work.

Abstract: The skin secretion of tree frogs contains a vast array of bioactive chemicals for repelling predators, but their structural and functional diversity is not fully understood. Paxilline (PAX), a compound synthesized by *Penicillium paxilli,* has been known as a specific antagonist of large conductance Ca^{2+}-activated K$^+$ Channels (BK$_{Ca}$). Here, we report the presence of PAX in the secretions of tree frogs (*Hyla japonica*) and that this compound has a novel function of inhibiting the potassium channel subfamily K member 18 (KCNK18) channels of their predators. The PAX-induced KCNK18 inhibition is sufficient to evoke Ca^{2+} influx in charybdotoxin-insensitive DRG neurons of rats. By forming π-π stacking interactions, four phenylalanines located in the central pore of KCNK18 stabilize PAX to block the ion permeation. For PAX-mediated toxicity, our results from animal assays suggest that the inhibition of KCNK18 likely acts synergistically with that of BK$_{Ca}$ to elicit tingling and buzzing sensations in predators or competitors. These results not only show the molecular mechanism of PAX-KCNK18 interaction, but also provide insights into the defensive effects of the enriched PAX.

Keywords: tree frogs; skin secretion; paxilline; KCNK18; defensive strategy

Key Contribution: This paper identified that KCNK18 serves as the target of paxilline, suggesting a defensive strategy of tree frogs by using exogenous chemicals.

Citation: Yin, C.; Zeng, F.; Huang, P.; Shi, Z.; Yang, Q.; Pei, Z.; Wang, X.; Chai, L.; Zhang, S.; Yang, S.; et al. The Bi-Functional Paxilline Enriched in Skin Secretion of Tree Frogs (*Hyla japonica*) Targets the KCNK18 and BK$_{Ca}$ Channels. *Toxins* **2023**, *15*, 70. https://doi.org/10.3390/toxins15010070

Received: 7 December 2022
Revised: 10 January 2023
Accepted: 11 January 2023
Published: 12 January 2023

1. Introduction

The skin of frogs is crucial to the survival and adaptability of these amphibians to various habitats and ecological conditions. Besides respiration, water regulation and antimicrobial and antifungal resistance, the skin of frogs serves many other functions. For example, several bioactive molecules have been identified from the secretion for deterring predators, such as toxic alkaloids [1–4] and peptide toxins [5–7]. Except for gene-encoding peptides, most chemical compounds are not directly synthesized by frogs. Alternatively, the frogs can either sequester these compounds unchanged from dietary sources into their skin glands or equip microbes with the necessary machinery to produce them [8,9]. Therefore, natural communities of microbes and fungi, such as those that live on amphibians' skin, are essential for enhancing chemical defenses [10]. Fungi produce various secondary metabolites with unique and complex structures. Among them, indolo-terpenoids of the paxilline (PAX) type belong to a large family of secondary metabolites that exhibit unique molecular architectures and diverse biological activities [11]. PAX has been known as a potent blocker of BK$_{Ca}$ (large-conductance voltage- and Ca^{2+}-activated K$^+$) channels, generally acting at low nanomolar concentrations [12]. Accordingly, the application of PAX induced tremors in rodents was sustained for several hours [13], suggesting the potent toxicity of this molecule.

Although with excellent jumping ability, some tree frogs with small body sizes are considered vulnerable prey in the arboreal habitat. In this case, several gene-encoded toxins have evolved to serve as chemical weapons to deter potential terrestrial predators [14]. Unlike the peptide toxins identified from typical venomous animals [15–18], most amphibia-derived peptides show less potency to their target [19]. Is there symbiotic fungus on the skin of tree frogs to facilitate this defensive strategy? In this study, we identified PAX from the skin secretion of tree frogs, suggesting the existence of symbiotic fungi that produce toxic chemicals. Furthermore, our functional tests demonstrated that KCNK18 acts as a novel target of PAX. By inhibiting both BK_{Ca} and KCNK18 channels of predators, PAX exhibits critical biological significance to evoke excitatory currents on sensory neurons and elicit tingling and buzzing sensations in predators or competitors. Therefore, our results show the existence of bi-functional PAX in the frog skin secretion, which may represent a unique defense mechanism in amphibian adaptation.

2. Results

2.1. PAX Evokes Ca^{2+} Signals on ChTx-Insensitive Neurons

By using calcium imaging, we assessed the effect that the collected secretions of tree frogs (*Hyla japonica*) would have on the dorsal root ganglion (DRG) of rats (*Rattus norvegicus*). Interestingly, we found that more than half of these neurons exhibited robust calcium signals in the presence of 20 mg/mL secretion (Figure 1A). To rule out the possibility that pore forming peptides may disrupt the membrane architecture to elicit the calcium influx [20], we used boiled secretion and again carried out the calcium imaging test. Similarly, the boiled secretion showed equal efficacy in evoking calcium signals (Figure 1B), suggesting that chemical compounds are likely to be responsible for this activity. We therefore employed high performance liquid chromatography (HPLC) to purify the active compound (Figure 1C). The results from LC-MS (Liquid Chromatography-Mass Spectrometry) and functional tracking indicated that PAX molecules bestowed the secretion with activity for eliciting calcium influx on neurons (Figure 1D). Although PAX and charybdotoxin (ChTx) have been known as the blockers of BK_{Ca} channels, we found that PAX is sufficient to evoke Ca^{2+} influx in ChTx-insensitive neurons (Figure 1E). These results suggest that PAX can target other receptors, except BK_{Ca} channels.

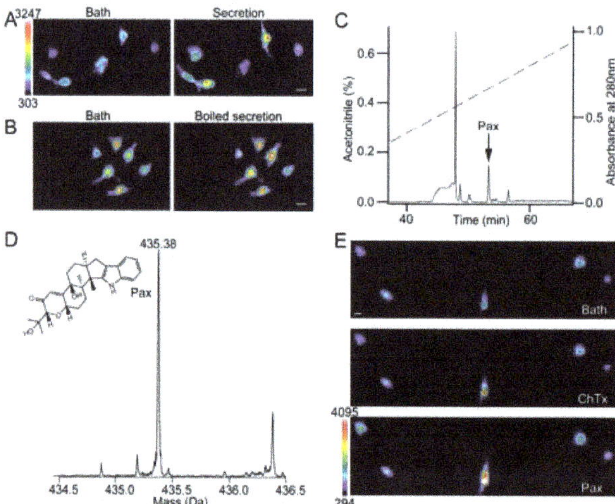

Figure 1. PAX induces Ca^{2+} signals on ChTx-insensitive neurons. (**A,B**) Representative calcium increase of rat DRG neurons in the presence of intact (**A**) or boiled (**B**) skin secretions of tree frogs. Scale bar, 50 μm (horizontal), 303 to 3247 AU (vertical). (**C**) Isolation of native PAX (black arrow) from frog skin secretions by a C_{18} RP-HPLC column. (**D**) The chemical structure and molecular weight of PAX. (**E**) Calcium imaging of rat DRG neurons in the presence of 100 nM ChTx and 20 μM PAX. Scale bar, 50 μm (horizontal), 294 to 4095 AU (vertical).

2.2. PAX Exhibits an Inhibitory Effect on Rat KCNK18

By using electrophysiological recordings, we screened several ion channels highly expressed in rat DRG neurons (Figure 2A). Among them, KCNK18 was potently inhibited by PAX (Figure 2B,C), yielding an IC_{50} value of 4.13 ± 0.04 μM. A kinetic analysis of PAX-induced KCNK18 inhibition revealed that the high affinity was due to a combination of rapid binding and very slow unbinding (Figure 2D,E). The washing out time course (τ_{off}) of PAX recorded from whole-cell configuration is 30.6 ± 2.1 s, which is much longer than that of sanshool (Figure 2F,G), a well-known KCNK18 inhibitor isolated from Szechuan peppercorns [21]. Furthermore, 100 μM PAX showed no effect on KCNK3, KCNK4, KCNK5 and KCNK9 (Figure 2H), demonstrating that PAX exhibits subtype selectivity among KCNK channels. We next used competing assays to investigate the interaction between PAX and KCNK18 to test whether PAX shares a similar binding pocket with sanshool. As illustrated in Figure 2I,J, the rate of PAX association was 11.4 ± 0.23 s, which was relatively slower than that of sanshool ($\tau_{on} = 1.15 \pm 0.12$ s). In addition, 20 μM PAX produced a blockage of KCNK18 currents after pretreatment with 50 μM sanshool (Figure 2I). However, the rate of blockage by PAX was intact (Figure 2J). This raised the possibility that PAX does not interact with the binding sites of sanshool, although none have yet been elucidated.

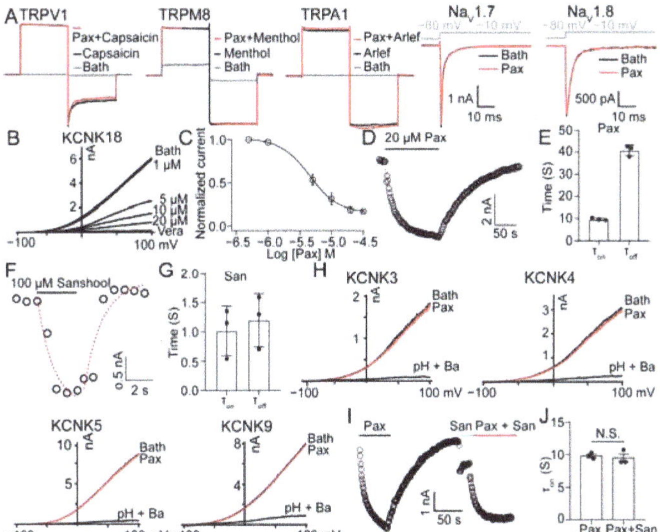

Figure 2. PAX selectively inhibits rat KCNK18. (**A**) 100 μM PAX had no inhibitory effect on TRPV1, TRPM8, TRPA1, $Na_V1.7$ and $Na_V1.8$. (**B**) Representative KCNK18 currents inhibited by PAX at different concentrations. 100 μM verapamil (vera) was used as a positive control. (**C**) Dose–response relationship of PAX inhibiting KCNK18. Data were fitted to a Hill equation (average $\pm$ SEM; $n = 3$ for each data point). (**D**) Representative wash-in and wash-out time course of 20 μM PAX on KCNK18 recorded at 100 mV, superimposed with fittings of a single-exponential function (red dotted curves). (**E**) The associated and dissociated time of 20 μM PAX on KCNK18 (average $\pm$ SEM; $n = 3$). (**F**) Representative wash-in and wash-out time course of 100 μM sanshool on KCNK18 recorded at 100 mV, superimposed with fittings of a single-exponential function (red dotted curves). (**G**) The associated and dissociated time of 100 μM sanshool on KCNK18 (average $\pm$ SEM; $n = 3$). (**H**) 100 μM PAX had no inhibitory effect on KCNK3, KCNK4, KCNK5 and KCNK9. (**I**) Representative wash-in time course of 20 μM PAX on KCNK18 recorded at 100 mV in the absence and presence of 50 μM sanshool, superimposed with fittings of a single-exponential function (red dotted curves). The solid black line indicates the presence of 20 μM PAX. The solid blue line indicates the presence of 50 μM sanshool. (**J**) The associated time of 20 μM PAX on KCNK18 in the absence and presence of 50 μM sanshool (average $\pm$ SEM; $n = 3$; N.S., no significance).

2.3. Key Residues for PAX-KCNK18 Interaction

To study the interaction between PAX and KCNK18, we screened the PAX effect on KCNK18 orthologs. Interestingly, panda KCNK18 was insensitive in the presence of PAX (Figure 3A,B). By using the chimeras constructed between rat and panda KCNK18 channels (Figure 3C), we found that the rat M3-M4 segment bestowed panda KCNK18 with PAX sensitivity, while rat KCNK18 containing panda M3-M4 segment had no response to PAX (Figure 3D,E). Therefore, the segment from M3 to M4 served as a transplantable domain to exhibit PAX sensitivity. Among the segment, only 14 residues differ between rat and panda KCNK18 channels (Figure 3F). Importantly, site 375 (in rat KCNK18 channel) is likely responsible for the interaction between KCNK18 and PAX, given that swapping this homologous site could either abolish or establish the PAX sensitivity on rat and panda KCNK18, respectively (Figure 3G–I). We therefore constructed the structural model of rat KCNK18 and tested residues nearby site 375 to screen other key amino acids that may also participate in the interaction between KCNK18 and PAX (Figure 3J). As shown in Figure 3K,L, the point mutant F167A lost the PAX sensitivity, suggesting that the side chains of F167 and F375 are likely to interact with PAX. Therefore, the interaction between PAX and KCNK18 is largely due to the π-π stacking interactions, which lead to the PAX blockage in the ion permeation pathway.

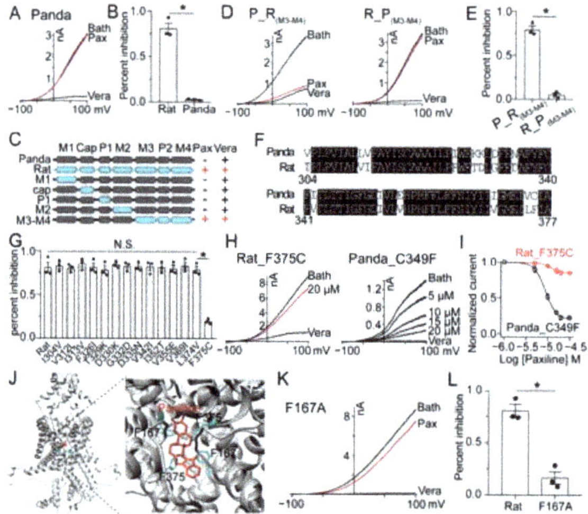

Figure 3. The binding pocket of PAX to KCNK18. (**A**) 20 μM PAX had no inhibitory effect on the panda KCNK18 channel. (**B**) The inhibitory rate of 20 μM PAX on rat and panda KCNK18 (average ± SEM; $n = 3$; * $p < 0.01$). (**C**) Schematic representation of the chimeras between panda (grey) and rat (cyan) KCNK18. The responses of wild-type and chimeric KCNK18 channels to 20 μM PAX and 100 μM vera are given. (**D**) Representative currents of chimeric KCNK18 inhibited by 20 μM PAX and 100 μM vera. P_R(M3-M4) represents that the M3-M4 region of panda KCNK18, was replaced by the homologous region of rat KCNK18, and vice versa R_P(M3-M4). (**E**) The inhibitory rate of 20 μM PAX on chimeric KCNK18 channels (average ± SEM; $n = 3$; * $p < 0.01$). (**F**) The sequence alignment of M3-M4 domain of panda and rat KCNK18. (**G**) The inhibitory rate of 20 μM PAX on rat KCNK18 mutants (average ± SEM; $n = 3$; N.S., no significance; * $p < 0.01$). (**H**) Representative currents of KCNK18 mutants inhibited by PAX and 100 μM vera. (**I**) Dose–response relationship of PAX inhibiting KCNK18 mutants. Data were fitted to a Hill equation (average ± SEM; $n = 3$). (**J**) The side view of PAX docking to rat KCNK18 model (left). The binding pocket of PAX was enlarged (right), and four phenylalanine are located near PAX. (**K**) Representative currents of rat KCNK18 mutants inhibited by 20 μM PAX and 100 μM vera. (**L**) The inhibitory rate of 20 μM PAX on rat KCNK18 and F167A mutant (average ± SEM; $n = 3$; * $p < 0.01$).

2.4. The Conserved Binding Pocket for PAX Binding

Among the KCNK family, none of the other subtypes possesses conserved phenylalanine at two key positions (Figure 4A), which may bestow PAX with subtype-selectivity. On the contrary, both phenylalanines are highly conserved in most of the avian and mammalian KCNK18 orthologs (Figure 4B), implying that the PAX equipped by tree frogs exerts general bioactivity against KCNK18 of potential predators or encounters. To evaluate the in vivo effect of the bi-functional PAX, we used ChTx, a BK$_{Ca}$ inhibitor without targeting KCNK18 (Figure 4C,D), as the control. To measure the pain reactions induced by KCNK dysfunction, the eye closure was monitored to assess the painful sensation of rat evoked by ChTx or PAX [22]. Compared with ChTx, the application of PAX elicited much more painful responses in the rat model (Figure 4E). Therefore, our results suggest that the inhibition of KCNK18 induced by PAX likely acts synergistically with its suppression of BK$_{Ca}$ to elicit tingling and buzzing sensations.

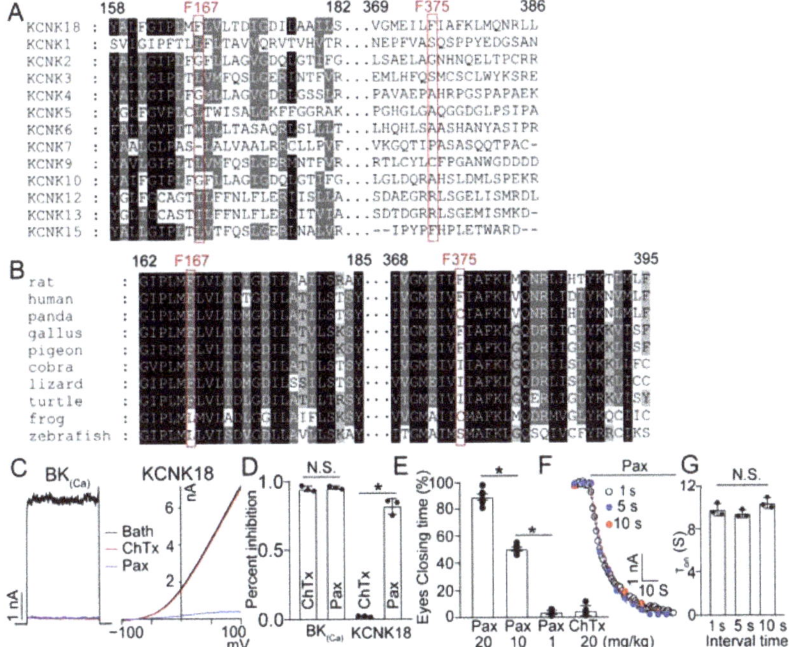

Figure 4. The sequential and functional comparison of KCNK families. (**A**) The sequence alignment of rat KCNK families. (**B**) The sequence alignment of KCNK18 orthologs. (**C**) Representative currents of BK$_{(Ca)}$ and KCNK18 inhibited by 1 μM ChTx (red) and 20 μM Pax (blue). (**D**) The inhibitory rate of 1 μM ChTx and 10 μM Pax on BK$_{(Ca)}$ and KCNK18 (average ± SEM; n = 3; * p < 0.01). (**E**) Eye closing response of rat after intravenous tail injection of PAX or ChTx. The time of both eyes closing was calculated (average ± SEM; n = 6; * p < 0.01). (**F**) Representative wash-in time course of 20 μM PAX on rat KCNK18. The perfusion of 20 μM PAX was constantly applied (indicated by a black bar). For evaluation of the inhibitory effect of PAX, the current was evoked from the holding potential (−80 mV) by the test pulse at +100 mV. The interval between each sweep was 1, 5 or 10 s, respectively. The currents were superimposed with fittings of a single-exponential function. (**G**) The associated time of 20 μM PAX on KCNK18 at different intervals of stimuli (average ± SEM; n = 3; N.S., no significance).

3. Discussion

Animals maintain a symbiotic relationship with commensal microbes, which allows them to regulate their biological operations by using functional molecules from these microbes. For instance, several marine protostomians and deuterostomians use symbionts to produce tetrodotoxin (TTX) from microbes [23]. Due to the absence of enzymes responsible for PAX biosynthesis, it is also likely that the PAX found in the skin secretions of tree frogs originates from symbiotic microbes, especially fungi. As a tremorgenic mycotoxin, PAX shows a robust inhibitory effect on BK_{Ca} channels [12]. In this work, our results demonstrate a novel target of PAX, which may promote the toxic effect induced by the blockage of BK_{Ca} channels. PAX-induced inhibition of KCNK18 currents persisted for minutes of washout (Figure 2D,E), which causes prolonged K^+ ions' retention in the cytoplasm of sensory neurons and continuous cell excitation [24]. The physiological effect of KCNK18 inhibition can be referred to as sanshool, the active compound from Szechuan peppers eliciting a unique sensation that is best described as tingling paresthesia or numbing [25]. It has been known that downregulated *kcnk18* mRNA or KCNK18 loss of function plays a crucial role in neuropathic pain [26], suggesting that the PAX-induced KCNK18 inhibition could evoke painful sensation in predators of the frogs. Due to the bifunction that targets BK_{Ca} and KCNK18 channels, we assumed that the defensive effect of PAX produced by symbiotic fungi was underestimated in previous reports.

The binding pocket of PAX on KCNK18 locates in the cavity region of the ion permeation pathway, which is likely responsible for the very slow dissociation (Figure 3J). Furthermore, our results emphasize the role of phenylalanines in PAX-KCNK18 interaction, given that mutation on either F167 or F375 (in rat KCNK18 channel) disrupted the inhibitory effect of PAX. In agreement with this, the binding models of other chemicals containing benzene rings have suggested a similar binding pocket [27], although the affinity of these chemicals is much lower than that of PAX. Interestingly, PAX did not respond to the state of KCNK18 (Figure 4F,G) in order to access this pocket, implying that the entry of PAX was directed from the bottom of this channel. According to the different binding mechanisms of PAX on KCNK18 and BK_{Ca} channels, PAX can be used as a template to develop selective KCNK18 inhibitors.

It is fascinating how animals are widely exposed to exogenous chemicals. In this case, the employment of PAX-producing microbes by tree frogs is expected to boost the toxicity of their skin secretion. One or a few transport proteins (toxin sponge molecules) in poison frogs acquired the ability to sequester toxic compounds during the toxin transportation from the gut to the skin [28]. Therefore, such a strategy may enable frogs to cope with various exogenous chemicals and dramatically elevate skin secretion's bioactivity by accumulating these compounds.

4. Materials and Methods

4.1. Animals, DRG Neurons and Skin Secretion Collection

Tree frogs (*Hyla japonica*, $n = 55$) used in this study were collected in Shangzhi, Heilongjiang province, China. The rats (*Rattus norvegicus*, $n = 3$) were purchased from Jiangsu province, China. As previously reported [29], a 3-V alternating current was used to stimulate the moistened skin manually, after which the secretion was washed using deionised water. For the preparation of the boiled secretion, the skin secretion underwent the water bath at 100 °C for 10 min. DRG neurons of rats were acutely dissociated and maintained in a short-term primary culture according to procedures as previously described [30]. All experiments involving animals conformed to the recommendations in the Guide for the Care and Use of Laboratory Animals of Northeast Forestry University. All experimental procedures were approved by the Institutional Animal Care and Use Committees at Northeast Forestry University (approval No: 2022070). All possible efforts were made to reduce the animals' sample size and minimize their suffering.

4.2. PAX Purification and Identification

The lyophilized frog secretion was dissolved in methanol (2 mg/mL) and filtered by 0.22 μm MF-Millipore filters. The filtrate was separated and purified using C_{18} reverse-phase high-performance liquid chromatography (RP-HPLC; XBrige C_{18} column, 5 μm particle size, 4.6 × 250 mm Column). PAX was analyzed using a Q-Exactive Hybrid Quadrupole-Orbitrap mass spectrometer (Thermo Scientific, Waltham, MA, USA) coupled with a Dionex UltiMate 3000 UHPLC system (Thermo Scientific, Waltham, MA, USA) and identified through searching ChemSpider database.

4.3. Plasmids and Mutagenesis

The corresponding NCBI codes of cDNA sequences of KCNK18 orthologues are 445,371 (*Rattus norvegicus*) and 100,471,846 (*Ailuropoda melanoleuca*). The corresponding NCBI code of cDNA sequences of rat BK_{Ca} is 83,731 (*Rattus norvegicus*). These coding sequences were synthesized by Tsingke (Beijing, China) and subcloned into the pCDNA3.1 vector. All KCNK18 chimeras and single-point mutants were constructed using Fast Mutagenesis Kit V2 (SBS Genetech) following the manufacturer's instructions. These channel chimeras and mutants were confirmed by DNA sequences.

4.4. Calcium Imaging

The dissociated rat DRG neurons were loaded with 3 μM Fluo-4 AM in Ringer's solution (140 mM NaCl, 5 mM KCl, 2 mM $MgCl_2$, 10 mM glucose, 2 mM $CaCl_2$ and 10 mM HEPES, pH 7.4) for 40–60 min. After incubation, the cells were washed by 2 mL Ringer's solution. The intact or boiled skin secretion (20 mg/mL) of tree frogs (Hyla japonica), 100 nM ChTx and 20 μM PAX were dissolved, respectively, in Ringer's solution to excite DRG neurons. The changes in calcium fluorescence intensity of DRG neurons were acquired with an Olympus IX71 microscope with Hamamatsu R2 charge-coupled device camera controlled by the MetaFluor Software (Molecular Devices, San Jose, CA, USA). Fluo-4 was excited by a LED light source (X-Cite 120LED, Lumen Dynamics, Mississauga, ON, Canada) with a 500/20 nm excitation filter and a 535/30 nm emission filter.

4.5. Cell Culture and Transient Transfection

HEK293 cells were cultured in Dulbecco's modified Eagle's medium with 10% fetal bovine serum and 1% penicillin/streptomycin at 37 °C with 5% CO_2. Cells were transiently transfected with DNA mixture (channel vectors and an enhanced green fluorescent protein (eGFP) using the Lipofectamine 2000 reagent (Invitrogen) following the instruction manual). Cells with fluorescence signals were selected for patch-clamp recordings 24 h after transfection. As expression of Nav1.8 is usually ineffective in HEK293 cells, rat Nav1.8 (3 μg) was transiently transfected into the neuroblastoma cell line N1E-115 by using a Nanofectin transfection kit (PAA Laboratories GMbH, Pasching, Australia). The culture condition of N1E-115 cells was the same as that of HEK 293 cells.

4.6. Electrophysiology

Whole-cell patches were recorded by using an EPC10 amplifier (HEKA) controlled by PatchMaster software (HEKA). Patch pipettes were made from borosilicate glass and fire-polished to a resistance of ~3 MΩ. For the potassium channel recording, the pipette solution contained 150 mM KCl, 3 mM $MgCl_2$, 10 mM HEPES and 5 mM EDTA, pH 7.4, and the bathing solution contained 145 mM NaCl, 2.5 mM KCl, 3 mM $MgCl_2$, 1 mM $CaCl_2$, 10 HEPES, pH 7.4. The membrane potential was held at -80 mV, and the currents were elicited by a ramp voltage from -100 mV to $+100$ mV for 400 ms. For the TRP channel recording, both the pipette solution and the bathing solution contained 130 mM NaCl, 0.2 mM EDTA, 3 mM HEPES (pH 7.2). The membrane potential was held at 0 mV and the currents were elicited by two steps, 300 ms to 80 mV followed by 300 ms to -80 mV. For the sodium channel recording, the pipette solution contained 110 mM CsF, 15 mM NaCl, 1 mM $CaCl_2$, 2 mM $MgCl_2$, 10 mM TEA-Cl, 10 mM EGTA, 2 mM ATP-Mg and 10 mM

HEPES (pH = 7.3). The bath solution contained 120 mM NaCl, 3 mM KCl, 2 mM MgCl$_2$, 2 mM CaCl$_2$, 10 mM HEPES, 30 mM glucose, 2 mM 4-aminopyridine (4-AP), 0.2 mM CdCl$_2$ and 0.5 µM tetrodotoxin (TTX) (pH = 7.3). The membrane potential was held at -80 mV, and the currents were elicited by a steady voltage to -10 mV for 100 ms. The current signals were filtered at 2.9 kHz and sampled at 10 kHz. A gravity-driven system (RSC-200, Bio-Logic) was performed to perfuse bath or stimulated solutions. The patched cells were placed at the perfusion tube outlet. The solutions were flowed through separated tubes to minimize the mixing of the solutions.

4.7. Structural Model Construction

The structure of rat KCNK18 was modeled by alpha fold molecular modeling suite version v2.0 [31]. Rosetta Ligand application from Rosetta program suite version 2020.27 was used to dock paxilline to KCNK18 [32,33]. The structural model of KCNK18 was relaxed in a membrane environment using the RosettaMembrane application [34], and the model with the lowest energy scores was used as the input structure for docking. Based on the chimera and mutant experiments, paxilline was first placed into the pocket nearby F375 and then the paxilline was docked to the optimal location and progressed from low-resolution conformational sampling and scoring to complete atom optimization using all-atom energy function. The generated 10,000 models were first screened with a total energy score. The top 1000 models with the lowest total energy score were selected and further scored with the binding energy between paxilline and KCNK18. The top 10 models with the lowest binding energy were chosen as the final docking model.

4.8. Animal Behavior

An equivalent number of male and female rats were used, between 6 and 8 weeks old weighing between 200–250 g. Rats were maintained in wired cages under a 12 h light/dark cycle at 24 °C and provided with free access to laboratory-standard food and water. All studies were approved by the Animal Care and Use Committees at Northeast Forestry University and were consistent with the guidelines.

All tail intravenous injections were performed with a 0.3 × 13 mm needle. Physiological saline was used as the diluent and vehicle. Mice were injected with 10 mL/kg physiological saline, ChTx (20 mg/kg), or PAX (1, 10, or 20 mg/kg), respectively, (n = 4–8 for each group). After injection, each rat was placed in a holding cage for 10 min to recover before detecting the painful response. After that, the time of both eyes closing was recorded for 30 min to assess the pain in rats [22].

4.9. Statistical Analysis

Igor Pro (WaveMatrix, version 6.37) and Prism (GraphPad version 8.0.1) were used to analyze the experimental data from electrophysiological recordings. All values are given as mean $\pm$ SEM for the number of measurements indicated (n). Statistical significance was determined using the Student's t-test and accepted at a level of $p < 0.01$. N.S. indicates no significance.

The currents of KCNK channel were measured at 100 mV, and normalized to the maximal currents in the absence of paxilline. The inhibition of PAX was calculated by the equation: Percent(inhibition) = $(I_{max} - I_x)/I_{max}$. I_X represents the KCNK18 current at 100 mV in the presence of concentration [x]. I_{max} represents the maximal current amplitude at 100 mV in the absence of paxilline.

EC_{50} values were calculated by fitting a Hill equation to the paxilline-induced dose-response relationship. Where n is an empirical Hill coefficient, EC_{50} is the concentration for the half-maximal effect of paxilline inhibition.

$$\frac{I_x}{I_{max}} = 1 - \frac{[X]^n}{EC_{50}^n + [X]^n}$$

τ_{on} and τ_{off} values were obtained from single exponential fits using the equations:

$$I_{(t)} = a_0 + a_1[1 - \exp(-t/\tau_{on})]$$

$$I_{(t)} = a_0 + a_1 \exp(-t/\tau_{off})$$

Author Contributions: Conceptualization, S.Y. and Y.W.; methodology, C.Y., F.Z., P.H., Z.S., L.C., S.Z. and W.D.; software, C.Y., W.D. and F.Z.; validation, P.H., Z.S. and Q.Y.; formal analysis, Z.P. and X.W.; investigation, C.Y., F.Z., P.H., X.L., Z.S., Q.Y., Z.P., X.W., L.C., S.Z. and W.D.; resources, L.C. and P.H.; data curation, S.Z. and Q.Y.; writing—original draft preparation, Y.W. and S.Y.; writing—review and editing, Y.W. and S.Y.; visualization, X.L. and C.Y.; supervision, Y.W.; project administration, Y.W.; funding acquisition, S.Y., Y.W. and X.L. All authors have read and agreed to the published version of the manuscript.

Funding: This research was funded by the National Natural Science Foundation of China (32000310 and 32030013), the Natural Science Foundation of Heilongjiang Province (YQ2021H001) and the Fundamental Research Funds for the Central Universities (2572020BE05) to Y.W; the National Natural Science Foundation of China (32022010 and 32170486), the Natural Science Foundation of Heilongjiang Province (JQ2021C001) and the National Forestry and Grassland Administration (2020132610) to S.Y.; and the National Natural Science Foundation of China (32100372) and the Natural Science Foundation of Heilongjiang Province (YQ2022H001) to X.L.

Institutional Review Board Statement: The animal study protocol was approved by the Institutional Review Board of Northeast Forestry University (2022070, issued on 04/07/2022) for studies involving animals.

Informed Consent Statement: Not applicable.

Data Availability Statement: All data generated or analysed during this study are included in this published article.

Acknowledgments: We are grateful to our laboratory members for assistance and insightful discussion.

Conflicts of Interest: The authors declare no conflict of interest.

References

1. Daly, J.W.; Myers, C.W. Toxicity of Panamanian poison frogs (Dendrobates): Some biological and chemical aspects. *Science* **1967**, *156*, 970–973. [CrossRef]
2. Daly, J.W.; Martin Garraffo, H.; Spande, T.F.; Jaramillo, C.; Stanley Rand, A. Dietary source for skin alkaloids of poison frogs (Dendrobatidae)? *J. Chem. Ecol.* **1994**, *20*, 943–955. [CrossRef]
3. Clarke, B.T. The natural history of amphibian skin secretions, their normal functioning and potential medical applications. *Biol. Rev. Camb. Philos. Soc.* **1997**, *72*, 365–379. [CrossRef]
4. Daly, J.W.; Myers, C.W.; Whittaker, N. Further classification of skin alkaloids from neotropical poison frogs (Dendrobatidae), with a general survey of toxic/noxious substances in the amphibia. *Toxicon* **1987**, *25*, 1023–1095. [CrossRef]
5. Lazarus, L.H.; Attila, M. The toad, ugly and venomous, wears yet a precious jewel in his skin. *Prog. Neurobiol.* **1993**, *41*, 473–507. [CrossRef]
6. Geller, R.G.; Govier, W.C.; Pisano, J.J.; Tanimura, T.; Van Clineschmidt, B. The action of ranatensin, a new polypeptide from amphibian skin, on the blood pressure of experimental animals. *Br. J. Pharmacol.* **1970**, *40*, 605–616. [CrossRef]
7. Endean, R.; Erspamer, V.; Falconieri Erspamer, G.; Improta, G.; Melchiorri, P.; Negri, L.; Sopranzi, N. Parallel bioassay of bombesin and litorin, a bombesin-like peptide from the skin of Litoria aurea. *Br. J. Pharmacol.* **1975**, *55*, 213–219. [CrossRef]
8. Clark, V.C.; Raxworthy, C.J.; Rakotomalala, V.; Sierwald, P.; Fisher, B.L. Convergent evolution of chemical defense in poison frogs and arthropod prey between Madagascar and the Neotropics. *Proc. Natl. Acad. Sci. USA* **2005**, *102*, 11617–11622. [CrossRef]
9. Vaelli, P.M.; Theis, K.R.; Williams, J.E.; O'Connell, L.A.; Foster, J.A.; Eisthen, H.L. The skin microbiome facilitates adaptive tetrodotoxin production in poisonous newts. *Elife* **2020**, *9*, e53898. [CrossRef]
10. Varga, J.F.A.; Bui-Marinos, M.P.; Katzenback, B.A. Frog Skin Innate Immune Defences: Sensing and Surviving Pathogens. *Front. Immunol.* **2018**, *9*, 3128. [CrossRef]
11. Thomas, W.P.; Pronin, S.V. New Methods and Strategies in the Synthesis of Terpenoid Natural Products. *Acc Chem Res* **2021**, *54*, 1347–1359. [CrossRef] [PubMed]
12. Gribkoff, V.K.; Lum-Ragan, J.T.; Boissard, C.G.; Post-Munson, D.J.; Meanwell, N.A.; Starrett, J.E., Jr.; Kozlowski, E.S.; Romine, J.L.; Trojnacki, J.T.; McKay, M.C.; et al. Effects of channel modulators on cloned large-conductance calcium-activated potassium channels. *Mol. Pharmacol.* **1996**, *50*, 206–217. [PubMed]

13. Tanner, M.R.; Pennington, M.W.; Chamberlain, B.H.; Huq, R.; Gehrmann, E.J.; Laragione, T.; Gulko, P.S.; Beeton, C. Targeting KCa1.1 Channels with a Scorpion Venom Peptide for the Therapy of Rat Models of Rheumatoid Arthritis. *J. Pharmacol. Exp. Ther.* **2018**, *365*, 227–236. [CrossRef] [PubMed]

14. Sani, M.A.; Separovic, F. How Membrane-Active Peptides Get into Lipid Membranes. *Acc. Chem. Res.* **2016**, *49*, 1130–1138. [CrossRef] [PubMed]

15. Yang, S.; Yang, F.; Wei, N.; Hong, J.; Li, B.; Luo, L.; Rong, M.; Yarov-Yarovoy, V.; Zheng, J.; Wang, K.; et al. A pain-inducing centipede toxin targets the heat activation machinery of nociceptor TRPV1. *Nat. Commun.* **2015**, *6*, 8297. [CrossRef] [PubMed]

16. Luo, L.; Li, B.; Wang, S.; Wu, F.; Wang, X.; Liang, P.; Ombati, R.; Chen, J.; Lu, X.; Cui, J.; et al. Centipedes subdue giant prey by blocking KCNQ channels. *Proc. Natl. Acad. Sci. USA* **2018**, *115*, 1646–1651. [CrossRef]

17. Li, B.; Silva, J.R.; Lu, X.; Luo, L.; Wang, Y.; Xu, L.; Aierken, A.; Shynykul, Z.; Kamau, P.M.; Luo, A.; et al. Molecular game theory for a toxin-dominant food chain model. *Natl. Sci. Rev.* **2019**, *6*, 1191–1200. [CrossRef]

18. Wang, Y.; Yin, C.; Zhang, H.; Kamau, P.M.; Dong, W.; Luo, A.; Chai, L.; Yang, S.; Lai, R. Venom resistance mechanisms in centipede show tissue specificity. *Curr. Biol.* **2022**, *32*, 3556–3563.e3553. [CrossRef]

19. You, D.; Hong, J.; Rong, M.; Yu, H.; Liang, S.; Ma, Y.; Yang, H.; Wu, J.; Lin, D.; Lai, R. The first gene-encoded amphibian neurotoxin. *J. Biol. Chem.* **2009**, *284*, 22079–22086. [CrossRef]

20. Zhao, Z.; Shi, Z.H.; Ye, C.J.; Zhang, Y. A pore-forming protein drives macropinocytosis to facilitate toad water maintaining. *Commun. Biol.* **2022**, *5*, 730. [CrossRef]

21. Gerhold, K.A.; Bautista, D.M. Molecular and cellular mechanisms of trigeminal chemosensation. *Ann. N. Y. Acad. Sci.* **2009**, *1170*, 184–189. [CrossRef] [PubMed]

22. Vuralli, D.; Wattiez, A.S.; Russo, A.F.; Bolay, H. Behavioral and cognitive animal models in headache research. *J. Headache Pain* **2019**, *20*, 11. [CrossRef] [PubMed]

23. Morita, M.; Schmidt, E.W. Parallel lives of symbionts and hosts: Chemical mutualism in marine animals. *Nat. Prod. Rep.* **2018**, *35*, 357–378. [CrossRef] [PubMed]

24. Bautista, D.M.; Sigal, Y.M.; Milstein, A.D.; Garrison, J.L.; Zorn, J.A.; Tsuruda, P.R.; Nicoll, R.A.; Julius, D. Pungent agents from Szechuan peppers excite sensory neurons by inhibiting two-pore potassium channels. *Nat. Neurosci.* **2008**, *11*, 772–779. [CrossRef] [PubMed]

25. Enyedi, P.; Braun, G.; Czirják, G. TRESK: The lone ranger of two-pore domain potassium channels. *Mol. Cell. Endocrinol.* **2012**, *353*, 75–81. [CrossRef]

26. Lafrenière, R.G.; Cader, M.Z.; Poulin, J.F.; Andres-Enguix, I.; Simoneau, M.; Gupta, N.; Boisvert, K.; Lafrenière, F.; McLaughlan, S.; Dubé, M.P.; et al. A dominant-negative mutation in the TRESK potassium channel is linked to familial migraine with aura. *Nat. Med.* **2010**, *16*, 1157–1160. [CrossRef]

27. Gao, Z.B.; Chen, X.Q.; Jiang, H.L.; Liu, H.; Hu, G.Y. Electrophysiological characterization of a novel Kv channel blocker N,N'-[oxybis(2,1-ethanediyloxy-2,1-ethanediyl)]bis(4-methyl)-benzenesulfonamide found in virtual screening. *Acta Pharmacol. Sin.* **2008**, *29*, 405–412. [CrossRef]

28. Abderemane-Ali, F.; Rossen, N.D.; Kobiela, M.E.; Craig, R.A.; Garrison, C.E.; Chen, Z.; Colleran, C.M.; O'Connell, L.A.; Du Bois, J.; Dumbacher, J.P.; et al. Evidence that toxin resistance in poison birds and frogs is not rooted in sodium channel mutations and may rely on "toxin sponge" proteins. *J. Gen. Physiol.* **2021**, *153*, e202112872. [CrossRef]

29. Chai, L.; Yin, C.; Kamau, P.M.; Luo, L.; Yang, S.; Lu, X.; Zheng, D.; Wang, Y. Toward an understanding of tree frog (Hyla japonica) for predator deterrence. *Amino Acids* **2021**, *53*, 1405–1413. [CrossRef]

30. Yang, S.; Lu, X.; Wang, Y.; Xu, L.; Chen, X.; Yang, F.; Lai, R. A paradigm of thermal adaptation in penguins and elephants by tuning cold activation in TRPM8. *Proc. Natl. Acad. Sci. USA* **2020**, *117*, 8633–8638. [CrossRef]

31. Jumper, J.; Evans, R.; Pritzel, A.; Green, T.; Figurnov, M.; Ronneberger, O.; Tunyasuvunakool, K.; Bates, R.; Žídek, A.; Potapenko, A.; et al. Highly accurate protein structure prediction with AlphaFold. *Nature* **2021**, *596*, 583–589. [CrossRef]

32. Davis, I.W.; Baker, D. RosettaLigand docking with full ligand and receptor flexibility. *J. Mol. Biol.* **2009**, *385*, 381–392. [CrossRef]

33. Meiler, J.; Baker, D. ROSETTALIGAND: Protein-small molecule docking with full side-chain flexibility. *Proteins* **2006**, *65*, 538–548. [CrossRef]

34. Yarov-Yarovoy, V.; Schonbrun, J.; Baker, D. Multipass membrane protein structure prediction using Rosetta. *Proteins* **2006**, *62*, 1010–1025. [CrossRef]

Article

A Novel Peptide from *Polypedates megacephalus* Promotes Wound Healing in Mice

Siqi Fu [1], Canwei Du [2], Qijian Zhang [3], Jiayu Liu [4], Xushuang Zhang [4] and Meichun Deng [4,5,6,*]

1   Hunan Key Laboratory of Medical Epigenomics, Department of Dermatology, The Second Xiangya Hospital, Central South University, Changsha 410013, China
2   Chengdu Pep Biomedical Co., Ltd., Chengdu 610041, China
3   Wound Center of Xiangya Hospital, Central South University, Changsha 410013, China
4   Hunan Province Key Laboratory of Basic and Applied Hematology, Department of Biochemistry and Molecular Biology, School of Life Sciences, Central South University, Changsha 410013, China
5   Hunan Key Laboratory of Animal Models for Human Diseases, School of Life Sciences, Central South University, Changsha 410013, China
6   Hunan Key Laboratory of Medical Genetics, School of Life Sciences, Central South University, Changsha 410013, China
*   Correspondence: dengmch@csu.edu.cn

Abstract: Amphibian skin contains wound-healing peptides, antimicrobial peptides, and insulin-releasing peptides, which give their skin a strong regeneration ability to adapt to a complex and harsh living environment. In the current research, a novel wound-healing promoting peptide, PM-7, was identified from the skin secretions of *Polypedates megacephalus*, which has an amino acid sequence of FLNWRRILFLKVVR and shares no structural similarity with any peptides described before. It displays the activity of promoting wound healing in mice. Moreover, PM-7 exhibits the function of enhancing proliferation and migration in HUVEC and HSF cells by affecting the MAPK signaling pathway. Considering its favorable traits as a novel peptide that significantly promotes wound healing, PM-7 can be a potential candidate in the development of novel wound-repairing drugs.

Keywords: frogs; peptides; skin wounds; HSF cells; HUVEC cells

Key Contribution: PM-7 purified from the skin of *Polypedates megacephalus* is a bioactive/effector compound with potential wound-healing ability through the MAPK signaling pathway.

Citation: Fu, S.; Du, C.; Zhang, Q.; Liu, J.; Zhang, X.; Deng, M. A Novel Peptide from *Polypedates megacephalus* Promotes Wound Healing in Mice. *Toxins* **2022**, *14*, 753. https://doi.org/10.3390/toxins14110753

Received: 3 August 2022
Accepted: 27 October 2022
Published: 2 November 2022

Publisher's Note: MDPI stays neutral with regard to jurisdictional claims in published maps and institutional affiliations.

1. Introduction

Skin, as a physical barrier, protects internal organs and tissues from external harm such as mechanical or physical damage, as well as pathogenic micro-organisms [1,2]. Skin includes the keratinized lamellar epidermis and the underlying dermal connective tissue, which is collagen-rich and able to provide support and nourishment [3,4]. Skin is vulnerable to accidental damage [5], metabolic dysfunction [6], and skin diseases [7]. Wound healing after injuries is important for human health and survival. Four stages are included in wound healing: hemorrhage, inflammation, proliferation, and tissue remodeling [8–10]; each of which play a vital role in ensuring the restoration of skin integrity and functional reconstruction [11]. Small-molecule chemicals from plants, and proteins represented by epidermal growth factors are the two main groups involved in wound healing [12–16]. Considering the unstable activity and high cost of these drugs, it is critical to develop new drugs with the function of wound healing.

Among vertebrates, amphibians live in a complicated environment. Their skins are more challenged by biotic or abiotic factors [17,18]. During physiological and pathological processes, amphibian skin, limbs, and tails can heal rapidly and show a strong regenerative ability [17,19]. In addition, it has been reported that fresh frog skin can effectively promote wound healing [19]. Bioactive compounds derived from amphibian skin have been used in traditional and folk medicine for hundreds of years [20]. Amphibians may be a rich resource pool of bioactive compounds and, up to now, more than 100 peptides from the skin of 2000 species have been reported, including antimicrobial peptides [21,22], neuropeptides [23–25], wound-healing peptides [26–28], and insulin-releasing peptides [29,30].

Herein, we reported a novel peptide PM-7 with an amino acid sequence of FLNWR-RILFLKVVR from the frog *Polypedates megacephalus*, which shows the activity of promoting wound healing in mice. Further, PM-7 can enhance cell proliferation and migration on Human Umbilical Vein Endothelial Cells (HUVEC) and Human skin fibroblast (HSF) cells and affect the MAPK signaling pathway, which exhibits an important function as an effective wound healing regulator.

2. Results

2.1. Purification and Identification of PM-7

The skin secretion of the frog *Polypedates megacephalus* was collected by anhydrous ether stimulation. The skin secretion was divided into four fractions using gel filtration (Figure 1A), and those fractions showing increased cell proliferation were purified with RP-HPLC C18 column (Figure 1B,C). The sample that displayed the activity of promoting cell proliferation was determined with a molecular weight of 1860.32 Da via a MALDI-TOF MS analysis (Figure 1D) and named PM-7. Using Edman degradation, PM-7 was identified as a new peptide containing 14 residues, with a sequence of FLNWRRILFLKVVR, which is consistent with the molecular mass.

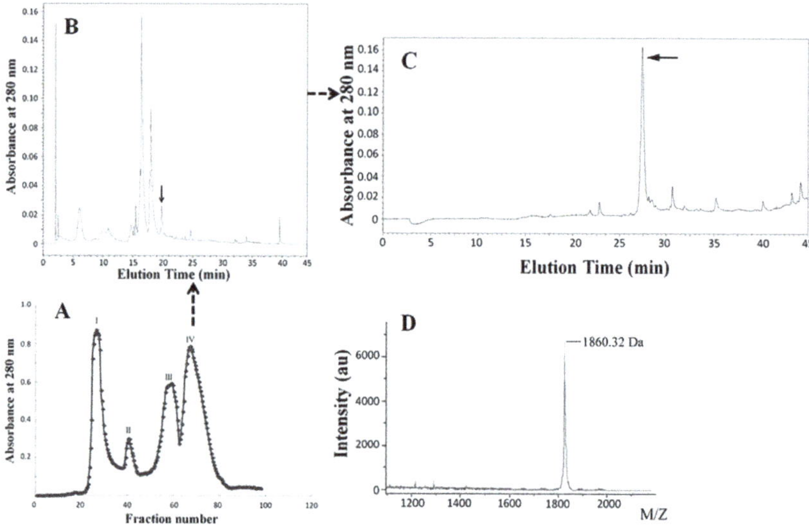

Figure 1. Isolation and purification of PM-7. (**A**) The skin secretion was isolated using G50 gel filtration and monitored at 280 nm, and the fraction IV displaying the activity of promoting cell proliferation is indicated with the arrow. (**B,C**) PM-7 was purified with RP-HPLC C18 and monitored at 280 nm. (**D**) The molecular weight of purified PM-7 was determined by mass spectrometry (1860.32 Da).

2.2. PM-7 Accelerated Cell Proliferation and Migration in HSF and HUVEC Cells

During wound healing, the proliferation and movement of fibroblasts can greatly promote the process of re-epithelialization and wound healing. Our results showed that PM-7 significantly enhances the proliferation of HSF cells and the effect displayed a significant correlation with the peptide concentration. The proliferation rates of HSF cells increased by 150% and 180% at 500 μg/mL and 1000 μg/mL of PM-7, respectively, but PM-7 at 2000 μg/mL reduced its activity of promoting cell proliferation compared to that at 1000 μg/mL (Figure 2A). For HUVEC cells, PM-7 at 500 μg/mL slightly increased the rate of cell proliferation and at 1000 µg/mL dramatically enhanced the cell proliferation by 80%. Similar to HSF cells, PM-7 at 2000 µg/mL displayed a moderate effect on the proliferation of HUVEC cells (Figure 2B). We also explored the effect of PM-7 on cell migration using a cell scratch assay in vitro. As shown in Figure 2C, after 16 h of treatment, PM-7 enhanced cells migration across the sound chasm compared with the control treatment. Therefore, PM-7 accelerated cell proliferation and migration of HSF and HUVEC cells.

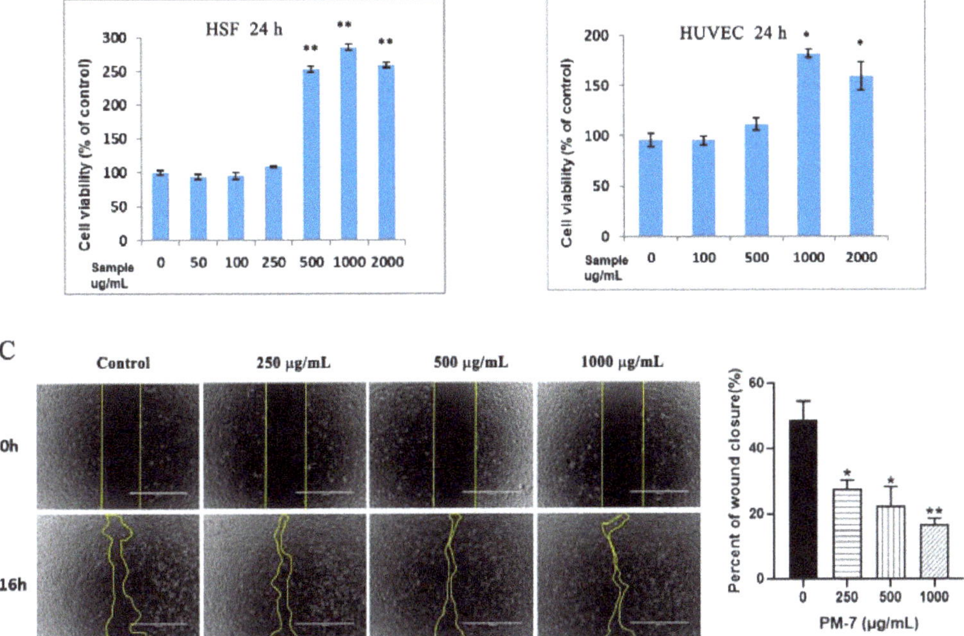

Figure 2. The effect of PM-7 on HSF and HUVEC cells. (**A,B**) Concentration-dependent effects of PM-7 on cell proliferation in HSF and HUVEC cells. (**C**) Representative images of PM-7 promoting healing of cell scratch on HSF cells. The line in each image represents the magnitude of 200 µm. Data represent the average of three independent experiments, expressed as a mean ± S.E.M., $n = 3$, ns: no statistical significance. * $p < 0.05$, ** $p < 0.01$. There were significant differences compared to the control group.

2.3. Effect of PM-7 on MAPK Signaling Pathway

The MAPK signaling pathway has been proven to be involved in promoting wound healing. We next carried out Western blot analyses to further investigate whether PM-7 could affect the MAPK signaling pathway in HUVEC cells in vitro or not. As shown in Figure 3, PM-7 significantly increased ERK and p38 phosphorylation after 12 h of 1000 µg/mL treatment. The results of Western blotting showed that the addition of PM-7 increased the expression of p-ERK1/2 and p-P38 that are the membranes of MAPK family.

2.4. PM-7 Enhanced the Healing of Full-Thickness Wounds in Mice

Since PM-7 showed activity in terms of promoting cell proliferation on HSF and HUVEC cells in vitro, the effect of PM-7 in vivo on the healing of full-thickness skin wounds was investigated by using an excision wound healing test in mice. The effect of PM-7 on full-thickness skin wounds in mice is shown in the figure. The application of PM-7 significantly accelerated wound closure in mice compared with that of treatment by PBS (Figure 4). On post-injury day 3, the wound area of mice treated with PM-7 was

50% smaller than that of the PBS-treated mice (Figure 4), suggesting that PM-7 accelerated wound healing. On post-injury day 5, the wounds treated with PM-7 and epidermal growth factor (EGF) were almost closed, while the wounds of control mice remained 45% open (Figure 4). After seven days, there was no obvious wound in PM-7-treated mice; however, the wounds in PBS-treated mice remained visible for several days (Figure 4). More importantly, there were no adverse effects on the body weight, overall health status, or behavior of the mice during the treatment with PM-7.

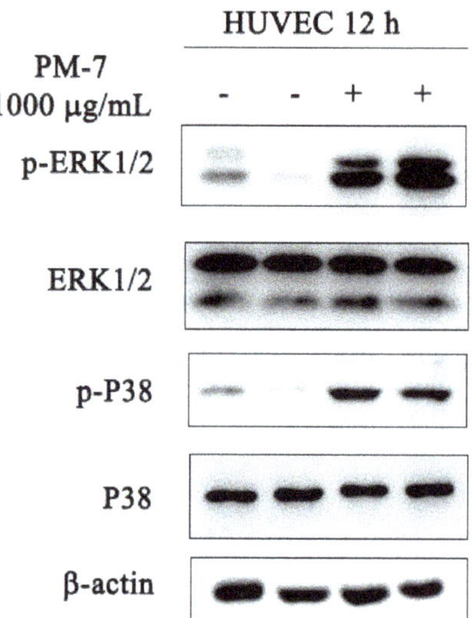

Figure 3. Effect of PM-7 on the MAPK signaling pathway in HUVEC cells using Western blot analyses. After a 12-h treatment with PM-7, the protein levels of p-ERK1/2 and p-P38 and the total ERK1/2 and p38 in HUVEC cells were detected using Western blot. β-actin was used to determine the amount of loaded protein.

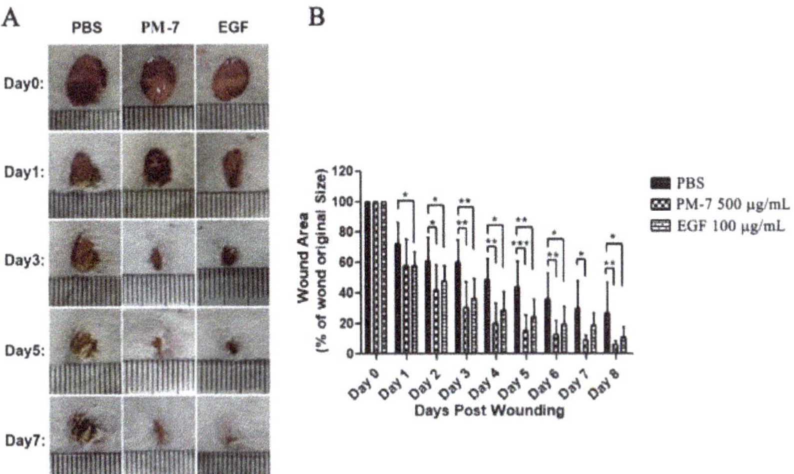

Figure 4. The effect of PM-7 on full-thickness skin wounds in mice. (**A**) Representative images of skin wounds on days 0, 1, 3, 5, and 7 with PM-7 treatment. PBS was a negative control and EGF was a positive control. A small cell on the ruler represents 1 mm. (**B**) Quantitative data showing the effect of PBS, PM-7, and EGF on full-thickness skin wounds in mice. Data are expressed as mean $\pm$ S.E.M., $n = 6$. * $p < 0.05$, ** $p < 0.01$, *** $p < 0.0001$.

3. Discussion

There are plentiful bioactive peptides in the skin secretions of amphibians, including antimicrobial peptides, wound-healing peptides, lectins, and protease inhibitors, as reported before [25,27,31,32]. It has also been reported that the skin secretions of amphibians could accelerate wound healing because the skin often suffers various injuries, and some peptides have been identified to be beneficial to wound healing, such as a short peptide, CW49, from the skin of the frog *Odorrana graham* [26]. Here, we discovered a novel peptide PM-7 from the skin secretions of *Polypedates megacephalus*. For the isolation of skin secretions, gel filtration chromatography was carried out first to separate the different fractions via molecular weight of the different fractions. Specifically, some peptides or small-molecule chemicals came out later, as fraction IV (Figure 1A). Following that, reverse-phase high-performance liquid chromatography (RP-HPLC) was performed to further purify the peptide PM-7 (Figure 1B,C). Through a sequence alignment, we found that PM-7 does not share a similar structure to any previously discovered peptide, suggesting that PM-7 may be a novel candidate for a wound-healing drug.

Keratinocytes and fibroblasts are circuits that repair the dominant cells involved in the proliferation phase of wound healing [17,33,34]. Here, we found that PM-7 could accelerate the proliferation of HSF cells and HUVEC cells in a dose-dependent manner (Figure 2). Notably, the acceleration of cell proliferation was delayed at a high concentration of PM-7, which may be due to the complicated mechanisms of wound healing. The MAPK signaling pathway is sensitized by different intracellular- and extracellular-related substances, including peptide growth factors, cytokines, and related hormones, and can regulate various cellular activities, including proliferation, differentiation, survival, and death [25,35]. We checked the effect of PM-7 on the signaling pathway and found that PM-7 enhanced ERK and p38 phosphorylation in HUVEC cells (Figure 3). According to previous reports, some other peptides from amphibian skin also exert significant effects on wound healing through the MAPK signaling pathway [19,35].

Some peptides have been found from the skin of frogs, such as Cathelicidin-NV [19] and OM-LV20 [36]. PM-7 had a shorter sequence than some other peptides which display wound healing activity, which indicates that PM-7 would cost less if obtained with solid state synthesis. Importantly, PM-7 clearly enhanced the healing of full-thickness dermal wounds in vivo (Figure 4), suggesting its possibility in clinical applications. Some growth factors used clinically, such as EGF, have been shown to enhance wound healing in a variety of tissues. The small peptide in this study, containing only 14 amino acid residues, might be a potential biomaterial or template for developing novel wound-healing agents.

In conclusion, PM-7 purified from the skin of *Polypedates megacephalus* is a bioactive/effector compound with potential wound-healing ability through the MAPK signaling pathway. It may facilitate the understanding of frog wound healing and skin regeneration. In addition, these properties make PM-7 a potent candidate for skin wound therapeutics.

4. Materials and Methods

4.1. Collection of Frog Skin Secretions and Isolation of Peptides

The frog skins were stimulated and purified as before [19], and the skin secretions were subjected to lyophilization and stored at $-80\,^{\circ}$C until use. Briefly, gel chromatography was carried out on a Sephadex G-50 gel filtration column (1.5×31 cm, superfine, GE Healthcare, Danderyd, Sweden) using 25 mM Tris-HCl buffer (pH 7.8) that contained 0.1 m NaCl. After pre-equilibrating the column, the skin secretions (500 µL, OD280 = 700 mAu), dissolved in a 25 mM Tris-HCl buffer, were eluted using the same buffer at a flow rate of 0.1 mL/min. A fraction collector (BSA-30A, HuXi Company, Shanghai, China) was used to automatically collect the fractions every 10 min and the fraction was checked at 280 nm. A reversed-phase high performance liquid chromatography (RP-HPLC) C18 column (Hypersil BDS C18, 4.0×300 mm, Elite, Shanghai, China) was pre-performed with water containing 0.1% (v/v) trifluoroacetic acid (TFA), followed by eluting along a linear gradient (0–70% acetonitrile, ACN in 70 min) with 0.1% (v/v) TFA and monitored at 280 nm. A mass spectrometer (MS) (Autoflex speed TOF/TOF, Bruker Daltonik GmbH, Leipzig, Germany) was used to analyze the purification of the crystallized samples and identify the molecular weight of the peptide. After lyophilization, the peptide was stored at $-20\,^{\circ}$C until use.

4.2. Edman Degradation Sequencing

Edman degradation was performed to determine the sequence of the peptide PM-7 using a PPSQ-31A protein sequencer (Shimadzu, Kyoto, Japan) according to the manufacturer's standard protocols for GFD.

4.3. Cell Culture

HSF and HUVEC cells were cultured in Dulbecco's Modified Eagle Medium and Ham's F12 Medium (DMEM/F12) (BI, Beit Haemek, Israel) with 10% (v/v) fetal bovine serum (FBS, Hyclone, Logan, UT, USA) and antibiotics (100 units/mL penicillin and 100 units/mL streptomycin) in an incubator of 5% CO_2 at 37 $^{\circ}$C.

4.4. Cell-Scratch Healing, Migration, and Proliferation

The cell-scratch healing and proliferation of HSF and HUVEC cells were carried out to assess the cellular pro-healing activities of PM-7. Briefly, HSF cells were cultured overnight in a 24-well plate at a density of 2×10^6 cells/well. Sterile 200-µL pipette tips were used to scratch the cell monolayer. Cells were washed with phosphate-buffered saline (PBS) solution three times before being cultured using a medium with PM-7 in different concentrations (0, 50, 100, 250, 500, 1000, and 2000 µg/mL). An inverted microscope (Motic, AE2000, Shenzhen, China) was used to obtain the images at different times. The methods for exploring the effect of PM-7 on the proliferation of HSF and HUVEC cells were the same as above. Briefly, 96-well plates were used to culture HSF and HUVEC cells at a density of 4×10^4 cells/well, and they were exposed to PM-7 at different concentrations for 24 h. An MTS cell proliferation assay kit from Promega (Madison, WI, USA) was used to assess the impact of PM-7 on the proliferation.

4.5. Full-Thickness Skin Wounds in Mice

Full-thickness skin wounds in mice were created to check the regenerative effects of PM-7 in vivo. Male Kunming mice were obtained from Hunan Slack Jingda Experimental Laboratory Animal Co., Ltd. (Chansha, Hunan, China). The mice were given an adaptive feeding with standard food and water for one week and anesthetized through the peritoneum using 1% pentobarbital sodium (0.1 mL/20 g body weight). The corresponding hairs were removed from the back surface and a full-thickness wound (8 mm in diameter) was made on the dorsal skin of the back of every mouse. PBS, PM-7 (20 μL, 500 μg/mL) or EGF (20 μL, 100 μg/mL) was applied directly to the wound site on mice twice a day from day 0 to day 8. A digital photograph was used to record the wound condition every two days. ImageJ software v1.51 was used to calculate the wound area (percentage of residual wound area to primitive area) and quantified using GraphPad Prism v.8 (GraphPad Software Inc., San Diego, CA, USA). All animal experiments were approved and implemented in accordance with the requirements of Central South University.

4.6. MAPKs Signaling Pathway

HUVEC cells (2×10^6) were cultured in six-well culture plates for 4 h. The adherent cells were seeded in a serum-free medium with the treatment of PM-7 for 12 h. After treatment with lysate (radio immunoprecipitation assay (RIPA)) containing phenylmethyl-sulphonyl fluoride (PMSF, Meilun Biotechnology, Dalian, China) and phosphatase inhibitor (Roche, Shanghai, China), cells were quantified using a BCA protein analysis kit (Meilun, Dalian, China). After separation of the cell samples using 12% SDS-PAGE, Western blotting was performed to analyze the protein samples and electroablation onto polyvinylidene fluoride (PVDF) membranes. ImageJ was used to check the grayscale and quantify the protein expression and degree of phosphorylation.

4.7. Statistical Analysis

Data are expressed as means $\pm$ S.E.M. The animals were randomly assigned (1:1:1 allocation) to receive PBS, EGF, or PM-7. For the cellular and animal experiments, two-tailed paired t-tests were used to compare two groups, and one-way ANOVA followed by Tukey's post hoc test was used to evaluate differences among more than two groups. Statistical analyses were performed using GraphPad Prism v.8 software (GraphPad Software Inc., San Diego, CA, USA). A value of $p < 0.05$ was considered to be statistically significant.

Author Contributions: Conceptualization, M.D.; data curation, S.F., C.D., Q.Z., J.L. and X.Z. All authors have read and agreed to the published version of the manuscript.

Funding: This work was supported by the National Natural Science Foundation of China under contract (Nos. 82071250, 31672290), Natural Science Foundation of Hunan Province (2021JJ30798, 2021JJ70150, 2022JJ30805), Innovation and entrepreneurship education reform research project of Central South University (2019CG052), Postgraduate education and teaching reform project of Central South University (2021JGB105), Biochemistry Maker Space of Central South University (2015CK009).

Institutional Review Board Statement: The study was conducted according to the guidelines of the Declaration of Helsinki, and approved by the Ethics Committee of School of Life Sciences, Central South University, Changsha, China (protocol code 2020-2-4 and date of approval 30 March 2020).

Informed Consent Statement: Not applicable.

Data Availability Statement: All data generated or analyzed during this study are included in this manuscript.

Conflicts of Interest: The authors declare no conflict of interest.

References

1. Kaplani, K.; Koutsi, S.; Armenis, V.; Skondra, F.G.; Karantzelis, N.; Champeris Tsaniras, S.; Taraviras, S. Wound healing related agents: Ongoing research and perspectives. *Adv. Drug Deliv. Rev.* **2018**, *129*, 242–253. [CrossRef] [PubMed]

2. Werner, S.; Grose, R. Regulation of wound healing by growth factors and cytokines. *Physiol. Rev.* **2003**, *83*, 835–870. [CrossRef] [PubMed]
3. Zhao, Y.; Wang, Q.; Jin, Y.; Li, Y.; Nie, C.; Huang, P.; Li, Z.; Zhang, B.; Su, Z.; Hong, A.; et al. Discovery and Characterization of a High-Affinity Small Peptide Ligand, H1, Targeting FGFR2IIIc for Skin Wound Healing. *Cell. Physiol. Biochem.* **2018**, *49*, 1033–1048. [CrossRef] [PubMed]
4. Rippa, A.L.; Kalabusheva, E.P.; Vorotelyak, E.A. Regeneration of Dermis: Scarring and Cells Involved. *Cells* **2019**, *8*, 607. [CrossRef]
5. Shiny, P.J.; Vimala Devi, M.; Felciya, S.J.G.; Ramanathan, G.; Fardim, P.; Sivagnanam, U.T. In vitro and in vivo evaluation of poly 3 hydroxybutyric acid sodium alginate as a core shell nanofibrous matrix with arginine and bacitracin nanoclay complex for dermal reconstruction of excision wound. *Int. J. Biol. Macromol.* **2021**, *168*, 46–58. [CrossRef]
6. Wang, Z.; Wang, Y.; Bradbury, N.; Gonzales Bravo, C.; Schnabl, B.; Di Nardo, A. Skin wound closure delay in metabolic syndrome correlates with SCF deficiency in keratinocytes. *Sci. Rep.* **2020**, *10*, 21732. [CrossRef]
7. Bagood, M.D.; Isseroff, R.R. TRPV1: Role in Skin and Skin Diseases and Potential Target for Improving Wound Healing. *Int. J. Mol. Sci.* **2021**, *22*, 6135. [CrossRef]
8. Diegelmann, R.F.; Evans, M.C. Wound healing: An overview of acute, fibrotic and delayed healing. *Front. Biosci.* **2004**, *9*, 283–289. [CrossRef]
9. Xie, Z.; Aphale, N.V.; Kadapure, T.D.; Wadajkar, A.S.; Orr, S.; Gyawali, D.; Qian, G.; Nguyen, K.T.; Yang, J. Design of antimicrobial peptides conjugated biodegradable citric acid derived hydrogels for wound healing. *J. Biomed. Mater. Res. A* **2015**, *103*, 3907–3918. [CrossRef]
10. Fu, S.; Panayi, A.; Fan, J.; Mayer, H.F.; Daya, M.; Khouri, R.K.; Gurtner, G.C.; Ogawa, R.; Orgill, D.P. Mechanotransduction in Wound Healing: From the Cellular and Molecular Level to the Clinic. *Adv. Skin Wound Care* **2021**, *34*, 1. [CrossRef]
11. Rujirachotiwat, A.; Suttamanatwong, S. Curcumin upregulates transforming growth factor-beta1, its receptors, and vascular endothelial growth factor expressions in an in vitro human gingival fibroblast wound healing model. *BMC Oral Health* **2021**, *21*, 535. [CrossRef] [PubMed]
12. Cao, X.; Wang, Y.; Wu, C.; Li, X.; Fu, Z.; Yang, M.; Bian, W.; Wang, S.; Song, Y.; Tang, J.; et al. Cathelicidin-OA1, a novel antioxidant peptide identified from an amphibian, accelerates skin wound healing. *Sci. Rep.* **2018**, *8*, 943. [CrossRef] [PubMed]
13. Lin, T.K.; Zhong, L.; Santiago, J.L. Anti-Inflammatory and Skin Barrier Repair Effects of Topical Application of Some Plant Oils. *Int. J. Mol. Sci.* **2017**, *19*, 70. [CrossRef]
14. Deng, Z.H.; Yin, J.J.; Luo, W.; Kotian, R.N.; Gao, S.S.; Yi, Z.Q.; Xiao, W.F.; Li, W.P.; Li, Y.S. The effect of earthworm extract on promoting skin wound healing. *Biosci. Rep.* **2018**, *38*, BSR20171366. [CrossRef]
15. Pazyar, N.; Yaghoobi, R.; Rafiee, E.; Mehrabian, A.; Feily, A. Skin wound healing and phytomedicine: A review. *Skin Pharmacol. Physiol.* **2014**, *27*, 303–310. [CrossRef] [PubMed]
16. Sui, H.; Wang, F.; Weng, Z.; Song, H.; Fang, Y.; Tang, X.; Shen, X. A wheat germ-derived peptide YDWPGGRN facilitates skin wound-healing processes. *Biochem. Biophys. Res. Commun.* **2020**, *524*, 943–950. [CrossRef]
17. Chang, J.; He, X.; Hu, J.; Kamau, P.M.; Lai, R.; Rao, D.; Luo, L. Bv8-Like Toxin from the Frog Venom of Amolops jingdongensis Promotes Wound Healing via the Interleukin-1 Signaling Pathway. *Toxins* **2019**, *12*, 15. [CrossRef]
18. Demori, I.; Rashed, Z.E.; Corradino, V.; Catalano, A.; Rovegno, L.; Queirolo, L.; Salvidio, S.; Biggi, E.; Zanotti-Russo, M.; Canesi, L.; et al. Peptides for Skin Protection and Healing in Amphibians. *Molecules* **2019**, *24*, 347. [CrossRef]
19. Wu, J.; Yang, J.; Wang, X.; Wei, L.; Mi, K.; Shen, Y.; Liu, T.; Yang, H.; Mu, L. A frog cathelicidin peptide effectively promotes cutaneous wound healing in mice. *Biochem. J.* **2018**, *475*, 2785–2799. [CrossRef]
20. Shi, Y.; Li, C.; Wang, M.; Chen, Z.; Luo, Y.; Xia, X.S.; Song, Y.; Sun, Y.; Zhang, A.M. Cathelicidin-DM is an Antimicrobial Peptide from Duttaphrynus melanostictus and Has Wound-Healing Therapeutic Potential. *ACS Omega* **2020**, *5*, 9301–9310. [CrossRef]
21. Zasloff, M. Magainins, a class of antimicrobial peptides from Xenopus skin: Isolation, characterization of two active forms, and partial cDNA sequence of a precursor. *Proc. Natl. Acad. Sci. USA* **1987**, *84*, 5449–5453. [CrossRef] [PubMed]
22. Casciaro, B.; Cappiello, F.; Loffredo, M.R.; Ghirga, F.; Mangoni, M.L. The Potential of Frog Skin Peptides for Anti-Infective Therapies: The Case of Esculentin-1a(1-21)NH2. *Curr. Med. Chem.* **2020**, *27*, 1405–1419. [CrossRef] [PubMed]
23. Baroni, A.; Perfetto, B.; Canozo, N.; Braca, A.; Farina, E.; Melito, A.; De Maria, S.; Carteni, M. Bombesin: A possible role in wound repair. *Peptides* **2008**, *29*, 1157–1166. [CrossRef] [PubMed]
24. Vethamany-Globus, S. Hormone action in newt limb regeneration: Insulin and endorphins. *Biochem. Cell Biol.* **1987**, *65*, 730–738. [CrossRef]
25. Luo, X.; Ouyang, J.; Wang, Y.; Zhang, M.; Fu, L.; Xiao, N.; Gao, L.; Zhang, P.; Zhou, J.; Wang, Y. A novel anionic cathelicidin lacking direct antimicrobial activity but with potent anti-inflammatory and wound healing activities from the salamander Tylototriton kweichowensis. *Biochimie* **2021**, *191*, 37–50. [CrossRef]
26. Liu, H.; Duan, Z.; Tang, J.; Lv, Q.; Rong, M.; Lai, R. A short peptide from frog skin accelerates diabetic wound healing. *FEBS J.* **2014**, *281*, 4633–4643. [CrossRef]
27. He, X.; Yang, Y.; Mu, L.; Zhou, Y.; Chen, Y.; Wu, J.; Wang, Y.; Yang, H.; Li, M.; Xu, W.; et al. A Frog-Derived Immunomodulatory Peptide Promotes Cutaneous Wound Healing by Regulating Cellular Response. *Front. Immunol.* **2019**, *10*, 2421. [CrossRef]
28. Thapa, R.K.; Diep, D.B.; Tonnesen, H.H. Topical antimicrobial peptide formulations for wound healing: Current developments and future prospects. *Acta Biomater.* **2020**, *103*, 52–67. [CrossRef]

29. Marenah, L.; Flatt, P.R.; Orr, D.F.; McClean, S.; Shaw, C.; Abdel-Wahab, Y.H. Brevinin-1 and multiple insulin-releasing peptides in the skin of the frog Rana palustris. *J. Endocrinol.* **2004**, *181*, 347–354. [CrossRef]
30. Zahid, O.K.; Mechkarska, M.; Ojo, O.O.; Abdel-Wahab, Y.H.; Flatt, P.R.; Meetani, M.A.; Conlon, J.M. Caerulein-and xenopsin-related peptides with insulin-releasing activities from skin secretions of the clawed frogs, Xenopus borealis and Xenopus amieti (Pipidae). *Gen. Comp. Endocrinol.* **2011**, *172*, 314–320. [CrossRef]
31. Cui, J.; Zheng, H.; Zhang, J.; Jia, L.; Feng, Y.; Wang, W.; Li, H.; Chen, F. Profiling of glycan alterations in regrowing limb tissues of Cynops orientalis. *Wound Repair Regen.* **2017**, *25*, 836–845. [CrossRef] [PubMed]
32. Rong, M.; Liu, J.; Liao, Q.; Lin, Z.; Wen, B.; Ren, Y.; Lai, R. The defensive system of tree frog skin identified by peptidomics and RNA sequencing analysis. *Amino Acids* **2019**, *51*, 345–353. [CrossRef] [PubMed]
33. Wang, S.; Feng, C.; Yin, S.; Feng, Z.; Tang, J.; Liu, N.; Yang, F.; Yang, X.; Wang, Y. A novel peptide from the skin of amphibian Rana limnocharis with potency to promote skin wound repair. *Nat. Prod. Res.* **2021**, *35*, 3514–3518. [CrossRef] [PubMed]
34. Liu, N.; Li, Z.; Meng, B.; Bian, W.; Li, X.; Wang, S.; Cao, X.; Song, Y.; Yang, M.; Wang, Y.; et al. Accelerated Wound Healing Induced by a Novel Amphibian Peptide (OA-FF10). *Protein Pept. Lett.* **2019**, *26*, 261–270. [CrossRef] [PubMed]
35. Wang, Y.; Feng, Z.; Yang, M.; Zeng, L.; Qi, B.; Yin, S.; Li, B.; Li, Y.; Fu, Z.; Shu, L.; et al. Discovery of a novel short peptide with efficacy in accelerating the healing of skin wounds. *Pharmacol. Res.* **2021**, *163*, 105296. [CrossRef]
36. Li, X.; Wang, Y.; Zou, Z.; Yang, M.; Wu, C.; Su, Y.; Tang, J.; Yang, X. OM-LV20, a novel peptide from odorous frog skin, accelerates wound healing in vitro and in vivo. *Chem. Biol. Drug Des.* **2018**, *91*, 126–136. [CrossRef] [PubMed]

Article

Venom Variation of Neonate and Adult Chinese Cobras in Captivity concerning Their Foraging Strategies

Xuekui Nie [1,†], Qianzi Chen [1,†], Chen Wang [1], Wangxiang Huang [1], Ren Lai [2], Qiumin Lu [2], Qiyi He [1,*] and Xiaodong Yu [1,*]

[1]   Animal Toxin Group, Engineering Research Center of Active Substance and Biotechnology, Ministry of Education, College of Life Science, Chongqing Normal University, Chongqing 401331, China
[2]   Key Laboratory of Animal Models and Human Disease Mechanisms of Chinese Academy of Sciences/Key Laboratory of Bioactive Peptides of Yunnan Province, Kunming Institute of Zoology, Kunming 650223, China
*   Correspondence: hqy7171@126.com (Q.H.); yxd@cqnu.edu.cn (X.Y.)
†   These authors contributed equally to this work.

Abstract: The venom and transcriptome profile of the captive Chinese cobra (*Naja atra*) is not characterized until now. Here, LC-MS/MS and illumine technology were used to unveil the venom and trascriptome of neonates and adults *N. atra* specimens. In captive Chinese cobra, 98 co-existing transcripts for venom-related proteins was contained. A total of 127 proteins belong to 21 protein families were found in the profile of venom. The main components of snake venom were three finger toxins (3-FTx), snake venom metalloproteinase (SVMP), cysteine-rich secretory protein (CRISP), cobra venom factor (CVF), and phosphodiesterase (PDE). During the ontogenesis of captive Chinese cobra, the rearrangement of snake venom composition occurred and with obscure gender difference. CVF, 3-FTx, PDE, phospholipase A$_2$ (PLA$_2$) in adults were more abundant than neonates, while SVMP and CRISP in the neonates was richer than the adults. Ontogenetic changes in the proteome of Chinese cobra venom reveals different strategies for handling prey. The levels of different types of toxin families were dramatically altered in the wild and captive specimens. Therefore, we speculate that the captive process could reshape the snake venom composition vigorously. The clear comprehension of the composition of Chinese cobra venom facilitates the understanding of the mechanism of snakebite intoxication and guides the preparation and administration of traditional antivenom and next-generation drugs for snakebite.

Keywords: Captive; Chinese cobra; Ontogeny; Snake venom; Proteomics; Transcriptomics

Key Contribution: This is the first time to combine proteomic and transcriptomic approaches to study the differences in venom composition between captive and wild individuals of Chinese cobras. The difference in venom composition between captive and wild individuals demonstrates the plasticity of venom composition under breeding conditions.

Citation: Nie, X.; Chen, Q.; Wang, C.; Huang, W.; Lai, R.; Lu, Q.; He, Q.; Yu, X. Venom Variation of Neonate and Adult Chinese Cobras in Captivity concerning Their Foraging Strategies. *Toxins* **2022**, *14*, 598. https://doi.org/10.3390/toxins14090598

Received: 17 July 2022
Accepted: 26 August 2022
Published: 29 August 2022

1. Introduction

On 8 April 2019, the WHO uploaded a report on its website about venomous snake bites, a neglected public health problem in many tropical and subtropical countries [1]. Approximately 5.4 million snake bites occur each year, resulting in 1.8 to 2.7 million envenomations from venomous snake bites, 81,410 to 137,880 deaths, and approximately three times as many amputations and permanent disabilities. Most of these bites occur in Africa, Asia, and Latin America. In Africa, approximately 435,000 to 580,000 snake bites require treatment each year. In Asia, 2 million people are bitten by venomous snakes each year. Women, children, and farmers in poor rural areas of low- and middle-income countries are the most affected. The burden is the heaviest in countries with the weakest health systems and scarce medical resources. Approximately 100–200,000 people were bitten each year in China, 70% of whom are young people, with a 5% mortality rate and 25–30% disability rate [2]. The Chinese cobra (*Naja atra*) is one of the major snake species in the epidemiology of snakebites in China [3]. Chinese cobra snakebites are usually accompanied by local swelling, pain, tissue necrosis, palpitations, arrhythmia, shock, cardiac arrest, severe coagulopathy, drooping eyelids, foaming at the mouth, slurred speech, respiratory failure, and skeletal muscle paralysis [4]. Gender [5–7], age [3,8–10], diet [11–14], and geographic distribution [15–17], may cause intraspecific variation in snake venom, including variations in the protein content and type of venom fractions. The plasticity of snake venom is a great challenge in the therapeutic process of snake bites.

Snake antivenom is the only treatment that is proven to be affective to treat snakebite envenoming. Antivenoms are made of antibodies that are harvested from the serum of hyperimmune animals, typically horse. Thus, the antivenom contains a large amount of heterologous proteins. After the administration of antivenom, heterologous protein induces life-threatening adverse effects such as serum sickness and shock [18]. Antivenom is not always effective against the various complications associated with snake bites [19]. After a snakebite occurs, it is often impossible to accurately determine how much venom of the snake has expelled into the patient's body through its fangs, or even the type of snake that bit the patient. In addition, there are cases where not all venom components are targeted by antivenom, especially the low molecular mass protein component, three-finger toxin (3-FTx), which is a major contributor to death in patients with snake envenomation [20]. During the treatment of a snakebite, the administration of the dosage and type of antivenom is extremely critical to reduce the possibility of side effects and to guarantee to completely neutralize the corresponding snake venom. To date, the impact of changes in venom components on the clinical symptoms of snakebites has been largely ignored in the treatment of snakebites [18]. At the same time, the standardized production of antivenom does not fully neutralize the venom component of snake venom, which can lead to increased mortality in snakebite patients [21]. Although the next generation of antivenom (recombinant antivenom) that is currently in development is independent of snake species, the formulation and dosing of antivenom are highly dependent on knowledge of the comprehensive venom profile [22]. Therefore, a clear understanding of the snake venom profile is essential for the effective treatment of snakebites.

In recent years, omics studies become popular among Elapidae, Viperidae, and Hydrophiidae [23]. The traditional strategy of snake venom research was to isolate and purify single snake venom proteins, following sequence determination, three-dimensional structural studies, and functional studies. Therefore, the traditional method is inefficient. Genomic, transcriptomic, and proteomic studies of Indian cobras (*N. naja*) have been completed [24,25]; transcriptomic studies of the venom glands of *N. kaothia* and *N. sumatrana* have been reported [15,26]; according to PubMed search results, most of the proteomic studies on the venom of the cobra genus have been reported. Currently, there is no report omics study on captive Chinese cobra. In this study, the venom and venom glands of captive Chinese cobras were studied using LC-MS/MS and Illumina sequencing technology separately, to investigate whether the captive breeding process could reshape the profile of snake venom under captive progress and ontogenesis.

2. Results

2.1. SDS-PAGE

The electrophoretic profiles of snake venom that were obtained by SDS-PAGE showed a similar pattern of approximately 10 protein bands for each group of snakes under reducing conditions and non-reducing conditions (Figure 1). However, venom from female and male snakes showed age-related variations in their electrophoretic profiles (differences in the intensity of bands). Regarding SDS-PAGE under reducing conditions, partial bands of the crude venom of adult and neonate snakes were distinctly varied. In neonate captive *N. atra* venom, the protein bands of approximately 26, 54, and 66 kDa densimetric more strongly than adult *N. atra* venom. In addition, four protein bands of about 14.4, 34, 45, and 100 kDa were more abundant in the venom of adult *N. atra* venom. This implies that there were differences in certain components of snake venom. However, there were no sex-related differences that showed up under reducing SDS-PAGE.

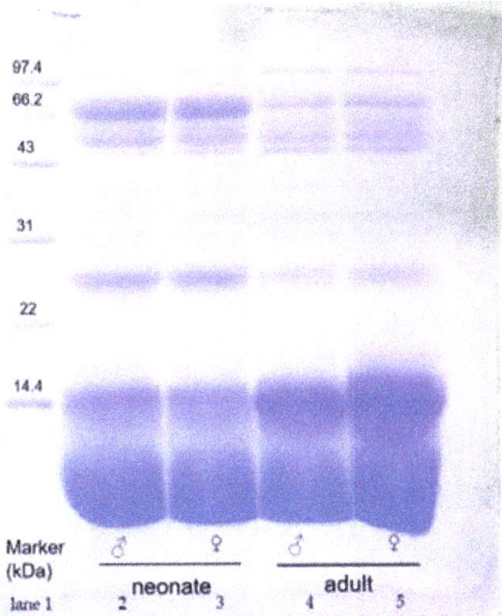

Figure 1. The 15% SDS-PAGE of *N. atra* crude venom (40 µg for each lane) under reducing. Lane 1: marker; Lane 2: neonate males; Lane 3: neonate females; Lane 4: adult males; Lane 5: adult females.

2.2. Venom Protein Concentration and Captive Snake Growth Status

The process of hatching and breeding of *N. atra* is shown in Figure 2A. The standard curve was obtained from the determination of BSA using a modified Bradford method protein concentration assay kit. According to the standard curve, the proportion of protein in each group of crude snake venom samples was calculated (Figure 2B). Indeed, there were no significant differences in the protein proportion of the different groups. There was no significant difference in the body total length between the diverse gender as well (Figure 2C). The average body weight of female neonate and adult *N. atra* was greater than males at the same age, respectively (Figure 2D).

2.3. Venom Proteomics

Accurate knowledge of snake venom proteomics contributes to an understanding of the poisoning mechanism of venom components in snakebites. In this study, a venom proteomics protocol was undertaken that included enzyme digestion, HPLC, LC-MS/MS, and sequence matching (Proteome Discoverer 2.4, PD 2.4, Thermo). In all the sample groups, a total of 127 proteins from 21 protein families were identified by MS analysis (Table 1). The proteins were identified by finding at least one unique peptide per protein (Table S1). The peptides that were used for sequencing after enzymatic hydrolysis ranged in length from 6 to 31 amino acids, with a predominant distribution of 7–14 amino acids (Figure S1). The most abundant protein component in snake venom was over 60% (Figure S2) and had a molecular mass of 0–14 kDa. Our results showed that captive *N. atra* venom mainly consisted of three-finger toxin (3-FTx, 60.87%, 62.22%, 67.99%, and 68.20% of total venom from the neonate male, neonate female, adult male, and adult female group, respectively); cysteine-rich secretory protein (CRISP, 11.39%, 9.37%, 7.08%, and 6.65%, respectively); snake venom metalloproteinase (SVMP, 10.22%, 10.13%, 4.70%, and 4.93%, respectively); cobra venom factor (CVF, 3.53%, 3.92%, 5.02%, and 4.78%, respectively); and phosphodiesterase (PDE, 3.49%, 3.30%, 4.35%, and 4.51%, respectively) (Table 2). The number of proteins that were jointly identified in the four groups was 119. Neonate female venom specimens had higher relative levels of the three finger toxins (neurotoxins and cytotoxins) snake venom proteins and lower relative levels of CVFs and PLA_2 compared to adult female venom specimens. The venom of the neonate males contained higher relative levels of SVMPs, CTXs, and neurotoxins than did the adult males. There were gender differences in the venom composition of snakes. Neonate male snake venom had higher relative levels of CTX, L-amino acid oxidase (LAAO), muscarinic tox-in-like protein (MTLP), kunitz-type serine protease inhibitor (KUN), but (SNT, CTX) were relatively low. The SVMP content was higher in the female adult specimens (Figure 3).

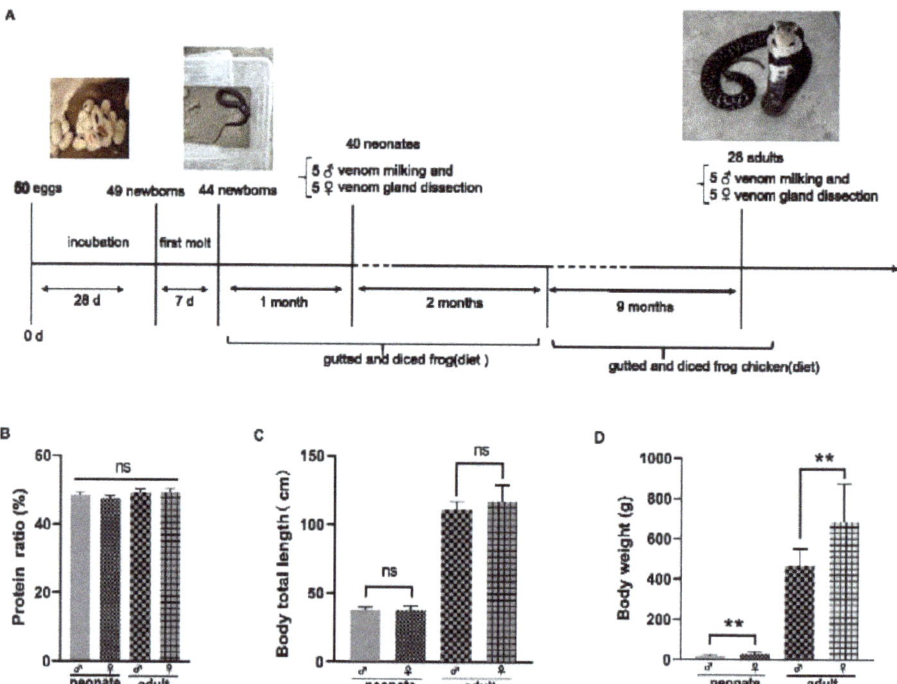

Figure 2. The breeding process of captive *N. atra* (**A**). Ratio of protein in crude venom (**B**). Snake body total length (**C**). Snake body weight (**D**). ♀: female; ♂: male. Significance analysis was performed by one-way ANOVA, $p > 0.05$ (ns), $p < 0.002$ (**).

Table 1. The list of proteins that were identified from *N. atra* snake venom by LC-MS/MS.

Protein Accessions	Description
V8N8G0	Alkaline phosphatase
ENSNNAP00000025215.1*	Acetylcholinesterase
Q9DF56	Acidic phospholipase A_2
P00598	Acidic phospholipase A_2 1
Q9DF33	Acidic phospholipase A_2 2
ENSNNAP00000015782.1*	Acidic phospholipase A_2 2
P60045	Acidic phospholipase A_2 3
Q6T179	Acidic phospholipase A_2 4
Q9I900	Acidic phospholipase A_2 D
P25498	Acidic phospholipase A_2 E
A4FS04	Acidic phospholipase A_2 natratoxin
P86542	Phospholipase A_2 3
P00599	Basic phospholipase A_2 1
P60043	Basic phospholipase A_2 1

Table 1. *Cont.*

Protein Accessions	Description
V8ND68	Phospholipase B-like
Q7LZI1	Phospholipase A$_2$ inhibitor 31 kDa subunit
ENSNNAP00000008460.1*	PLIalpha-like protein
V8NQ76	Atriopeptidase
V8N495	Carboxypeptidase
ENSNNAP00000003269.1*	Carboxypeptidase D
I2C090	Ophiophagus venom factor
Q91132	Cobra venom factor
Q91132	Cobra venom factor
ENSNNAP00000021869.1*	Cobra venom factor
ENSNNAP00000023282.1*	Cobra venom factor
ENSNNAP00000010931.1*	Complement C3
ENSNNAP00000013063.1*	Complement C3
Q01833	Complement C3
Q90WI6	C-type lectin BML-1
E3P6P4	Cystatin
P86543	Cysteine-rich venom protein
P84807	Cysteine-rich venom protein 25-A
P84808	Cysteine-rich venom protein kaouthin-2
Q7T1K6	Cysteine-rich venom protein natrin-1
Q53B46	Beta-cardiotoxin CTX15
P60305	Cytotoxin 1
P86541	Cytotoxin 10
Q9W6W6	Cytotoxin 10
ENSNNAP00000013731.1*	Cytotoxin 11
Q98956	Cytotoxin 1b
P85429	Cytotoxin 1f
P01442	Cytotoxin 2
Q9DGH9	Cytotoxin 2
P01445	Cytotoxin 2
P01463	Cytotoxin 2
O93472	Cytotoxin 2c
P01452	Cytotoxin 4
O93473	Cytotoxin 4a
O73856	Cytotoxin 4b
P07525	Cytotoxin 5
P01457	Cytotoxin 5
P24779	Cytotoxin 5
P25517	Cytotoxin 5
P24779	Cytotoxin 5
Q98965	Cytotoxin 6
P01465	Cytotoxin 6
P49122	Cytotoxin 7
O73859	Cytotoxin 7
P86540	Cytotoxin 8
P60311	Cytotoxin KJC3
P60308	Cytotoxin SP15c
P62377	Cytotoxin-like basic protein
P18328	Muscarinic toxin 2
P82462	Muscarinic toxin-like protein 1
P82463	Muscarinic toxin-like protein 2
P82464	Muscarinic toxin-like protein 3
Q9PSN6	Neurotoxin 3
Q9DEQ3	Neurotoxin homolog NL1
Q9W717	Neurotoxin-like protein NTL2
P14613	Short neurotoxin 1
P01431	Short neurotoxin 1

Table 1. *Cont.*

Protein Accessions	Description
P68418	Short neurotoxin 1
P01424	Short neurotoxin 1
P01432	Short neurotoxin 3
E2IU01	Long neurotoxin 7
E2ITZ9	Cobrotoxin-b
P85092	Toxin AdTx1
P29180	Weak neurotoxin 6
O42256	Weak neurotoxin 6
Q2VBN2	Weak neurotoxin WNTX34
P01401	Weak toxin CM-11
P25679	Weak toxin CM-9a
P85520	Oxiana weak toxin
Q9YGI4	Probable weak neurotoxin NNAM2
P25669	Long neurotoxin 2
Q2VBP3	Long neurotoxin LNTX37
P01391	Alpha-cobratoxin
ENSNNAP00000012975.1*	Cobrotoxin
P82849	Cobrotoxin II
P59275	Cobrotoxin-b
D9IX97	Natriuretic peptide Na-NP
V8NF35	Dipeptidyl peptidase 2
V8P9G9	Venom dipeptidylpeptidase IV
ENSNNAP00000025792.1*	Dipeptidyl peptidase 7
V8P395	Glutathione peroxidase
ENSNNAP00000008637.1*	Hyaluronidase-2-like isoform X1
ENSNNAP00000024830.1*	Insulin like growth factor 1
ENSNNAP00000006918.1*	Insulin like growth factor binding protein 3
V8N7H9	BPTI/Kunitz domain-containing protein-like isoform X2
P20229	Kunitz-type serine protease inhibitor
ENSNNAP00000023266.1*	Kunitz-type serine protease inhibitor 4-like
A8QL51	L-amino acid oxidase
ENSNNAP00000002373.1*	L-amino-acid oxidase
A8QL58	L-amino-acid oxidase
V8P5W4	Putative serine carboxypeptidase CPVL
ENSNNAP00000011792.1*	Serpin family E member 2
A0A2I4HXH5	5′-nucleotidase
V8NYW9	5′-nucleotidase
A8QL53	Snake venom serine protease NaSP
P86545	Thrombin-like enzyme TLP
V8NCP7	Vascular endothelial growth factor C
P61899	Venom nerve growth factor
Q5YF90	Venom nerve growth factor 1
A0A2D0TC04	Venom phosphodiesterase
P83234	Vespryn-21
D3TTC2	Zinc metalloproteinase-disintegrin-like atragin
ENSNNAP00000015130.1*	Zinc metalloproteinase-disintegrin-like atragin
D5LMJ3	Zinc metalloproteinase-disintegrin-like atrase-A
ENSNNAP00000015250.1*	Zinc metalloproteinase-disintegrin-like atrase-A
ENSNNAP00000015377.1*	Zinc metalloproteinase-disintegrin-like atrase-A
D6PXE8	Zinc metalloproteinase-disintegrin-like atrase-B
Q9PVK7	Zinc metalloproteinase-disintegrin-like cobrin
A8QL59	Zinc metalloproteinase-disintegrin-like NaMP

Table 1. *Cont.*

Protein Accessions	Description
Q10749	Snake venom metalloproteinase-disintegrin-like mocarhagin
P82942	Hemorrhagic metalloproteinase-disintegrin-like kaouthiagin

""* Matched with *N. naja* snake veonom reference proteome (ftp://ftp.ensembl.org/pub/release-101/fasta/naja_naja/pep/, accessed on 22 January 2021) and the others matched with Uniprot (https://www.uniprot.org/). Accessed on 22 January 2021.

Table 2. The relative abundance of venom components in *N. atra* venom and quantification of the relative abundance of venom components using TMT label.

% Proteome	Neonate Males	Neonate Females	Adult Males	Adult Females
PLB	0.07%	0.06%	0.10%	0.10%
HAase	0.09%	0.10%	0.11%	0.11%
DPP-IV	0.14%	0.14%	0.18%	0.14%
VEGF	0.22%	0.22%	0.23%	0.21%
PLI	0.19%	0.28%	0.26%	0.30%
CTL	0.35%	0.37%	0.47%	0.51%
LAAO	0.38%	0.32%	0.24%	0.27%
5'-NT	0.43%	0.39%	0.40%	0.45%
QC	0.49%	0.50%	0.40%	0.39%
SVSP	0.50%	0.50%	0.26%	0.27%
Acidic PLA$_2$	0.53%	0.52%	0.94%	1.3%
Basic PLA$_2$	0.02%	0.02%	0.07%	0.1%
KUN	0.95%	0.70%	0.71%	0.73%
NP	1.05%	1.83%	1.03%	1.08%
VESP	1.12%	1.18%	0.72%	0.86%
AChE	1.44%	1.23%	1.80%	1.41%
NGF	2.52%	2.70%	2.82%	2.94%
PDE	3.49%	3.30%	4.35%	4.51%
CVF	3.53%	3.92%	4.70%	4.78%
SVMP(P-III)	10.23%	10.13%	5.02%	4.93%
CRISP	11.39%	9.37%	7.08%	6.65%
LNX(3-FTx)	0.10%	0.09%	0.13%	0.14%
SNX(3-FTx)	15.67%	15.73%	11.00%	13.65%
WNX(3-FTx)	5.08%	6.18%	5.09%	5.79%
CTX(3-FTx)	40.00%	40.22%	51.77%	48.63%

2.4. Protein Sequence Alignment

3-FTx are the main components that make up the snake venom proteins and are responsible for snakebite lethality [27]. According to our results, 3-FTx was most abundant component in *N. atra* snake venom and can be classified into CTX and neurotoxin based on their biological targets of action. There are two subfamilies, CTX and neurotoxin, that were high in content and had massive members in each subfamily. Our results revealed that 3-FTx contains 8–10 cysteine residues for disulfide bond formation and further leads to the formation of the large family of 3-FTx. 3-FTx signal peptides were extremely conserved (Figure 4). The neurotoxin long neurotoxin (LNX) had five pairs of disulfide bonds (Figure 4A), weak neurotoxin (WNX) had five pairs of disulfide bonds (Figure 4B), short neurotoxin (SNX) had four pairs of disulfide bonds (Figure 4C), and CTX had four pairs of disulfide bonds (Figure 3D). LNX binds with high affinity to muscular (alpha-1/CHRNA1) and neuronal (alpha-7/CHRNA7) nicotinic acetylcholine receptor (nAChR) and inhibits acetylcholine from binding to the receptor, thereby impairing neuromuscular and neuronal transmission. P01391was well studied and there were four forms of existence.

In Figure 4C (SNX), the muscarinic acetylcholine receptor antagonists consisted of P18328, P82462, and P82463. The mechanism of toxin-receptor interaction was comprised of at least two steps. The first step is fast with no competition between the toxin and the antagonist [28]. The second step is slow with the formation of a more stable toxin-receptor complex and the inhibition of the antagonist binding. Other SNX bound to muscle nicotinic acetylcholine receptor (nAChR) and inhibit acetylcholine from binding to the receptor, thereby impairing neuromuscular transmission.

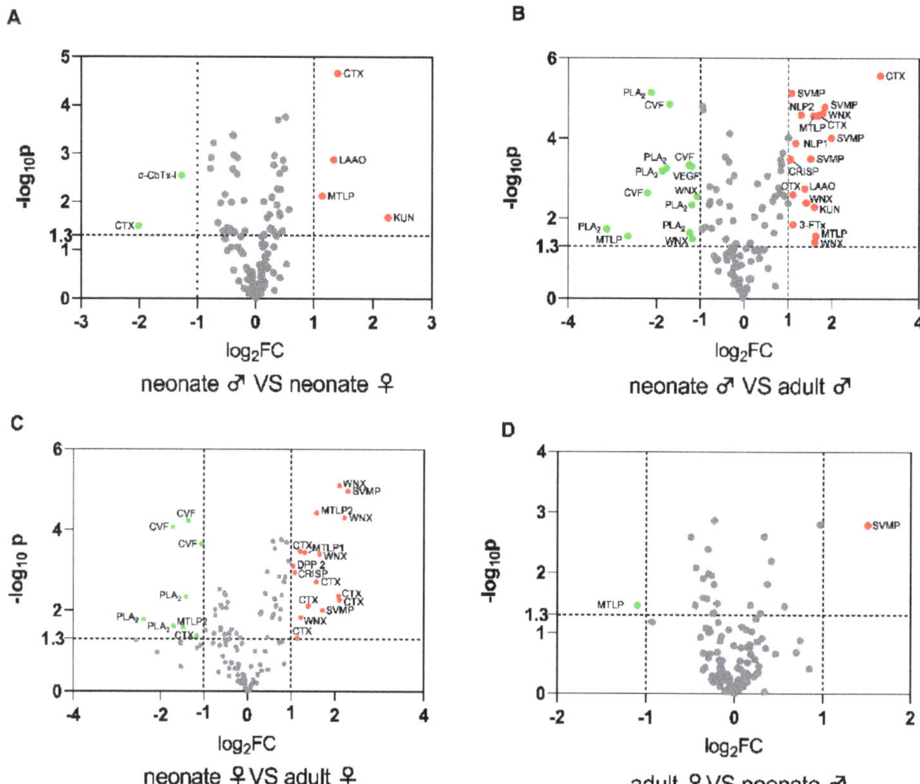

Figure 3. The relative differences of *N. atra* venomic groups. (**A**) Neonate males versus neonate females. (**B**) Neonate males versus adult males. (**C**) Neonate females versus adult females. (**D**) Adult females versus adult males. The red dots represent the upregulated and green dots present the downregulated proteins. $p < 0.05 (-\log 10\, p > 1.3)$, $|\log2FoldChange| > 1$. Abbreviations: 3-FTX, three-finger toxins; CRISP, cysteine-rich secretory protein; SVMP, snake venom metalloproteinase; CVF, cobra venom factor; PDE, phosphodiesterase; NGF, nerve growth factor; AChE, acetylcholinesterase; PLA$_2$, phospholipase A$_2$; NP, natriuretic peptide; VESP, vespryns; KUN, kunitz-type inhibitor; CTL, C-type lectin; 5′-NT, 5′-nucleotidase; QC, glutaminyl-peptide cyclotransferases; PLI, phospholipase A$_2$ inhibitor; SVSP, snake venom serine protease; LAAO, L-amino acid oxidase; VEGF, vascular endothelial growth factor; DPP, dipeptidylpeptidase; HAase, hyaluronidase; PLB, phospholipase B; CTX, cytotoxin; WNX, weak neurotoxin; α-CbTx-I, type I alpha-neurotoxin (short neurotoxin, SNX); MTLP, Muscarinic toxin-like protein; NLP, neurotoxin like protein.

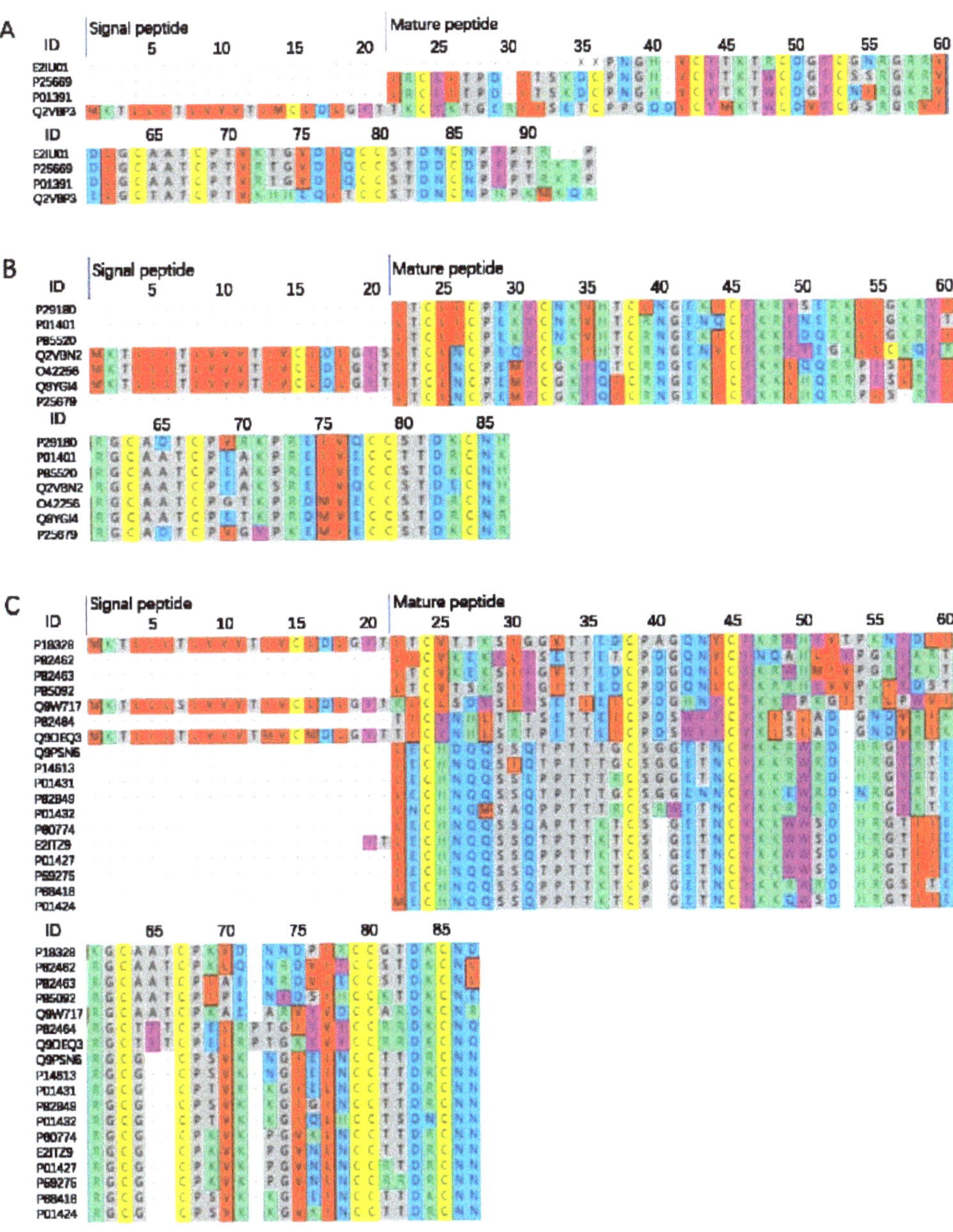

Figure 4. *Cont.*

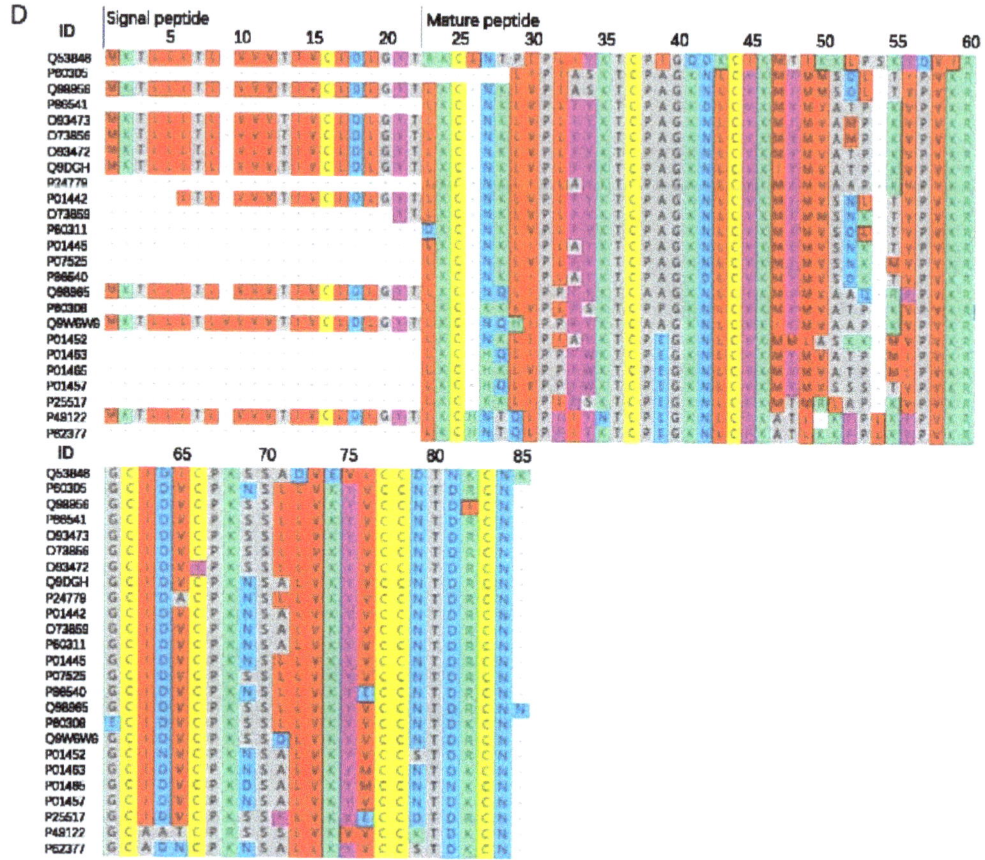

Figure 4. *N. atra* 3-FTx sequence alignment analysis. LNX (long neurotoxin) (**A**). WNX (weak neurotoxin) (**B**). SNX (short neurotoxin) (**C**). CTX (cytotoxin) (**D**). ID: the ID number of the protein in Uniprot. Online multi-sequence alignment by MAFFT Version 7, parameter default value.

In this study, more than 60% of the three-finger toxins were CTX, and they usually induced apoptosis of multiple classes of cells. The main CTX shows cytolytic activity on many different cells by forming a pore in the lipid membranes. In vivo, they increase heart rate or kill the animal by cardiac arrest. In addition, it binds to heparin with high affinity, interacts with Kv channel-interacting protein 1 (KCNIP1) in a calcium-independent manner, and binds to integrin alpha-V/beta-3 (ITGAV/ITGB3) with moderate affinity [29]. However, there were some special CTX. Q53B46 Acts as a beta-blocker by binding to beta-1 and beta-2 adrenergic receptors (ADRB1 and ADRB2). It dose-dependently decreases the heart rate (bradycardia), whereas conventional cardiotoxins increase it [30]. P60305 shows cytolytic activity (apoptosis is induced in C2C12 cells). Basic protein (P01442) that binds to the cell membrane and depolarizes cardiomyocytes (may interact with sulfatides in the cell membrane which induces pore formation and cell internalization) and it also may target the mitochondrial membrane and induce mitochondrial swelling and fragmentation [31]. P07525 functions as same as P01442 and also inhibits protein kinases C [32]. 3-FTx is abundant in Chinese cobras with structural and functional diversity.

Snake venom metalloproteinases (SVMP) are part of the cobra venom. This class of proteins consists of a combination of structural domains, forming assemblies of different structural domains, leading to the formation of structurally and functionally diverse

SVMP families. SVMP possesses substantial biological activity. It hydrolyzes human fibrin (pro), acts as a prothrombin activator (activates blood coagulation factor X) [33], inhibits platelet aggregation [34], promotes the inflammatory response of the body [35], inhibits the activity of serine protease inhibitors in the blood [36], and hydrolyzes basement membrane components (laminin, nestin, Type IV collagen) leading to bleeding. SVMP also has pro-inflammatory effects [18]. Cofactors of the SVMP family members (Ca^{2+}, Zn^{2+}) and glycosylation modifications of proteins (metalloproteases, disintegrin, cysteine-rich structural domains on peptide chains asparagine residues that are linked to glycosyl groups). Glycosylation sites are present on the structural domains of the basal genus protease, disintegrin, and cysteine-rich, but not all three structural domains of the same protein have glycosylation modifications (Table 3). The results show (Figure 5) that the described SVMPs are all P-III metalloproteases, except for P82942 and D6PXE8, which only have 14 pairs of disulfide bonds within the molecule, and the remaining SVMPs, which form 17 pairs of disulfide bonds within the molecule. The active center of all the SVMPs is the amino acid glutamate residue at position 343, only Q10749 at position 343 is a glutamine residue, with opposite acidity.

Table 3. Distribution of the SVMP glycosylation modification sites.

ID	Distribution of Glycosylation Sites (N-linked Asparagine)		
	Metalloprotease Structural Domains	Disintegrin Structural Domain	Cysteine-Rich Structural Domains
P82942	304	-	-
D6PXE8	320	-	509
D3TTC2	-	438	-
Q9PVK7	-	438	-
Q10749	303	-	509
D5LMJ3	221, 273, 304	438	527
A8QL59	225, 268, 319	-	551

- No data available. ID: ID number of the protein in Uniprot.

The PLA_2 family of snake venom proteins are widely distributed in the Elapidae, Viperidae, and Colubridae. PLA_2, active forms include dimers and monomers, has a highly conserved Ca^{2+} binding loop (XCGXGG), and active center (DXCCXXHD) in both major structures, corresponding to amino acid numbers 55–60 and 69–76, respectively, in Figure 6. All PLA_2s contain seven pairs of disulfide bonds according to the sequence comparison in Figure 6. P00598 inhibits G protein-coupled muscarinic acetylcholine receptors and A4FS04 effectively inhibits A-type K^+ currents (Kv/KCN) in acutely dissociated rat dorsal root ganglion (DRG) neurons. This inhibitory effect is independent of its enzymatic activity. The anticoagulant effect of Q6T179, Q9DF33, and P00598 is through interference with the coagulation cascade. All are acidic PLA_2s except for P60043 and P00599, which are basic PLA_2s. The active sites are located at positions 75 (Histidine) and 121 (Aspartic acid).

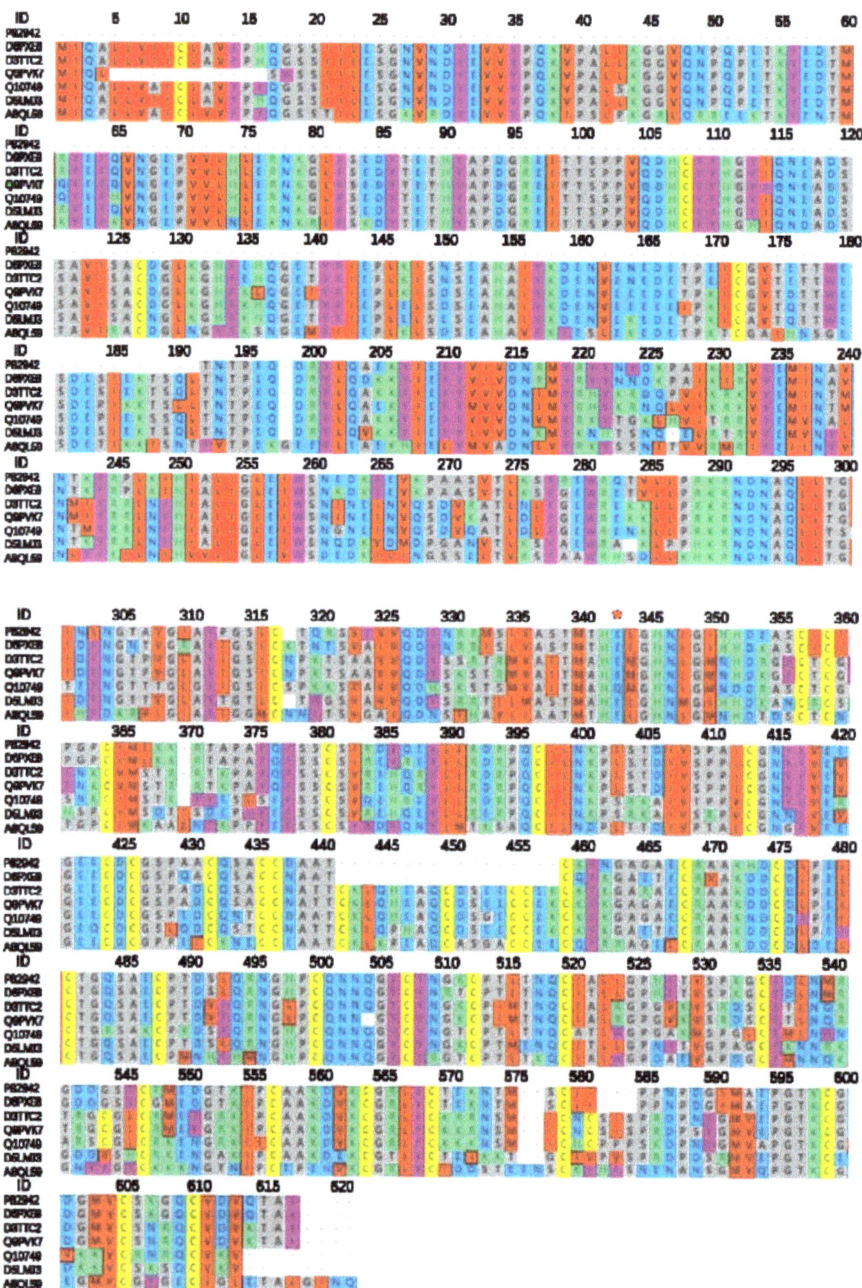

Figure 5. SVMP of *N. atra* sequence alignment analysis. This was compared online by MAFFT Version, parameter default value. ID: the ID number of the protein in Uniprot., *: active site.

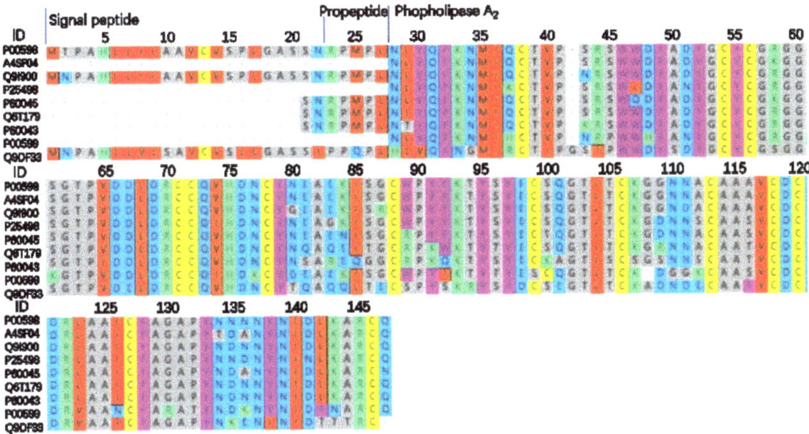

Figure 6. *N. atra* PLA$_2$ sequence alignment analysis. This was compared online by MAFFT Version 7, parameter default value. ID: the ID number of the protein in Uniprot.

2.5. Venom Gland Transcriptome

Revealing the transcriptome profile of snake venom glands is essential to decipher the functional role of snake venom, which was performed by next-generation sequencing. In our results, from neonate male, neonate female, adult male, and adult female groups, clean reads and unique gene (FPKM > 1) were identified, 43,700,115 and 11,153, 42,159,554 and 10,872, 42,965,160 and 10,456, 45,060,349, and 10,600, respectively. In the neonate male, neonate female, adult male, adult female groups, 70% of the paired regions were exons, the rest of the paired regions were intron and intergenic. However, we found 98 shared snake venom unique genes, among these unique genes were contained 10 new transcripts and 88 unique transcripts (FPKM > 1) (Table 4). Our results showed that the *N. atra* venom gland transcriptome mainly consists of 3-FTx (85.31, 88.22, 87.82, and 80.40% of the total transcriptome from the neonate male, neonate female, adult male, and adult female group, respectively); PLA$_2$ (1.31, 1.25, 3.41, and 1.83%, respectively); natriuretic peptide (NP, 4.74, 3.81, 2.56, and 6.93%, respectively); snake toxin and toxin-like protein (STLK, 1.67%, 2.18%, 1.85%, and 3.06%, respectively); and glutathione peroxidase (GPX, 2.35, 1.52, 1.39, and 2.59%, respectively) (Table 5). Based on the results of bioinformatics analysis, the total percentage of low-abundance expression of the transcripts of venom components in the four groups was less than 4% (Table 5).

Table 4. The list of genes that were obtained from the identification of the *N. atra* venom gland transcriptome. All the genes co-existed in all the groups (FPKM > 1).

Gene ID	Gene Chromosome	Gene Coding Protein	Abbreviation
ENSNNAG00000009518	3	Alpha-elapitoxin-Nk2a	α-CbTx(3-FTx)
ENSNNAG00000009534	3	Muscarinic toxin-like protein 2	MTLP-2(3-FTx)
ENSNNAG00000011050	3	Muscarinic toxin-like protein 3 homolog	MTLP-3(3-FTx)
ENSNNAG00000009404	3	Cytotoxin 3a	CTX(3-FTx)
ENSNNAG00000009217	3	Cytotoxin 4N	CTX(3-FTx)
ENSNNAG00000011032	3	Cytotoxin 2	CTX(3-FTx)
ENSNNAG00000011010	3	Cytotoxin 5	CTX(3-FTx)
ENSNNAG00000008699	3	Probable weak neurotoxin NNAM1	WNX(3-FTx)
ENSNNAG00000008719	3	Tryptophan-containing weak neurotoxin	WNX(3-FTx)
ENSNNAG00000009048	3	Neurotoxin-like protein NTL2	NTL2(3-FTx)

Table 4. *Cont.*

Gene ID	Gene Chromosome	Gene Coding Protein	Abbreviation
ENSNNAG00000011613	SOZL01001066.1	Neurotoxin homolog NL1	NL1(3-FTx)
ENSNNAG00000011709	SOZL01001066.1	Cardiotoxin 7	CTX(3-FTx)
ENSNNAG00000008731	3	Cobrotoxin	CBT(3-FTx)
ENSNNAG00000007474	1	Snake venom 5′-nucleotidase	5′ NT
ENSNNAG00000007520	1	Snake venom 5′-nucleotidase	5′ NT
novel.2091	SOZL01000663.1	5′-nucleotidase	5′ NT
ENSNNAG00000018750	2	5′-nucleotidase	5′ NT
ENSNNAG00000016482	SOZL01001525.1	Acetylcholinesterase	AchE
ENSNNAG00000016482	SOZL01001525.1	Acetylcholinesterase	AchE
ENSNNAG00000015104	MIC_6	Aminopeptidase	AP
ENSNNAG00000008828	2	Complement C3	C3
ENSNNAG00000008657	2	Complement C3	C3
ENSNNAG00000012844	SOZL01001814.1	Cathelicidin-related peptide	CATH
ENSNNAG00000014282	1	Cysteine-rich venom protein ophanin	CRISP
ENSNNAG00000014151	1	Cysteine-rich venom protein ophanin	CRISP
ENSNNAG00000018595	MIC_4	Cysteine-rich secretory protein	CRISP
ENSNNAG00000000526	2	C-type lectin	CTL
ENSNNAG00000018045	7	C-type lectin lectoxin-Thr1	CTL
ENSNNAG00000000430	2	C-type lectin	CTL
ENSNNAG00000009410	1	C-type lectin	CTL
ENSNNAG00000018053	7	C-type lectin	CTL
ENSNNAG00000016651	MIC_7	C-type lectin	CTL
novel.1968	MIC_9	Cystatin (cysteine proteinase inhibitor)	CYS
ENSNNAG00000010061	2	Cystatin-B	CYS
ENSNNAG00000005504	MIC_9	Cystatin	CYS
ENSNNAG00000012223	3	Cystatin-2	CYS
novel.1967	MIC_9	Cystatin	CYS
ENSNNAG00000006541	1	Dipeptidyl peptidase IV	DPP-IV
ENSNNAG00000013472	2	Glutathione peroxidase 3	GPX
ENSNNAG00000013545	1	Glutathione peroxidase 2	GPX
ENSNNAG00000013509	2	Glutathione peroxidase 6	GPX
ENSNNAG00000008046	2	Glutathione peroxidase 1	GPX
ENSNNAG00000002583	2	Hyaluronidase	HAase
ENSNNAG00000005815	MIC_2	Hyaluronidase	HAase
ENSNNAG00000018577	MIC_3	Kunitz-type serine protease inhibitor	KSPI
ENSNNAG00000018577	MIC_3	Kunitz-type protease inhibitor	KSPI
ENSNNAG00000015199	3	Kunitz-type protease inhibitor	KSPI
ENSNNAG00000001233	1	Kunitz-type protease inhibitor	KSPI
ENSNNAG00000001705	2	L-amino-acid oxidase	LAAO
ENSNNAG00000001808	2	L-amino-acid oxidase	LAAO
ENSNNAG00000001642	2	L-amino-acid oxidase	LAAO
ENSNNAG00000001705	2	L-amino-acid oxidase	LAAO
ENSNNAG00000013519	MIC_2	Nerve growth factor	NGF
ENSNNAG00000004808	3	Venom nerve growth factor	NGF
ENSNNAG00000004800	3	Venom nerve growth factor1	NGF
ENSNNAG00000004808	3	Venom nerve growth factor	NGF
ENSNNAG00000006727	4	Natriuretic peptide Na-NP	NP
ENSNNAG00000006779	4	Natriuretic peptide Na-NP	NP
ENSNNAG00000006809	4	Natriuretic peptide Na-NP	NP
ENSNNAG00000006727	4	Natriuretic peptide Na-NP	NP
ENSNNAG00000000097	1	Venom phosphodiesterase 1	PDE
ENSNNAG00000013762	MIC_11	Group IIE secretory phospholipase A_2	PLA$_2$
ENSNNAG00000003407	1	Phospholipase A_2 crotoxin basic subunit	PLA$_2$
ENSNNAG00000010611	SOZL01000342.1	Acidic phospholipase A_2	PLA$_2$
ENSNNAG00000003618	4	Phospholipase A_2	PLA$_2$
ENSNNAG00000015696	7	Phospholipase B	PLB

Table 4. *Cont.*

Gene ID	Gene Chromosome	Gene Coding Protein	Abbreviation
novel.1890	MIC_8	Phospholipase B	PLB
novel.1889	MIC_8	Phospholipase B	PLB
ENSNNAG00000005694	4	PLIalpha-like protein	PLI
ENSNNAG00000003592	MIC_3	Phospholipase A_2 inhibitor	PLI
ENSNNAG00000003672	MIC_3	Phospholipase A_2 inhibitor	PLI
ENSNNAG00000013551	2	Snake venom serine protease NaSP	SVSP
ENSNNAG00000015301	MIC_3	Snake venom serine protease	SVSP
ENSNNAG00000015396	MIC_3	Snake venom serine protease	SVSP
ENSNNAG00000017712	6	Venom prothrombin activator	VPA
ENSNNAG00000017242	1	Vascular endothelial growth factor A	VEGF
ENSNNAG00000011309	MIC_5	Vascular endothelial growth factor	VEGF
ENSNNAG00000008620	MIC_9	Vascular endothelial growth factor A	VEGF
ENSNNAG00000004103	MIC_2	Cysteine-rich with EGF-like domain	CREGF
ENSNNAG00000001343	2	Cysteine-rich with EGF-like domain protein	CREGF
ENSNNAG00000007968	4	Vascular endothelial growth factor C	VEGF
ENSNNAG00000017242	1	Vascular endothelial growth factor A	VEGF
ENSNNAG00000013375	SOZL01001403.1	Cobra venom factor	CVF
ENSNNAG00000008178	2	Cobra venom factor	CVF
ENSNNAG00000013375	SOZL01001403.1	Cobra venom factor	CVF
ENSNNAG00000008178	2	Cobra venom factor	CVF
novel.1103	3	Snake toxin and toxin-like protein	STLK
novel.1007	3	Snake toxin and toxin-like protein	STLK
novel.1104	3	Snake toxin and toxin-like protein	STLK
novel.2094	SOZL01000688.1	Snake toxin and toxin-like protein	STLK
ENSNNAG00000009894	MIC_1	Zinc metalloproteinase-disintegrin-like NaMP	SVMP
ENSNNAG00000017863	MIC_8	Disintegrin and metalloproteinase domain-containing protein 9	SVMP
ENSNNAG00000013917	2	A disintegrin and metalloproteinase with thrombospondin motifs 12	SVMP
ENSNNAG00000011023	Z	Disintegrin and metalloproteinase domain-containing protein 11	SVMP
ENSNNAG00000010003	MIC_1	Zinc metalloproteinase-disintegrin-like atragin	SVMP
ENSNNAG00000008597	MIC_6	A disintegrin and metalloproteinase with thrombospondin motifs 17	SVMP
ENSNNAG00000005308	MIC_1	Disintegrin and metalloproteinase domain-containing protein 9	SVMP
novel.1209	4	Zinc metalloprotease	SVMP

Table 5. The relative abundance of venom components of *N. atra* gland venom transcriptomes. Quantification of relative abundance of venom components by TMT label.

FPKM%	Neonate Males	Neonate Females	Adult Males	Adult Females
PLI	0.001%	0.001%	0.001%	0.004%
C3	0.005%	0.002%	0.007%	0.008%
CREGF	0.015%	0.012%	0.006%	0.053%
AP	0.015%	0.013%	0.007%	0.062%
DPP-IV	0.017%	0.007%	0.014%	0.022%
VEGF	0.027%	0.015%	0.013%	0.059%
PLB	0.039%	0.037%	0.053%	0.083%
HAase	0.057%	0.031%	0.067%	0.067%
VPA	0.069%	0.025%	0.010%	0.063%

Table 5. *Cont.*

FPKM%	Neonate Males	Neonate Females	Adult Males	Adult Females
CATH	0.072%	0.006%	0.011%	0.022%
AChE	0.091%	0.072%	0.049%	0.167%
CYS	0.107%	0.081%	0.064%	0.172%
PDE	0.112%	0.059%	0.134%	0.136%
SVSP	0.168%	0.124%	0.145%	0.210%
LAAO	0.213%	0.121%	0.055%	0.207%
KUN	0.221%	0.151%	0.203%	0.221%
CVF	0.223%	0.096%	0.213%	0.179%
CTL	0.253%	0.212%	0.177%	0.456%
5'-NT	0.347%	0.211%	0.154%	0.400%
CRISP	0.615%	0.346%	0.135%	0.497%
NGF	0.972%	0.887%	1.386%	1.450%
SVMP	0.985%	0.515%	0.078%	0.653%
Basic PLA$_2$	1.305%	1.244%	3.405%	1.829%
Acidic PLA$_2$	0.001%	0.001%	0.001%	0.001%
STLK	1.673%	2.182%	1.853%	3.056%
GPX	2.352%	1.521%	1.386%	2.593%
NP	4.738%	3.806%	2.556%	6.932%
LNX(3-FTx)	0.751%	2.630%	1.918%	7.968%
CTX(3-FTx)	37.825%	38.359%	26.955%	29.696%
WNX(3-FTx)	12.815%	9.078%	3.417%	3.429%
SNX(3-FTx)	33.919%	38.158%	55.530%	39.307%

Abbreviations: 3-FTx, three-finger toxins; NP, natriuretic peptide; STLK, snake toxin and toxin-like protein; GPX, glutathione peroxidase; PLA$_2$, phospholipase A$_2$; NGF, nerve growth factor; SVMP, snake venom metalloproteinase; CRISP, cysteine-rich secretory protein; CTL, C-type lectin; 5'-NT, 5'-nucleotidase; KUN, kunitz-type inhibitor; SVSP, snake venom serine protease; LAAO, L-amino acid oxidase; CVF, cobra venom factor; CYS, cystatin; AChE, acetylcholinesterase; PDE, phosphodiesterase; PLB, phospholipase B; HAase, hyaluronidase; VPA, venom prothrombin activator; AP, aminopeptidase; CREGF, cysteine-rich with EGF-like domain protein; VEGF, vascular endothelial growth factor; DPP IV, dipeptidylpeptidase IV; CATH, cathelicidin-related peptide; C3, complement C3; PLI, phospholipase A$_2$ inhibitor; CTX, cytotoxin; WNX, weak toxin; SNX, short neurotoxin; LNX, long neurotoxin; MTLP, muscarinic toxin-like protein.

2.6. Cooperative Proteomic and Transcriptomic Analysis

Among all the captive *N. atra* (neonate male, neonate female, adult male, and adult female groups), we found that high correlations only for the transcriptomic and proteomic statistical analysis of 3-FTx among dominant snake venom families, but significantly different levels for other snake venom families.

3-FTx, a non-enzymatic snake venom protein, is present in the majority of Elapidae snake venoms and contains a variety of biological activities. [37]. Based on the length of the peptide chain and the biological target of the 3-FTx, 3-FTx can be classified into five subclasses, cytotoxins (CTX, 40.00% neonate male group, 40.22% neonate female group, 51.77% adult male group, and 48.63% adult female group, respectively) including 26 proteins; weak neurotoxins (WNX, 5.08, 6.18, 5.09, and 5.79%, respectively) including 7 proteins; long-neurotoxins (LNX, 0.10, 0.09, 0.13, and 0.14%, respectively) including 3 proteins; and short-neurotoxins (SNX, 15.67, 15.73, 11.00, and 13.65%, respectively) including 20 proteins (Tables 1 and 2). The above results verified the 3-FTx plays an important role in the envenomation of *N. atra* that is responsible for the death of victims. Our results showed that the correlation between the changes in the proteome and transcriptome of cultured Chinese cobra snake venom was not significant ($p > 0.05$), except for neonate females versus adult females (Figure 7). Indeed, many previous studies lack concordance between transcriptome and proteome, including *Boiga irregularis* [38], *N. kaouthia* [39], *Crotalus simus* [40].

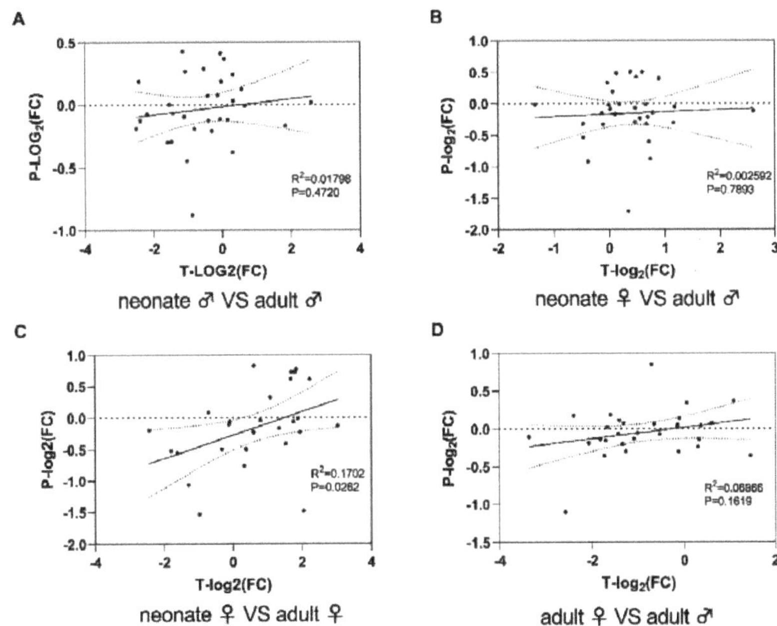

Figure 7. The correlation analysis of the proteome and venom gland transcriptome of *N. atra* venom. Neonate males versus adult males (**A**); neonate females versus neonate males (**B**); neonate females versus adult females (**C**); adult females versus adult males (**D**). T: transcript; P: protein; FC: fold change; R^2: Pearson R squared.

A total of 37 proteins of venom were matched to one or more transcript sequences using BLAST, while the vast majority of other proteins did not match the corresponding transcripts (Table 6).

Table 6. Matched *N. atra* snake venom proteome and venom gland transcriptome.

Protein Accessions	Gene	Protein Name
A0A2D0TC04	ENSNNAG00000000097	Venom phosphodiesterase
A8QL53	ENSNNAG 00000013551	Snake venom serine protease NaSP
	ENSNNAG00000015396	Snake venom serine protease NaSP
A8QL58	ENSNNAG00000001808	L-amino-acid oxidase
A8QL59	ENSNNAG00000009894	Zinc metalloproteinase-disintegrin-like NaMP
	ENSNNAG00000008597	Zinc metalloproteinase-disintegrin-like NaMP
D5LMJ3	ENSNNAG00000005308	Zinc metalloproteinase-disintegrin-like atrase-A
	ENSNNAG00000013917	Zinc metalloproteinase-disintegrin-like atrase-A

Protein Accessions	Gene	Protein Name
D9IX97	ENSNNAG00000006809	Natriuretic peptide Na-NP
	ENSNNAG00000006727	Natriuretic peptide Na-NP
	ENSNNAG00000006779	Natriuretic peptide Na-NP

Table 6. *Cont.*

E3P6P4	ENSNNAG00000005504	Cystatin
ENSNNAP00000002373.1	ENSNNAG00000001642	L-amino-acid oxidase (Fragment chain B)
	ENSNNAG00000001705	L-amino-acid oxidase
ENSNNAP00000008460.1	ENSNNAG00000005694	PLI alpha-like protein
ENSNNAP00000008637.1	ENSNNAG00000005815	Hyaluronidase-2-like isoform X1
	ENSNNAG00000002583	Hyaluronidase-2-like isoform X1
ENSNNAP00000009791.1	ENSNNAG00000006541	Venom dipeptidylpeptidase IV
ENSNNAP00000011851.1	ENSNNAG00000008620	Vascular endothelial growth factor C
	ENSNNAG00000007968	Vascular endothelial growth factor C
	ENSNNAG00000017242	Vascular endothelial growth factor C
ENSNNAP00000012520.1	ENSNNAG00000008178	Cobra venom factor
ENSNNAP00000012975.1	ENSNNAG00000008731	Cobrotoxin
ENSNNAP00000013063.1	ENSNNAG00000008657	Complement C3
ENSNNAP00000014120.1	ENSNNAG00000009518	Alpha-cobratoxin
ENSNNAP00000015130.1	ENSNNAG00000010003	Zinc metalloproteinase-disintegrin-like atragin
ENSNNAP00000021869.1	ENSNNAG00000013375	Cobra venom factor
ENSNNAP00000023044.1	ENSNNAG00000001233	BPTI/Kunitz domain-containing protein-like isoform X2
	ENSNNAG00000015199	BPTI/Kunitz domain-containing protein-like isoform X2
ENSNNAP00000025215.1	ENSNNAG00000016482	Acetylcholinesterase
P00598	ENSNNAG00000010611	Acidic phospholipase A2 1
P01442	ENSNNAG00000009217	Cytotoxin 2
	ENSNNAG00000011010	Cytotoxin 2
P01445	ENSNNAG00000011032	Cytotoxin 2
P29180	ENSNNAG00000008699	Weak neurotoxin 6
P61899	ENSNNAG00000004808	Venom nerve growth factor
P62377	ENSNNAG00000011709	Cytotoxin-like basic protein
P86540	ENSNNAG00000009404	Cytotoxin 8
Q01833	ENSNNAG00000008828	Complement C3
Q10749	ENSNNAG00000017863	Snake venom metalloproteinase-disintegrin-like mocarhagin
	ENSNNAG00000011023	Snake venom metalloproteinase-disintegrin-like mocarhagin
Q5YF90	ENSNNAG00000004800	Venom nerve growth factor 1
	ENSNNAG00000013519	Venom nerve growth factor 1
Q6T179	ENSNNAG00000013762	Acidic phospholipase A2 4
	ENSNNAG00000003407	Acidic phospholipase A2 4
Q7LZI1	ENSNNAG00000003592	Phospholipase A2 inhibitor 31 kDa subunit
Q7T1K6	ENSNNAG00000014151	Cysteine-rich venom protein natrin-1
Q9DEQ3	ENSNNAG00000011613	Neurotoxin homolog NL1
	ENSNNAG00000011050	Neurotoxin homolog NL1
Q9W717	ENSNNAG00000009048	Neurotoxin-like protein NTL2
Q9YGI4	ENSNNAG00000008719	Probable weak neurotoxin NNAM2
V8P395	ENSNNAG00000013472	Glutathione peroxidase
	ENSNNAG00000008046	Glutathione peroxidase
	ENSNNAG00000013545	Glutathione peroxidase

2.7. Toxicological and Enzymatic Activity of N. atra Venom

To characterize the relationship of the dose-effects between the snake venom protein families and the corresponding functional assays, we performed toxicological and enzymatic activities of *N. atra* crude venom. 3-FTx is thought to act on the neuromuscular junction, acetylcholine receptors, and muscarinic receptors to achieve a blocking effect on nerve signaling transmission. The activity of SVMP, LAAO, and 5′-nucleotidase (5′-NT) of neonate *N. atra* was higher than the adult and without a gender difference (Figure 8A,B,D). There was a higher toxicity in the adult specimens. There were age differences (adult > neonate) but no sex differences in PLA2 activity in *N. atra* crude toxin (Figure 8C); there were neither sex nor age differences in AChE activity (Figure 8E). The LD_{50} of crude venom of the neonate male, neonate female, adult male, and adult female groups were 0.73, 0.74, 0.82, and 0.82 µg/g (intraperitoneal injection), respectively (Figure 8F). All of the tested enzymes indicated a good correlation ($p < 0.05$) between enzyme activity and enzyme content in each group (Figure 9A–E).

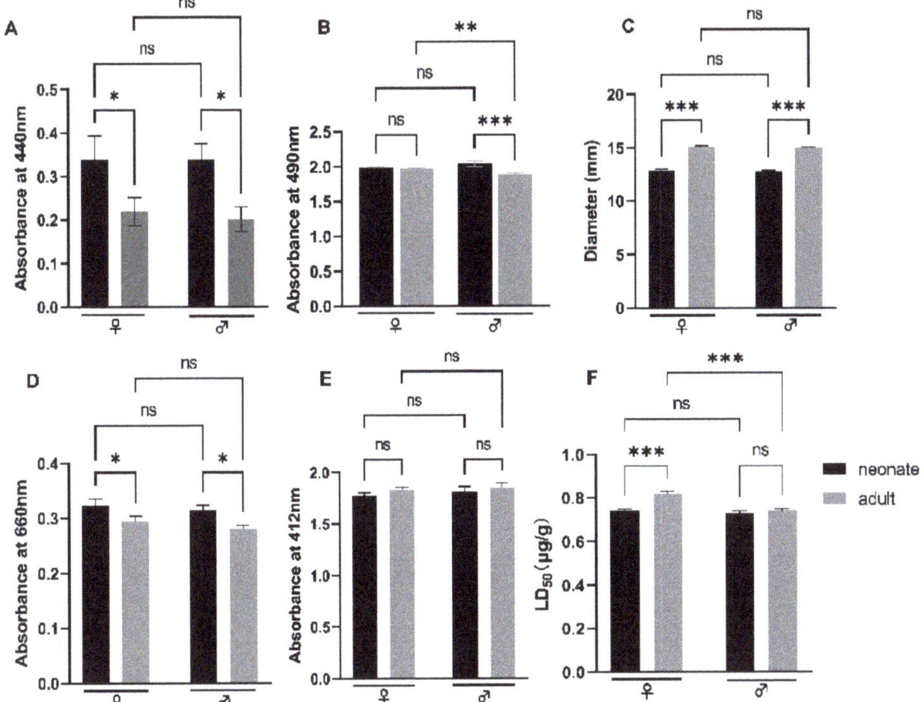

Figure 8. The enzyme activities of *N. atra* crude venom. SVMP (**A**); LAAO (**B**); PLA$_2$ (**C**); 5′-NT (**D**); AChE (**E**); LD$_{50}$ (**F**). ♀: female; ♂: male. Significance analyzed by Tukey's multiple comparison test, the value of $p > 0.05$ (ns), $p < 0.05$ (*), $p < 0.002$ (**), $p < 0.001$ (***).

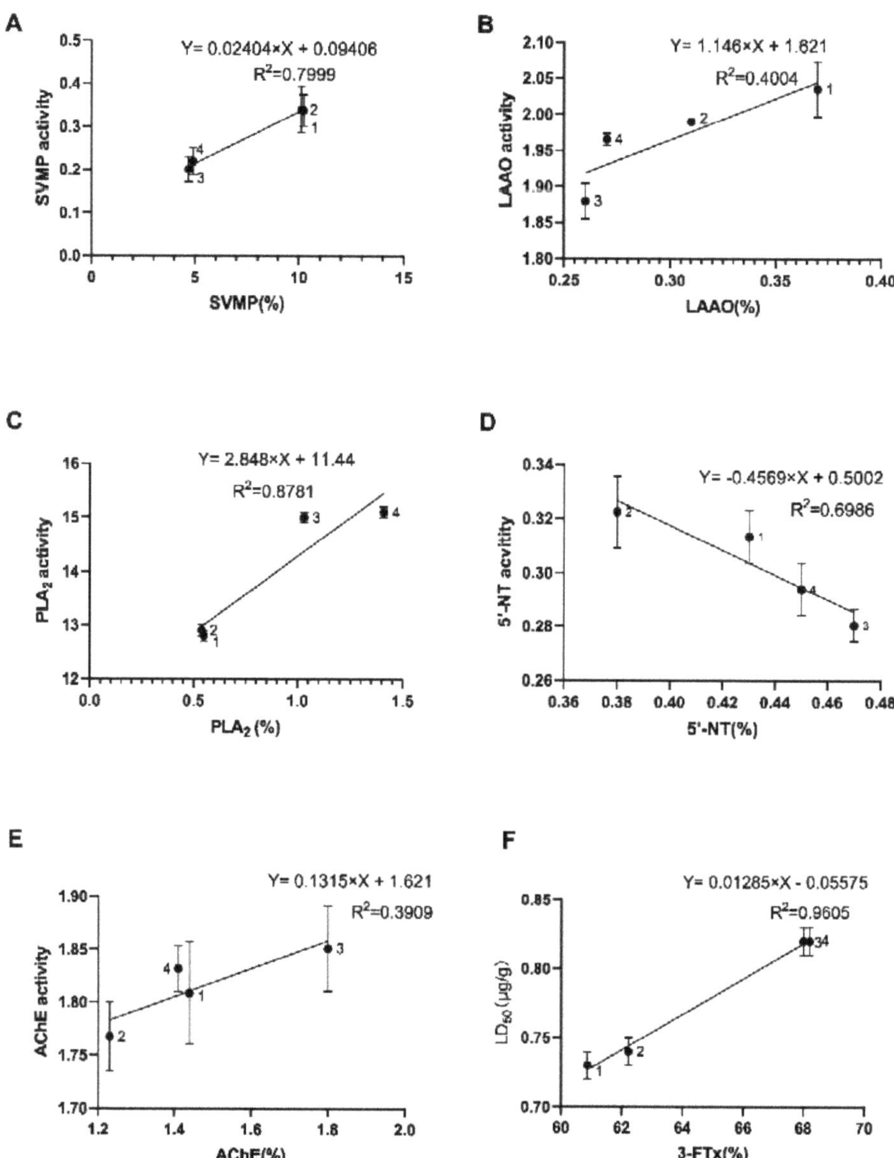

Figure 9. The correlation of *N. atra* venom protein ratios with corresponding enzyme activities. SVMP (**A**). LAAO (**B**). PLA$_2$ (**C**). 5′-NT (**D**). AChE (**E**). LD$_{50}$ (**F**). 1: neonate males; 2: neonate females; 3: adult males; 4: adult females. Significance analyzed by Simple linear regression analysis, the value of *p* < 0.05 considered significant.

3. Discussion

3.1. Ontogenetic Variation Predominates in the Snake Venom of Captive Species

Our results indicated that there were age differences in the venom composition of captive Chinese cobras. The relative contents of SVMP and CRISP were higher in the neonates; the relative contents of 3-FTX, CVF, and PLA2 were higher in the adults. This result is generally consistent with the variation of venom composition in neonate and adult specimens of wild cobras in Guangxi Province [3]. The results of SDS-PAGE electrophoresis of the crude venom of the snakes (Figure 1) indicated age differences in the venom content. Furthermore, the results of enzyme activity and semi-lethal concentration assay further confirmed this (Figures 8 and 9). It is noteworthy that the changes in the relative content of the transcripts and proteins were consistent only for PLA_2. In addition, the snake venom components that showed gender differences included AChE (male specimens > female specimens), NGF (male specimens < female specimens), and CRISP (male specimens > female specimens). In our study, only AChE was more abundant in male specimens than that in the female specimens, and both adult specimens and neonate specimens. Based on the transcript of venom toxin distribution of toxin genes on chromosomes, no toxin-related genes were found on the W and Z sex chromosomes. Therefore, the genes for these sexually distinct snake venom components are not distributed on the sex chromosomes and the cause of the appearance of sex differences is not companionship. Other non-genetic factors may be suspected to cause the sex differences in the toxin.

Previous studies have shown that macroscopic factors contributing to changes in snake venom composition during species ontogeny include prey species (ectothermic prey or endothermic prey) [41], surface area to volume ratio (same prey species during ontogeny) [8], ontogenetic shifts in dietary habits, competition, predation pressure [3], and microscopic factors, such as miRNA [42,43]. In captivity, the environmental factors for neonate and adult snakes were the same except for the food, which was thought to be the same, and the adult and neonate snakes were fed block chicks and frogs (*Pelophylax nigromaculatus*), respectively. Amphibian reptiles are variable temperature animals, while chicks are constant temperature animals. The prey for wild Chinese cobra neonates is mainly amphibious reptiles and small-sized fish, while the prey for adult snakes are mainly birds and mice. Therefore, the difference in venom fractions during ontogenesis may be due to the type of food (ectothermic prey or endothermic prey).

According to the results in Figure 6, the relative changes of snake venom proteins during the growth and development of individuals in captive-bred Chinese cobras were poorly correlated with the transcripts of the corresponding snake venom proteins. There are several possible reasons for this result. First, the limited number of samples that were taken and the variability of the sampled individuals. Second, the venom glands were sampled at an inappropriate time. Third, inaccuracies in the proteomic or transcriptional sequencing process. Fourth, improper handling and storage of proteomic and transcriptomic samples resulted in the degradation or contamination of the samples.

We confirmed at the transcriptomic and proteomic levels that all snake venom metalloproteases in Chinese cobra venom were Class III SVMPs. Class III SVMPs exhibit complex biological activities due to the diversity of molecular structures (metalloproteinase, deintegrin, cysteine-rich domain). Another source of Class I and Class II SVMPs in snake venom, in addition to the corresponding gene expression, is the self-hydrolysis process of Class III [44]. However, the disappearance of metalloproteinase self-hydrolysis may be due to its own glycosylation modifications (Table 3). The snake venom proteases are associated with prey digestion (cytotoxins and hydrolases) [45]. For example, SVMP takes responsibility for the local and systemic hemorrhage by hydrolyzing basement membranes, extracellular matrix, promoting apoptosis, inhibiting platelet aggregation, hydrolyzing collagen, fibrinogen, and also inducing leukocyte rolling [44,46]. The destruction of blood vessels may promote the penetration and diffusion of other toxins and accelerate the digestion process of prey. It is assumed that the neonate snake's strategy to ensure survival is rapid growth. What was puzzling was that females weighed significantly more than the males at the same growth period, but there were no gender differences in the protein hydrolase relative content levels. Moreover, there was no significant difference in the body lengths between the sexes. Therefore, the levels of cytotoxic and hydrolytic enzymes might not be directly related to the body length and weight.

The relative content of CRISP protein in snake venom of captive Chinese cobras is rich. CRISP is widely distributed in snake venom and acts on ion channels, the inflammatory response, and smooth muscle contraction [47]. Snake venom CRISPs with a molecular weight between 20 and 30 kDa are non-enzymatic proteins, containing 16 highly conserved cysteine residues forming eight disulfide bonds, four of these residues form a cysteine-rich domain (CRD region) at the C-terminus. Despite the richness of the biological targets of action of CRISP, its specific function remains unclear. Since CRISP can promote smooth muscle contraction, it is presumed that CRISP may act on the circulatory system, to raise blood pressure and promote increased hemorrhage. The silica molecular docking study of cobra CRISP in complex with TLR4-MD2 receptor shows that CRISP is involved in its cysteine-rich structural domain (CRD) interacting with the complex [48]. The rich diversity of CRISP in intra- and inter-species snake venoms suggests functional diversity. Dissecting these rich structural and functional CRISPs promotes an understanding of the mechanism of action of snakebites and exploits resources for drug development. Surprisingly, the relative levels of CRISP in the venom of captive-bred individuals increased dramatically compared to the wild individuals, but the exact reason for this is not clear.

CVF leads to a greater understanding of complement C3. CVF is a spherical glycoprotein, a homologous molecule that is formed in evolution with complement C3, similar in structure to C3a, and reversibly bound to factor B in the presence of Mg^{2+}. The complex CVF-B is then hydrolyzed by factor D in serum into two fragments, CVF-Bb and Ba. CVF-Bb can be present in the blood for 6–7 h and the small fragment Ba is released. The product CVF-Bb is a complement-activated alternative pathway C3/C5 convertase with serine protease activity that hydrolyzes the peptide bond at position 77 of the alpha chain of C3 (Arg-Ser) and the peptide bond at position 74 of the alpha chain of C5 (Arg-Leu) to form C3b and C5b, releasing the allergenic toxins C3a and C5a. C5b contributes to the formation of the membrane attack complex, leading to cell membrane perforation and cell lysis. It also caused the depletion of C3 and C5–C9 components, and C3 was largely depleted in the blood of rats after 2 h of intraperitoneal injection of CVF (1 mg/kg) [49]. Finally, CVF weakens complement resistance to other snake venom proteins and assists other toxins in their toxicity.

Snake venom PDEs are among the least studied snake venom enzymes. During envenomation, they exhibit various pathological effects, ranging from the induction of hypotension, to the inhibition of platelet aggregation, edema, and paralysis [50]. In the present study, only one phosphodiesterase was obtained and was not an abundant component. In addition, no PDE transcripts were found in the venom gland transcriptome of wild *N. atra* that were collected from Zhejiang Province [51].

PLA2 is distributed among various families of snakes. Generally, PLA_2 is exceptionally plentiful in biological activities, including neurotoxicity, hemorrhagic toxicity, myotoxicity, cardiotoxicity, and the induction or inhibition of platelet aggregation. In our study, the proportion of acid PLA_2 content was significantly higher than that of alkaline PLA_2. The active catalytic center of all PLA_2 matched was His48-Asp49-Tyr52, the motif underlying the high activity of PLA_2. There are two major amino acids, Lys115 and Arg117, that are thought to underlie the neurotoxicity of these enzymes and indicated that 3 out of 11 PLA_2 has neurotoxicity among all the groups. The conserved structures of these enzymes are essentially the same in different snake species they have the same evolutionary origin. They both have 14 cysteine residues and form seven disulfide bonds. According to the results of previous studies, *N. atra* expressed up to 12.2% PLA_2 in their venom in the wild (n = 13, all adults), while the expression of PLA_2 in the venom of the artificially hatched individuals in this study was less than 1% (among all groups) [52].

3.2. Neonate and Adult Snakes have Different Strategies for Handling Prey

The biological functions of snake venom include predation, defense, and digestion [53]. In the process of prey predation, snake venom neurotoxins play a key role in immobilizing and then killing the prey, increasing the efficiency of prey predation, which likewise is the synergistic effect of 3-FTx and fewer abundance components in the venom [54,55]. Different components of snake venom play different biological functions, among which enzymes have the role of helping to digest prey, and 3-FTx is mainly used to kill prey. In the present study, the content of 3FTx was lower in the neonates than in the adults, while the content of enzymes was higher in the neonates than in the adults. The predation strategy of the cobra family is to violently control the movement of the prey after completing the injection of venom into the prey until the toxicity kicks in and the prey is incapacitated. In addition, the prey of neonate Chinese cobras is mainly small-sized amphibians, while the prey of adults is mainly rodents and reptiles [56]. Therefore, the venom pattern of neonate and adult Chinese cobras differs, with neonates having high enzymatic activity and low toxicity and adults having low enzymatic activity and high toxicity. High doses of neurotoxin are incompatible with high protein hydrolytic activity [57]. The LD_{50} of venom from adult Chinese cobras was significantly lower than that of juveniles when measured in mice. In conclusion, the predation strategies of adult and neonate captive *N. atra* were different.

3.3. The Hatchery Breeding Process Rearranges Snake Venom

Snake venom composition is regulated by intrinsic factors and also environmental factors. Differences in the venom fractions between captive-bred and wild individuals have been reported for both cobras and vipers [58–62]. However, a comparison of the proteome of venom from long-term captive and recently wild-caught eastern brown snakes (*Pseudonaja textilis*) showed that the venom was not altered by captivity. In this study, the humidity (70%) and temperature (28 °C) were constant during captive breeding, food was the same, and the snakes went without hibernation. The habitat of wild Chinese cobras is more complex and variable. The protein number of wild Chinese cobra venom that was determined by Shotgun-LC-MS/MS, 1DE-LC-MS/MS, GF-LC-MS/MS, and GF-2DE-MALDI-TOF-MS were 78, 65, 77, and 16, respectively [63]. However, this study lacked the description of protein sequence and protein function annotation. In addition, the venomic profile of Chinese cobra that separately inhabited eastern and western Taiwan were described and the results indicated that dominating component contents both were the same and but the proportion of protein families were not equal. The most components of the eastern inhabitants are 3-FTx, phospholipase A_2 (PLA_2), and cysteine-rich secretory protein (CRISP) (76.35%, 16.82%, and 2.41%, respectively) and the western inhabitants were 3-FTx, PLA_2, and CRISP (79.96%, 13.95%, and 2.16%, respectively) [16]. The main components of adult *N. atra* (Zhoushan, Zhejiang province, China) venom was 3-FTx, PLA_2, CRISP, SVMP, and nerve growth factor (NGF) (84.3, 12.2, 1.8, 1.6, and 0.1%, respectively) [52]. A similar proteome profile was discovered in *N. kouthia* that was caught in China [39]. The transcriptome profile of the Chinese Cobra showed that the 3-FTx was 95.80%, followed by PLA_2 1.20%, CRISP 0.70%, C-type lectins 0.60%, and vespryn 0.50% [51]. In our study, the consequence of the venomic (Table 2) and transcriptomes (Table 5) of captive Chinese cobras were distinct from the wild ones about the previous description of snake venom composition. The main components of wild Chinese cobra venom are 3-FTx, PLA_2, and CRISP. Although 3-FTx was still the main component of captive Chinese cobra venom, its content was significantly lower than that of wild specimens; however, PLA_2 was not the main component, and its content was nearly ten times lower. Although CRISP was the main component of both cultured and wild specimens, the content of CRISP in captive specimens increased by more than two times compared with that of the wild specimens. Therefore, the captive breeding process with strictly controlled conditions may alter snake venom partial composition.

3.4. Asynchronous Snake Venom Synthesis Process

The process of snake venom secretion involves the secretion of a complex mixture of peptides and proteins by secretory columnar cells of the glandular epithelium and storage in the lumen of the venom gland. The process of venom release involves muscle contraction of the venom gland to expel the venom through the fangs. However, there are different views on the origin of the venomous gland apparatus. The venom gland of the king cobra is rich in vertebrate pancreas-specific miRNA (miR-375), confirming that its venom gland originates from the pancreas [64]. In contrast, Kochva proved that the snake venom gland has been adapted from the salivary gland during vertebrate evolution [65]. Heterogeneity of venom gland cells in snakes was observed in king cobra (*Ophiophagus annah*) and horned coral snake (*Aspidelaps lubricus cowlesi*) [64,66]. In addition, the secretory processes of both the snake venom gland and the pancreas are under hormonal and neurological control [67–69]. The snake venom that is stored in the lumen is secreted by the venom gland secretory cells after a process of transcription, translation, and modification. The transcript directs the synthesis and modification of the toxin, and the transcript is hydrolyzed by nucleic acid endonucleases once the toxin synthesis process is completed [70]. By comparing the protein sequences of snake venom with the transcript sequences, we found that some of the transcripts of the toxins were absent from the secretory cells of the venom glands and only the corresponding proteins remained, or the transcripts were present but the proteins were missing (Tables 1, 4 and 6). These results suggest that there was a sequential order in the synthesis and secretion of toxins. It is hypothesized that the response times of different toxic gland secreting cells to toxic gland milking stimuli differ.

3.5. Widespread Distribution of Toxin Gene on Chromosomes

Snake venom, which contains more than 20 well recognized protein families (also including many non-toxic proteins), is a result of different protein families being recruited to the venom gland at different times in the evolution of the snake. The generation and diversity of snake venoms are attributed to gene recruitment, doubling, and neogenization [71,72]. The vast majority of snake venom genes have been shown to originate from gene recruitment during venom evolution, and toxin recruitment events have been found to have occurred at least 24 times [73,74].In addition, the process of snake venom recruitment seems to be uninterrupted, with new members of the snake venom metalloproteinase family recently reported in rear-fanged snake venom [72].

Genomic data and karyotype maps of Indian cobras were used as a reference for our study. We found that the majority of Chinese cobra snake venom genes were distributed on macro-chromosomes (2n = 38) 1–4, 6, and 7, micro-chromosomes (11 pairs) 1–9 and 11, and without toxin genes were found on the sex chromosomes (Table 4). The gene distribution of the 3-FTx, DPP-IV, PDE, VPA, NP, LAAO, C3, and AP family is located on macro-chromosome 3, 1, 1, 6, 4, 2, 2, and micro-chromosome 6, respectively. Since all 3-FTx genes are localized on the same chromosome, it is speculated that this gene family is amplified by tandem duplication [46]. The gene distribution of the other protein families is scattered across multiple chromosomes, for instance, CTL (macro-chromosome 1, 2, and 7, micro-chromosome 7), VEGF (macro-chromosome 1, 2, 4, micro-chromosome 5, and 9), SVMP (macro-chromosome 4, micro-chromosome 1, 6, and 8), and PLA$_2$ (macro-chromosome 1 and 4, micro-chromosome 11). The snake venom gene family expands diversely. These data on the genetic distribution of snake venom protein genes on chromosomes provide the basis for the theoretical analysis of the evolutionary study of toxins and evidence that toxin recruitment and toxin family expansion have occurred over a long period and through a large number of evolutionary events.

To date, transcriptomics and proteomics technology have been widely used in snake venom research. In the past decades, a large number of toxic or non-toxic venom proteins have been sequenced and characterized for physicochemical properties and function. With the popularity of toxin proteomics, a large number of proteins regarding snake venom components have been identified at a very fast rate [75]. In snake venom and venom gland studies, the combination of venomics and venom gland transcriptomics is increasingly being applied jointly to find evidence of intra- and inter-specific individual differences in snake venom secretion species, and the combination of these two omics approaches helps to explain the differences that already exist more scientifically and rigorously. All snakebite treatments rely on accurate snake venom protein profiles, including antivenom, next-generation treatments (including monoclonal antibodies and antibody fragments, nanobodies, small molecule inhibitors, aptamers, and peptides, metal ion chelators, and antivenoms that are manufactured using synthetic immunogens) [76]. In the present study, a combined transcriptomic and proteomic approach was adopted to study the corresponding changes of captive snakes in ontogenesis. Based on the results of our study, we found that 3-FTx relative content varied in a discrete and uncorrelated manner between 3-FTx transcriptional and proteomic levels among different ages and gender specimens. Moreover, the same trend of variation was not synchronous in the relative content of 3-FTx subclasses. The same was true for several other major components of snake venom, the CRISP, CVF, SVMP, PDE, and NGF. The relative content of the venom protein family varied considerably among the captive *N. atra* individuals of different ages, and secondly, the venom composition between the captive and wild individuals also differed significantly.

4. Conclusions

In the present research, the transcriptome, proteome, and functional assays of captive *N. atra* venom was depicted. Under the captive conditions, females grew faster than males during ontogenesis. Ontogenesis has an overwhelming effect on the venom protein profile of captive *N. atra*, while there is an obscure gender difference. The main components of captive *N. atra* venom was very different from the wild specimens, although they were obtained from the same geographical distribution area, Guangxi province, China. Presumably, the loss of transcripts of some toxin proteins indicates that the expression of different snake venom families was asynchronous under the same toxin extraction stimulation conditions. An analysis of the proteomic and transcriptomic results revealed significant effects of ontogenesis and captive breeding on the relative content of the snake venom protein, demonstrating the plasticity of snake venom expression and secretion. The protein profile of snake venom provides the foundation for the formulation and administration of traditional antivenom and next-generation antivenom.

5. Materials and Methods

5.1. Animals Keeping and Feeding

All the experimental procedures involving animals were carried out under the Chinese Animal Welfare Act and our protocol (20090302 on 10 January 2021) was approved by Chongqing Municipal Public Health Bureau. The healthy female Kunming mice (20 ± 2 g of body weight) were obtained from the Laboratory Animal Center of the Army Medical University. They were housed in temperature-controlled rooms and received water and food ad libitum until use.

Snake eggs were obtained from Guangxi province, China. All the eggs were incubated and all snakes were captive in our laboratory without hibernation, Chongqing, China. Newborn snakes were fed with a starter diet (gutted and chopped frog) for three months and a gradual increase in the feeding of dehaired, frozen chicks (gutted and chopped). The feeding frequency was once every two days. The constant ambient temperature was 28 °C, the humidity was 70%, and a supply of sterilized water. An average of 50 snakes in a four-square meter enclosure, which were enclosed in nylon netting.

5.2. Snake Venom and Gland

A total of 10 snakes in each group (neonate male, neonate female, adult male, and adult female group), were randomly selected to measure the body total length and body weight. The venom milking process was that snake bites on parafilm-wrapped jars along with gently massaging the venomous glands or aspirate snake venom with a pipette aimed at the fangs. A total of five snakes were randomly selected from each group (neonate male, neonate female, adult male, and adult female group) to collect venom and venom glands. The fresh venom was centrifuged to remove impurities for 15 min at 10,000 g 4 °C, lyophilized, marked, and stored at -80 °C until use. During the lyophilization process, the lyophilizer that was used was a SCIENTZ-18N (SCIENTZ, Ningbo, China) with the condition of $-70\,^\circ$C sample temperature and $-45\,^\circ$C cold trap temperature. The venom gland was dissected four days after the venom was milked and stored in RNA later solution at $-80\,^\circ$C until use.

5.3. Library Preparation for Transcriptome Sequencing

A total amount of 1 µg RNA per sample was used as the input material for the RNA sample preparations. Sequencing libraries were generated using NEBNext® UltraTM RNA Library Prep Kit for Illumina® (NEB, San Diego, CA, USA) following the manufacturer's recommendations and index codes were added to attribute sequences to each sample. Briefly, mRNA was purified from the total RNA using poly-T oligo-attached magnetic beads. Fragmentation was carried out using divalent cations under elevated temperature in NEBNext First Strand Synthesis Reaction Buffer ($5\times$). First-strand cDNA was synthesized using random hexamer primer and M-MuLV Reverse Transcriptase. Second strand cDNA synthesis was subsequently performed using DNA Polymerase I and RNase H. The remaining overhangs were converted into blunt ends via exonuclease/polymerase activities. After adenylation of $3'$ ends of DNA fragments, NEBNext Adaptor with hairpin loop structure was ligated to prepare for hybridization. To select cDNA fragments of preferentially 250–300 bp in length, the library fragments were purified with the AMPure XP system (Beckman Coulter, Beverly, CA, USA). Then, 3 µL USER Enzyme (NEB, San Diego, CA, USA) was used with size-selected, adaptor-ligated cDNA at 37 °C for 15 min followed by 5 min at 95 °C before PCR. Then PCR was performed with Phusion High -Fidelity DNA polymerase, Universal PCR primers, and Index (X) Primer. At last, the PCR products were purified by AMPure XP system(Beckman Coulter, Beverly, USA) and the library quality was assessed on the Agilent Bioanalyzer 2100 system (Agilent, Santa Clara, CA, USA).

5.4. cDNA Sequencing and Data Analysis

The clustering of the index-coded samples was performed on a cBot Cluster Generation System using TruSeq PE Cluster Kit v3-cBot-HS (Illumia, San Diego, CA, USA) according to the manufacturer's instructions. After cluster generation, the library preparations were sequenced on an Illumina Novaseq platform and 150 bp paired-end reads were generated. At the same time, the Q20, Q30, and GC content of the clean data were calculated. All the downstream analyses were based on clean data with high quality. Reference genome and gene model annotation files were downloaded from the genome website directly. Index of the reference genome was built using Hisat2 v2.0.5 and paired-end clean reads were aligned to the reference genome (GenBank GCA_009733165.1) using Hisat2 v2.0.5. StringTie uses a novel network flow algorithm as well as an optional de novo assembly step to assemble and quantitate the full-length transcripts representing multiple splice variants for each gene locus. FeatureCounts v1.5.0-p3 was used to count the reads numbers that are mapped to each gene. The FPKM of each gene was calculated based on the length of the gene and the read counts that were mapped to this gene. Differential expression analysis of two conditions/groups (two biological replicates per condition) was performed using the DESeq2 R package (1.16.1, Bioconductor, Seattle, DC, USA, 2005). The resulting p-values were adjusted using Benjamini and Hochberg's approach for controlling the false discovery rate. Genes with an adjusted $p < 0.05$ found by DESeq2 were assigned as differentially expressed.

5.5. TMT Labeling of Peptides

Each equivalent snake protein sample was taken and the volume was made up to 100 µL with DB dissolution buffer (8 M Urea, 100 mM TEAB, pH 8.5). Trypsin (Promega/V5280, Madison, WI, USA) and 100 mM TEAB buffer (Sigma/T7408-500ML, St. Louis, MO, USA) were added, the sample was mixed and digested at 37 °C for 4 h. Then, trypsin and $CaCl_2$ were added and the sample was digested overnight. Formic acid was mixed with the digested sample, adjusted pH under 3, and centrifuged at 12,000 g for 5 min at room temperature. The supernatant was slowly loaded to the C18 desalting column (Waters BEH C18, 4.6 × 250 mm, 5 μm, Milford, MA, USA), washed with washing buffer, including 0.1% formic acid (Thermo Fisher Scientific/A117-50, Waltham, MA, USA), 3% acetonitrile (Thermo Fisher Chemical/A955-4, Waltham, MA, USA), 3 times, then eluted by some elution buffer (0.1% formic acid, 70% acetonitrile). The eluents of each sample were collected and lyophilized. A total of 100 μL of 0.1 M TEAB buffer was added to reconstitute, and 41 μL of acetonitrile-dissolved TMT labeling reagent was added, the sample was mixed with shaking for 2 h at room temperature. Then, the reaction was stopped by adding 8% ammonia. All the labeling samples were mixed with equal volume, desalted, and lyophilized.

5.6. Separation of Fractions

Mobile phase A (2% acetonitrile, adjusted pH to 10.0 using ammonium hydroxide) and B (98% acetonitrile) were used to develop a gradient elution. The lyophilized powder was dissolved in solution A and centrifuged at 12,000 g for 10 min at room temperature. The sample was fractionated using a C18 column (Waters BEH C18, 4.6 × 250 mm, 5 μm, Milford, MA, USA) on a Rigol L3000 HPLC system (Rigol, Shanghai, China), the column oven was set as 45 °C. The detail of the elution gradient is shown in Table S2. The eluates were monitored at UV 214 nm, collected for a tube per minute, and combined into 10 fractions finally. All the fractions were dried under vacuum, and then, reconstituted in 0.1% (v/v) formic acid in water. During the lyophilization process, the lyophilizer that was used was a SCIENTZ-18N (SCIENTZ, Ningbo, China) with a sample temperature of -70 °C and a cold trap temperature of -45 °C.

5.7. LC-MS/MS Analysis

For transition library construction, shotgun proteomics analyses were performed using an EASY-nLCTM 1200 UHPLC system (Thermo Fisher, Waltham, MA, USA) coupled with a Q ExactiveTM HF-X mass spectrometer (Thermo Fisher, Waltham, MA, USA) operating in the data-dependent acquisition (DDA) mode. A total of 1 μg sample was injected into a homemade C18 Nano-Trap column (4.5 cm × 75 μm, 3 μm). The peptides were separated in a homemade analytical column (15 cm × 150 μm, 1.9 μm), using a linear gradient elution as listed in Table S3. The separated peptides were analyzed by Q ExactiveTM HF-X mass spectrometer (Thermo Fisher, Waltham, MA, USA), with ion source of Nanospray Flex™ (ESI), spray voltage of 2.3 kV, and ion transport capillary temperature of 320 °C. Full scan ranges from m/z 350 to 1500 with the resolution of 60,000 (at m/z 200), an automatic gain control (AGC) target value was 3×10^6, and a maximum ion injection time was 20 ms. The top 40 precursors of the highest abundant in the full scan were selected and fragmented by higher-energy collisional dissociation (HCD) and analyzed in MS/MS, where the resolution was 45,000 (at m/z 200) for 10 plex, the automatic gain control (AGC) target value was 5×104 the maximum ion injection time was 86 ms, the normalized collision energy was set as 32%, the intensity threshold was 1.2×10^5, and the dynamic exclusion parameter was 20 s.

5.8. The Identification and Quantitation of Protein

The resulting spectra from each run were searched separately against the Uniprot database by Proteome Discoverer 2.4 (Thermo, Waltham, MA, USA). The searched parameters are set as follows: mass tolerance for precursor ion was 10 ppm and mass tolerance for production was 0.02 Da. Carbamidomethyl was specified as fixed modifications, oxidation of methionine (M) and TMT plex were specified as dynamic modification, acetylation, and TMT plex were specified as N-terminal modification in PD 2.4. A maximum of 2 miscleavage sites was allowed. To improve the quality of analysis results, the software PD 2.4 further filtered the retrieval results and peptide spectrum matches (PSMs) with the credibility of more than 99% were identified PSMs. The identified protein contained at least 1 unique peptide. The identified PSMs and protein were retained and performed with FDR no more than 1.0%. The protein quantitation results were statistically analyzed by t-test.

5.9. Protein and Transcript Alignment

The protein family members sequence alignment was performed using online multiple sequence alignment analysis (MAFFT version 7) with default parameters. The protein sequence and transcriptome sequence alignment analysis were performed using BLASTx, E-value $< 1 \times 10^{-5}$, with the other parameters set as default.

5.10. Snake Protein Concentration Determination

The protein concentration of each snake venom was obtained using the Pierce® Bicinchoninic Acid (BCA) Protein Assay (Thermo Scientific, Rockford, IL, USA) following the manufacturer's instructions and using bovine serum albumin (BSA) as a standard.

5.11. SDS-PAGE

A total of 40 µg of 15 µL snake protein sample was loaded to 15% SDS-PAGE gel electrophoresis, wherein the concentrated gel was performed at 80 V for 20 min, and the separation gel was performed at 120 V for 90 min. Electrophoresis instrument (BIO-RAD/PowerPac Basic, Hercules, CA, USA) and electrophoresis tank (Bei-jing JUNYI DONGFANG/JY-SCZ2+, Beijing, China) were used in this research. The gel was stained by Coomassie brilliant blue R-250 (Beyotime Biotechnology, Jiangsu, China) and decolored with elution solution until the bands were visualized clearly by a gel reader (Beijing Liuyi/WD-9406, Beijing, China).

5.12. Median Lethal Dose (LD$_{50}$) Determination

Groups of 5 healthy female Kunming mice (20 ± 2 g) from the Laboratory Animal Center of Army Medical University were intraperitoneally injected with various doses of captive crude *N. atra* venom, which was dissolved in 100 µL sterile saline, while the control test with an equal volume of sterile saline injection only. The deaths were recorded 6 h after injection, and the LD$_{50}$ was calculated using the Spearman–Karber method.

5.13. Enzyme Activity Assay

For the SVMP activity assay [77], 0.5 mL of substrate (0.2 M Tris-HCl, pH 8.5, containing 1% azo-casein) was added to the snake venom solution (30 µg dissolved in 10 µL of dd water), mixed well, and incubated at 37 °C for 1 h. A total of 75 µL of 10% TCA(Sigma/T6508-100ML, St. Louis, MO, USA) was added and incubated at 37 °C for 30 min to complete the reaction. An equal amount (100 µL) of supernatant was collected, 50 µL of 2 M Na$_2$CO$_3$ (Sigma/5330050050, St. Louis, MO, USA)was added and mixed, and the absorbance values of the samples were measured at 440 nm (Flash® SP-Max2300A2, Shanghai, China).

For the PLA2 activity assay [78], the preparation of egg yolk solution was conducted as follows: the egg yolk was mixed in 0.85% NaCI (SCR/10019318-500g, Shanghai, China) solution at a ratio of 1:3, centrifuged at 3000 r/min for 5 min (IKA® VORTEX, Guangzhou, China), the supernatant was removed, and left to stand at -20 °C. For the plates preparation: add 0.05 M pH 6.5 NaAC (SCR/10018818-500g, Shanghai, China) solution, add 1% agarose (SCR/10000561-250g, Shanghai, China) without shaking and microwave for 90 s. Reduce the temperature of the agarose to 50 °C, add 4% egg yolk solution and 2% 0.01 M CaCl$_2$ (SCR/20011160-500g, Shanghai, China), mix the mixture well, and spread it on the plates. The mixture was allowed to cool to room temperature in the plate, and then the plates were punched and 30 µg venom was loaded per well. After incubation at 37 °C for 8 h, the diameter of the transparent circle was measured using vernier calipers.

For the LAAO activity assay [79], a total of 30 µg venom was added to 90 µL of the substrate system, containing 50 mM Tris-HCl (Sigma/T5941-100g, St. Louis, MO, USA), pH 8.0, 5 mM L-leucine (Sigma/L8000-25g, St. Louis, MO, USA), 2 mM O-phenylenediamine (Sigma/P23938-5g, St. Louis, MO, USA), and 0.81 U/mL horseradish peroxidase (Sigma/P8375-5KU, St. Louis, MO, USA), mixed, and incubated at 37 °C. After one hour of incubation, the reaction was stopped by adding 50 µL of 2 M sulfuric acid. The absorbance was recorded at 490 nm (Flash®

SP-Max2300A2, Shanghai, China). The change in the absorbance at 0.3 units/min/µg venom was used to express the activity.

For the 5′-NT activity assay [80], a total of 40 µg venom was added to 90 µL of substrate solution, containing 50 mM Tris-HCl, pH 7.4, containing 10 mM $MgCl_2$ (Sigma/M8266-100g, St. Louis, MO, USA), 50 mM NaCl, 10 mM KCl (Sigma/P3911-25g, St. Louis, MO, USA), and 10 mM 5′-AMP (Sigma/01930-5g, St. Louis, MO, USA), incubated at 37 °C for 60 min, then 20 µL of 2.5 M sulfuric acid was added and mixed and stood for 5 min, and 80 µL of chromogenic solution (8% The absorbance values were measured at 660 nm (Flash ® SP-Max2300A2, Shanghai, China). The H2PO4- activity was expressed as nmol /min/mg of venom that was produced, using kH2PO4 (SCR/10017605-250g, Shanghai, China) as the standard.

For the AChE activity assay [81], the substrate solution contained 75 mM acetylcholine iodide and 10 mM DTNB (Sigma/D8130-1g, St. Louis, MO, USA) that was dissolved in 0.1 M PBS (pH 8.0) buffer containing 0.1 M NaCl and 0.02 M $MgCl_2 \cdot 6H_2O$, respectively. A total of 100 µL of substrate was added to each 96-well plate, 30 µg of snake venom was added immediately, and the reaction was carried out at 37 °C for 30 min. The samples were vortexed gently for 15 s (IKA ® VORTEX, Guangzhou, China) and the absorbance values were measured at 412 nm (Flash ® SP-Max2300A2, Shanghai, China).

Each of the enzyme activity enzyme activity test methods were simply modified.

5.14. Statistical Analyses

There were three replicates per treatment that were used in all the studies. The means were obtained by taking the average of three measurements for each experiment with the standard error of the means (±SE; standard error) obtained. An analysis of variance (ANOVA) was applied to analyze the variation of the means with a 95% confidence interval. With $p < 0.05$ as a significant difference, using the same statistical software GraphPad prism 9.0.0.

Supplementary Materials: The following supporting information can be downloaded at: https://www.mdpi.com/article/10.3390/toxins14090598/s1, Figure S1: Distribution of peptides in enzymatic hydrolysis of *N. atra* snake venom; Figure S2: Molecular weight distribution of *N. atra* venom protein; Table S1: Protein matches that were obtained by tandem MS/MS of tryptic peptides from captive *N. atra* venom in different ages and genders; Table S2: Peptide fraction separation liquid chromatography elution gradient table; Table S3: Liquid chromatography elution gradient table.

Author Contributions: Conceptualization, X.N. and Q.C.; methodology, X.N. and Q.C.; software, X.N.; validation, Q.C., C.W. and W.H.; formal analysis, X.N. and Q.C.; investigation, Q.C.; resources, R.L.; data curation, X.N. and Q.C.; writing—original draft preparation, X.N. and Q.C.; writing—review and editing, R.L. and Q.L.; visualization, X.N.; supervision, Q.H. and X.Y.; project administration, X.Y.; funding acquisition, X.Y. and Q.H. All authors have read and agreed to the published version of the manuscript.

Funding: This work was supported by grants from the Chongqing forestry reform and development fund (No. Yu-lin-ke-tui -2019-2) and a project for public well-being funded by Chongqing Science and Technology Bureau (cstc2017chmsxdny0246).

Institutional Review Board Statement: Not applicable.

Informed Consent Statement: Not applicable.

Data Availability Statement: The raw data that were used in this study and the statistical analyses are available via the corresponding author at Chongqing Normal University.

Acknowledgments: The authors are grateful to Chongqing forestry reform and development fund (No. Yu-lin-ke-tui -2019-2) and Chongqing Science and Technology Bureau (cstc2017chmsxdny0246).

Conflicts of Interest: The authors declare no conflict of interest.

References

1. WHO. Snakebite-Envenoming. Available online: https://www.who.int/zh/news-room/fact-sheets/detail/snakebite-envenoming (accessed on 8 April 2019).
2. Pin, L.; Wang, D.; Lutao, X.; Linjie, X. 2019 China Medical Rescue Association Animal Injury Treatment Branch Group Standard "Snake Bite Treatment Specification" Interpretation. In Proceedings of the 2020 China Animal Injury Diagnosis and Treatment Summit, Zhengzhou, China; 2020; p. 35. [CrossRef]
3. Ying, H.; Jianfang, G.; Longhui, L.; Xiaomei, M.; Xiang, J. Age-related Variation in Snake Venom: Evidence from Two Snakes (*Naja atra* and *Deinagkistrodon acutus*) in Southeastern China. *Asian Herpetol. Res.* **2014**, *5*, 119–127. [CrossRef]
4. Wang, W.; Chen, Q.-F.; Yin, R.-X.; Zhu, J.-J.; Li, Q.-B.; Chang, H.-H.; Wu, Y.-B.; Michelson, E. Clinical features and treatment experience: A review of 292 Chinese cobra snakebites. *Environ. Toxicol. Pharmacol.* **2014**, *37*, 648–655. [CrossRef] [PubMed]
5. Machado Braga, J.R.; de Morais-Zani, K.; Pereira, D.D.S.; Sant'Anna, S.S.; da Costa Galizio, N.; Tanaka-Azevedo, A.M.; Gomes Vilarinho, A.R.; Rodrigues, J.L.; da Rocha, M.M.T. Sexual and ontogenetic variation of *Bothrops leucurus* venom. *Toxicon* **2020**, *184*, 127–135. [CrossRef] [PubMed]
6. Amorim, F.G.; Costa, T.R.; Baiwir, D.; De Pauw, E.; Quinton, L.; Sampaio, S.V. Proteopeptidomic, Functional and Immunoreactivity Characterization of *Bothrops moojeni* Snake Venom: Influence of Snake Gender on Venom Composition. *Toxins* **2018**, *10*, 177. [CrossRef] [PubMed]
7. Zelanis, A.; Menezes, M.C.; Kitano, E.S.; Liberato, T.; Tashima, A.K.; Pinto, A.F.; Sherman, N.E.; Ho, P.L.; Fox, J.W.; Serrano, S.M. Proteomic identification of gender molecular markers in *Bothrops jararaca* venom. *J. Proteom.* **2016**, *139*, 26–37. [CrossRef] [PubMed]
8. Mackessy, S.P.; Leroy, J.; Mociño-Deloya, E.; Setser, K.; Bryson, R.W.; Saviola, A.J. Venom Ontogeny in the Mexican Lance-Headed Rattlesnake (*Crotalus polystictus*). *Toxins* **2018**, *10*, 271. [CrossRef] [PubMed]
9. Castro, E.N.; Lomonte, B.; del Carmen Gutierrez, M.; Alagon, A.; Gutierrez, J.M. Intraspecies variation in the venom of the rattlesnake *Crotalus simus* from Mexico: Different expression of crotoxin results in highly variable toxicity in the venoms of three subspecies. *J. Proteom.* **2013**, *87*, 103–121. [CrossRef]
10. Zelanis, A.; Travaglia-Cardoso, S.R.; De Fátima Domingues Furtado, M. Ontogenetic changes in the venom of *Bothrops insularis* (Serpentes: Viperidae) and its biological implication. *S. Am. J. Herpetol.* **2008**, *3*, 43–50. [CrossRef]
11. Cipriani, V.; Debono, J.; Goldenberg, J.; Jackson, T.N.W.; Arbuckle, K.; Dobson, J.; Koludarov, I.; Li, B.; Hay, C.; Dunstan, N.; et al. Correlation between ontogenetic dietary shifts and venom variation in Australian brown snakes (Pseudonaja). *Comp. Biochem. Physiol. C Toxicol. Pharmacol.* **2017**, *197*, 53–60. [CrossRef]
12. Gibbs, H.L.; Sanz, L.; Chiucchi, J.E.; Farrell, T.M.; Calvete, J.J. Proteomic analysis of ontogenetic and diet-related changes in venom composition of juvenile and adult Dusky Pigmy rattlesnakes (*Sistrurus miliarius barbouri*). *J. Proteom.* **2011**, *74*, 2169–2179. [CrossRef]
13. Sasa, M. Diet and snake venom evolution: Can local selection alone explain intraspecific venom bvariation? *Toxicon* **1999**, *37*, 249–252. [PubMed]
14. Daltry, J.C.; Wüster, W.; Thorpe, R.S. Diet and snake venom evolution. *Nature* **1996**, *379*, 537–540. [CrossRef] [PubMed]
15. Chanda, A.; Patra, A.; Kalita, B.; Mukherjee, A.K. Proteomics analysis to compare the venom composition between *Naja naja* and *Naja kaouthia* from the same geographical location of eastern India: Correlation with pathophysiology of envenomation and immunological cross-reactivity towards commercial polyantivenom. *Expert Rev. Proteom.* **2018**, *15*, 949–961. [CrossRef]
16. Huang, H.-W.; Liu, B.-S.; Chien, K.-Y.; Chiang, L.-C.; Huang, S.-Y.; Sung, W.-C.; Wu, W.-G. Cobra venom proteome and glycome determined from individual snakes of *Naja atra* reveal medically important dynamic range and systematic geographic variation. *J. Proteom.* **2015**, *128*, 92–104. [CrossRef]
17. Alape-Girón, A.; Sanz, L.; Escolano, J.; Flores-Díaz, M.; Madrigal, M.; Sasa, M.; Calvete, J.J. Snake venomics of the lancehead pitviper *Bothrops asper* geographic, individual, and ontogenetic variations. *J. Proteome Res.* **2008**, *7*, 3556–3571. [CrossRef]
18. Warrell, D.A. Snake bite. *Lancet* **2010**, *375*, 77–88. [CrossRef]
19. Alangode, A.; Rajan, K.; Nair, B.G. Snake antivenom: Challenges and alternate approaches. *Biochem. Pharmacol.* **2020**, *181*, 114135. [CrossRef]
20. Liu, C.C.; You, C.H.; Wang, P.J.; Yu, J.S.; Huang, G.J.; Liu, C.H.; Hsieh, W.C.; Lin, C.C. Analysis of the efficacy of Taiwanese freeze-dried neurotoxic antivenom against *Naja kaouthia*, *Naja siamensis* and *Ophiophagus hannah* through proteomics and animal model approaches. *PLoS Negl. Trop. Dis.* **2017**, *11*, e0006138. [CrossRef]
21. Casewell, N.R.; Jackson, T.N.W.; Laustsen, A.H.; Sunagar, K. Causes and Consequences of Snake Venom Variation. *Trends Pharmacol. Sci.* **2020**, *41*, 570–581. [CrossRef]
22. Knudsen, C.; Laustsen, A.H. Recent Advances in Next Generation Snakebite Antivenoms. *Trop. Med. Infect. Dis.* **2018**, *3*, 42. [CrossRef]
23. Utkin, Y.N. Modern trends in animal venom research-omics and nanomaterials. *World J. Biol. Chem.* **2017**, *8*, 4–12. [CrossRef] [PubMed]
24. Suryamohan, K.; Krishnankutty, S.P.; Guillory, J.; Jevit, M.; Schröder, M.S.; Wu, M.; Kuriakose, B.; Mathew, O.K.; Perumal, R.C.; Koludarov, I.; et al. The Indian cobra reference genome and transcriptome enables comprehensive identification of venom toxins. *Nat. Genet.* **2020**, *52*, 106–117. [CrossRef] [PubMed]

25. Chanda, A.; Mukherjee, A.K. Quantitative proteomics to reveal the composition of Southern India spectacled cobra (*Naja naja*) venom and its immunological cross-reactivity towards commercial antivenom. *Int. J. Biol. Macromol.* **2020**, *160*, 224–232. [CrossRef]

26. Yap, M.K.K.; Fung, S.Y.; Tan, K.Y.; Tan, N.H. Proteomic characterization of venom of the medically important Southeast Asian Naja sumatrana (Equatorial spitting cobra). *Acta Trop.* **2014**, *133*, 15–25. [CrossRef] [PubMed]

27. Liu, B.-S.; Jiang, B.-R.; Hu, K.-C.; Liu, C.-H.; Hsieh, W.-C.; Lin, M.-H.; Sung, W.-C. Development of a Broad-Spectrum Antiserum against Cobra Venoms Using Recombinant Three-Finger Toxins. *Toxins* **2021**, *13*, 556. [CrossRef] [PubMed]

28. Kornisiuk, E.; Jerusalinsky, D.; Cerveñansky, C.; Harvey, A.L. Binding of muscarinic toxins MT × 1 and MT × 2 from the venom of the green mamba *Dendroaspis angusticeps* to cloned human muscarinic cholinoceptors. *Toxicon* **1995**, *33*, 11–18. [CrossRef]

29. Chien, K.Y.; Chiang, C.M.; Hseu, Y.C.; Vyas, A.A.; Rule, G.S.; Wu, W. Two distinct types of cardiotoxin as revealed by the structure and activity relationship of their interaction with zwitterionic phospholipid dispersions. *J. Biol. Chem.* **1994**, *269*, 14473–14483. [CrossRef]

30. Rajagopalan, N.; Pung, Y.F.; Zhu, Y.Z.; Wong, P.T.H.; Kumar, P.P.; Kini, R.M. β-cardiotoxin: A new three-finger toxin from *Ophiophagus hannah* (king cobra) venom with beta-blocker activity. *FASEB J. Off. Publ. Fed. Am. Soc. Exp. Biol.* **2007**, *21*, 3685–3695. [CrossRef]

31. Jang, J.-Y.; Kumar, T.K.S.; Jayaraman, G.; Yang, P.-W.; Yu, C. Comparison of the hemolytic activity and solution structures of two snake venom cardiotoxin analogues which only differ in their N-terminal amino acid. *Biochemistry* **1997**, *36*, 14635–14641. [CrossRef]

32. Chiou, S.H.; Raynor, R.L.; Zheng, B.; Chambers, T.C.; Kuo, J.F. Cobra venom cardiotoxin (cytotoxin) isoforms and neurotoxin: Comparative potency of protein kinase C inhibition and cancer cell cytotoxicity and modes of enzyme inhibition. *Biochemistry* **1993**, *32*, 2062–2067. [CrossRef]

33. Tans, G.; Rosing, J. Snake venom activators of factor X: An overview. *Pathophysiol. Haemost. Thromb.* **2001**, *31*, 225–233. [CrossRef] [PubMed]

34. Hsu, C.C.; Wu, W.B.; Huang, T.F. A snake venom metalloproteinase, kistomin, cleaves platelet glycoprotein VI and impairs platelet functions. *J. Thromb. Haemost.* **2008**, *6*, 1578–1585. [CrossRef] [PubMed]

35. Menezes, M.C.; Kitano, E.S.; Bauer, V.C.; Oliveira, A.K.; Cararo-Lopes, E.; Nishiyama, M.Y., Jr.; Zelanis, A.; Serrano, S.M.T. Early response of C2C12 myotubes to a sub-cytotoxic dose of hemorrhagic metalloproteinase HF3 from Bothrops jararaca venom. *J. Proteom.* **2019**, *198*, 163–176. [CrossRef] [PubMed]

36. Rau, J.C.; Beaulieu, L.M.; Huntington, J.A.; Church, F.C. Serpins in thrombosis, hemostasis and fibrinolysis. *J. Thromb. Haemost.* **2007**, *5*, 102–115. [CrossRef] [PubMed]

37. Sala, A.; Cabassi, C.S.; Santospirito, D.; Polverini, E.; Flisi, S.; Cavirani, S.; Taddei, S. Novel *Naja atra* cardiotoxin 1 (CTX-1) derived antimicrobial peptides with broad spectrum activity. *PLoS ONE* **2018**, *13*, e0190778. [CrossRef]

38. Pla, D.; Petras, D.; Saviola, A.J.; Modahl, C.M.; Sanz, L.; Pérez, A.; Juárez, E.; Frietze, S.; Dorrestein, P.C.; Mackessy, S.P.; et al. Transcriptomics-guided bottom-up and top-down venomics of neonate and adult specimens of the arboreal rear-fanged Brown Treesnake, *Boiga irregularis*, from Guam. *J. Proteom.* **2018**, *174*, 71–84. [CrossRef]

39. Xu, N.; Zhao, H.-Y.; Yin, Y.; Shen, S.-S.; Shan, L.-L.; Chen, C.-X.; Zhang, Y.-X.; Gao, J.-F.; Ji, X. Combined venomics, antivenomics and venom gland transcriptome analysis of the monocoled cobra (*Naja kaouthia*) from China. *J. Proteom.* **2017**, *159*, 19–31. [CrossRef]

40. Durban, J.; Sanz, L.; Trevisan-Silva, D.; Neri-Castro, E.; Alagón, A.; Calvete, J.J. Integrated Venomics and Venom Gland Transcriptome Analysis of Juvenile and Adult Mexican Rattlesnakes *Crotalus simus*, *C. tzabcan*, and *C. culminatus* Revealed miRNA-modulated Ontogenetic Shifts. *J. Proteome Res.* **2017**, *16*, 3370–3390. [CrossRef]

41. Zelanis, A.; Tashima, A.K.; Rocha, M.M.T.; Furtado, M.F.; Camargo, A.C.M.; Ho, P.L.; Serrano, S.M.T. Analysis of the Ontogenetic Variation in the Venom Proteome/Peptidome of *Bothrops jararaca* Reveals Different Strategies to Deal with Prey. *J. Proteome Res.* **2010**, *9*, 2278–2291. [CrossRef]

42. Durban, J.; Pérez, A.; Sanz, L.; Gómez, A.; Bonilla, F.; Rodríguez, S.; Chacón, D.; Sasa, M.; Angulo, Y.; Gutiérrez, J.M.; et al. Integrated "omics" profiling indicates that miRNAs are modulators of the ontogenetic venom composition shift in the Central American rattlesnake, *Crotalus simus simus*. *BMC Genom.* **2013**, *14*, 234. [CrossRef]

43. Bartel, D.P. MicroRNAs: Target recognition and regulatory functions. *Cell* **2009**, *136*, 215–233. [CrossRef]

44. Markland, F.S., Jr.; Swenson, S. Snake venom metalloproteinases. *Toxicon* **2013**, *62*, 3–18. [CrossRef]

45. Nicholson, J.; Mirtschin, P.; Madaras, F.; Venning, M.; Kokkinn, M. Digestive properties of the venom of the Australian Coastal Taipan, *Oxyuranus scutellatus* (Peters, 1867). *Toxicon* **2006**, *48*, 422–428. [CrossRef] [PubMed]

46. Holub, E.B. The arms race is ancient history in Arabidopsis, the wildflower. *Nat. Rev. Genet.* **2001**, *2*, 516–527. [CrossRef] [PubMed]

47. Bernardes, C.P.; Menaldo, D.L.; Zoccal, K.F.; Boldrini-França, J.; Peigneur, S.; Arantes, E.C.; Rosa, J.C.; Faccioli, L.H.; Tytgat, J.; Sampaio, S.V. First report on BaltCRP, a cysteine-rich secretory protein (CRISP) from *Bothrops alternatus* venom: Effects on potassium channels and inflammatory processes. *Int. J. Biol. Macromol.* **2019**, *140*, 556–567. [CrossRef] [PubMed]

48. Deka, A.; Sharma, M.; Mukhopadhyay, R.; Devi, A.; Doley, R. *Naja kaouthia* venom protein, Nk-CRISP, upregulates inflammatory gene expression in human macrophages. *Int. J. Biol. Macromol.* **2020**, *160*, 602–611. [CrossRef]

49. Sun, Q.-Y.; Chen, G.; Guo, H.; Chen, S.; Wang, W.-Y.; Xiong, Y.-L. Prolonged cardiac xenograft survival in guinea pig-to-rat model by a highly active cobra venom factor. *Toxicon* **2003**, *42*, 257–262. [CrossRef]
50. Ullah, A.; Ullah, K.; Ali, H.; Betzel, C.; ur Rehman, S. The Sequence and a Three-Dimensional Structural Analysis Reveal Substrate Specificity Among Snake Venom Phosphodiesterases. *Toxins* **2019**, *11*, 625. [CrossRef]
51. Jiang, Y.; Yan, L.; Lee, W.; Xu, X.; Zhang, Y.; Zhao, R.; Zhang, Y.; Wang, W. Venom gland transcriptomes of two elapid snakes (*Bungarus multicinctus* and *Naja atra*) and evolution of toxin genes. *BMC Genom.* **2011**, *12*, 1. [CrossRef]
52. Shan, L.-L.; Gao, J.-F.; Zhang, Y.-X.; Shen, S.-S.; He, Y.; Wang, J.; Ma, X.-M.; Ji, X. Proteomic characterization and comparison of venoms from two elapid snakes (*Bungarus multicinctus* and *Naja atra*) from China. *J. Proteom.* **2016**, *138*, 83–94. [CrossRef]
53. El-Aziz, T.M.A.; Soares, A.G.; Stockand, J.D. Snake Venoms in Drug Discovery: Valuable Therapeutic Tools for Life Saving. *Toxins* **2019**, *11*, 564. [CrossRef] [PubMed]
54. Tan, C.H.; Tan, K.Y.; Fung, S.Y.; Tan, N.H. Venom-gland transcriptome and venom proteome of the Malaysian king cobra (*Ophiophagus hannah*). *BMC Genom.* **2015**, *16*, 687. [CrossRef] [PubMed]
55. Koh, D.C.I.; Armugam, A.; Jeyaseelan, K. Snake venom components and their applications in biomedicine. *Cell. Mol. Life. Sci.* **2006**, *63*, 3030–3041. [CrossRef] [PubMed]
56. Ji, X.; Li, P. The IUCN Red List of Threatened Species 2014: E.T192109A2040894. Available online: https://doi.org/10.2305/IUCN.UK.2014-3.RLTS.T192109A2040894.en (accessed on 28 April 2019).
57. Mackessy, S.P.; Williams, K.; Ashton, K.G.; Lannoo, M.J. Ontogenetic Variation in Venom Composition and Diet of *Crotalus oreganus concolor*: A Case of Venom Paedomorphosis? *Copeia* **2003**, *2003*, 769–782. [CrossRef]
58. Modahl, C.M.; Doley, R.; Kini, R.M. Venom analysis of long-term captive Pakistan cobra (*Naja naja*) populations. *Toxicon* **2010**, *55*, 612–618. [CrossRef]
59. Amazonas, D.R.; Freitas-De-Sousa, L.A.; Orefice, D.P.; de Sousa, L.F.; Martinez, M.G.; Mourão, R.H.V.; Chalkidis, H.M.; Camargo, P.B.; Moura-Da-Silva, A.M. Evidence for Snake Venom Plasticity in a Long-Term Study with Individual Captive *Bothrops atrox*. *Toxins* **2019**, *11*, 294. [CrossRef]
60. De Farias, I.B.; de Morais-Zani, K.; Serino-Silva, C.; Sant'Anna, S.S.; da Rocha, M.M.; Grego, K.F.; Andrade-Silva, D.; Serrano, S.M.T.; Tanaka-Azevedo, A.M. Functional and proteomic comparison of *Bothrops jararaca* venom from captive specimens and the Brazilian Bothropic Reference Venom. *J. Proteom.* **2018**, *174*, 36–46. [CrossRef]
61. Freitas-de-Sousa, L.A.; Amazonas, D.R.; Sousa, L.F.; Sant'Anna, S.S.; Nishiyama, M.Y., Jr.; Serrano, S.M.T.; Junqueira-de-Azevedo, I.L.M.; Chalkidis, H.M.; Moura-da-Silva, A.M.; Mourão, R.H.V. Comparison of venoms from wild and long-term captive *Bothrops atrox* snakes and characterization of Batroxrhagin, the predominant class PIII metalloproteinase from the venom of this species. *Biochimie* **2015**, *118*, 60–70. [CrossRef]
62. Saad, E.; Curtolo Barros, L.; Biscola, N.; Pimenta, D.C.; Barraviera, S.R.C.S.; Barraviera, B.; Seabra Ferreira, R.S., Jr. Intraspecific variation of biological activities in venoms from wild and captive Bothrops jararaca. *J. Toxicol. Environ. Health. Part A* **2012**, *75*, 1081–1090. [CrossRef]
63. Li, S.; Wang, J.; Zhang, X.; Ren, Y.; Wang, N.; Zhao, K.; Chen, X.; Zhao, C.; Li, X.; Shao, J.; et al. Proteomic characterization of two snake venoms: *Naja naja atra* and *Agkistrodon halys*. *Biochem. J.* **2004**, *384*, 119–127. [CrossRef]
64. Vonk, F.J.; Casewell, N.R.; Henkel, C.V.; Heimberg, A.M.; Jansen, H.J.; McCleary, R.J.R.; Kerkkamp, H.M.E.; Vos, R.A.; Guerreiro, I.; Calvete, J.J.; et al. The king cobra genome reveals dynamic gene evolution and adaptation in the snake venom system. *Proc. Natl. Acad. Sci. USA* **2013**, *110*, 20651–20656. [CrossRef] [PubMed]
65. Kochva, E. The origin of snakes and evolution of the venom apparatus. *Toxicon* **1987**, *25*, 65–106. [CrossRef]
66. Post, Y.; Puschhof, J.; Beumer, J.; Kerkkamp, H.M.; de Bakker, M.A.G.; Slagboom, J.; de Barbanson, B.; Wevers, N.R.; Spijkers, X.M.; Olivier, T.; et al. Snake Venom Gland Organoids. *Cell* **2020**, *180*, 233–247.e21. [CrossRef] [PubMed]
67. Yamanouye, N.; Britto, L.R.; Carneiro, S.M.; Markus, R.P. Control of venom production and secretion by sympathetic outflow in the snake *Bothrops jararaca*. *J. Exp. Biol.* **1997**, *200*, 2547–2556. [CrossRef]
68. Chey, W.Y. Regulation of pancreatic exocrine secretion. *Int. J. Pancreatol. Off. J. Int. Assoc. Pancreatol.* **1991**, *9*, 7–20. [CrossRef]
69. Chandra, R.; Liddle, R.A. Modulation of pancreatic exocrine and endocrine secretion. *Curr. Opin. Gastroenterol.* **2013**, *29*, 517–522. [CrossRef]
70. Beelman, C.A.; Parker, R. Degradation of mRNA in eukaryotes. *Cell* **1995**, *81*, 179–183. [CrossRef]
71. Giorgianni, M.W.; Dowell, N.L.; Griffin, S.; Kassner, V.A.; Selegue, J.E.; Carroll, S.B. The origin and diversification of a novel protein family in venomous snakes. *Proc. Natl. Acad. Sci. USA* **2020**, *117*, 10911–10920. [CrossRef]
72. Bayona-Serrano, J.D.; Viala, V.L.; Rautsaw, R.M.; Schramer, T.D.; Barros-Carvalho, G.A.; Nishiyama, M.Y.; Freitas-de-Sousa, L.A.; Moura-da-Silva, A.M.; Parkinson, C.L.; Grazziotin, F.G.; et al. Replacement and Parallel Simplification of Nonhomologous Proteinases Maintain Venom Phenotypes in Rear-Fanged Snakes. *Mol. Biol. Evol.* **2020**, *37*, 3563–3575. [CrossRef]
73. Fry, B.G. From genome to "venome": Molecular origin and evolution of the snake venom proteome inferred from phylogenetic analysis of toxin sequences and related body proteins. *Genome Res.* **2005**, *15*, 403–420. [CrossRef]
74. Fry, B.G.; Wüster, W. Assembling an Arsenal: Origin and Evolution of the Snake Venom Proteome Inferred from Phylogenetic Analysis of Toxin Sequences. *Mol. Biol. Evol.* **2004**, *21*, 870–883. [CrossRef] [PubMed]
75. Zelanis, A.; Tashima, A.K. Unraveling snake venom complexity with 'omics' approaches: Challenges and perspectives. *Toxicon* **2014**, *87*, 131–134. [CrossRef] [PubMed]

76. Laustsen, A.H.; Engmark, M.; Milbo, C.; Johannesen, J.; Lomonte, B.; Gutiérrez, J.M.; Lohse, B. From Fangs to Pharmacology: The Future of Snakebite Envenoming Therapy. *Curr. Pharm. Des.* **2016**, *22*, 5270–5293. [CrossRef] [PubMed]
77. Garcia, E.S.; Guimarães, J.A.; Prado, J.L. Purification and characterization of a sulfhydryl-dependent protease from Rhodnius prolixus midgut. *Arch. Biochem. Biophys.* **1978**, *188*, 315–322. [CrossRef]
78. Habermann, E.; Hardt, K.L. A sensitive and specific plate test for the quantitation of phospholipases. *Anal. Biochem.* **1972**, *50*, 163–173. [CrossRef]
79. Kishimoto, M.; Takahashi, T. A spectrophotometric microplate assay for L-amino acid oxidase. *Anal. Biochem.* **2001**, *298*, 136–139. [CrossRef] [PubMed]
80. Ouyang, C.; Huang, T.-F. Inhibition of platelet aggregation by 5′-nucleotidase purified from *Trimeresurus gramineus* snake venom. *Toxicon* **1983**, *21*, 491–501. [CrossRef]
81. Ellman, G.L.; Courtney, K.D.; Andres, V., Jr.; Feather-Stone, R.M. A new and rapid colorimetric determination of acetyl-cholinesterase activity. *Biochem. Pharmacol.* **1961**, *7*, 88–95. [CrossRef]

Article

In Vitro Efficacy of Antivenom and Varespladib in Neutralising Chinese Russell's Viper (*Daboia siamensis*) Venom Toxicity

Mimi Lay [1], Qing Liang [1,2], Geoffrey K. Isbister [1,3] and Wayne C. Hodgson [1,*]

1 Monash Venom Group, Department of Pharmacology, Biomedical Discovery Institute, Monash University, Clayton, VIC 3800, Australia
2 Department of Emergency Medicine, The First Affiliated Hospital of Guangzhou Medical University, 151 Yanjiang Rd, Guangzhou 510120, China
3 Clinical Toxicology Research Group, University of Newcastle, Callaghan, NSW 2308, Australia
* Correspondence: wayne.hodgson@monash.edu

Abstract: The venom of the Russell's viper (*Daboia siamensis*) contains neurotoxic and myotoxic phospholipase A_2 toxins which can cause irreversible damage to motor nerve terminals. Due to the time delay between envenoming and antivenom administration, antivenoms may have limited efficacy against some of these venom components. Hence, there is a need for adjunct treatments to circumvent these limitations. In this study, we examined the efficacy of Chinese *D. siamensis* antivenom alone, and in combination with a PLA_2 inhibitor, Varespladib, in reversing the in vitro neuromuscular blockade in the chick biventer cervicis nerve-muscle preparation. Pre-synaptic neurotoxicity and myotoxicity were not reversed by the addition of Chinese *D. siamensis* antivenom 30 or 60 min after venom (10 µg/mL). The prior addition of Varespladib prevented the neurotoxic and myotoxic activity of venom (10 µg/mL) and was also able to prevent further reductions in neuromuscular block and muscle twitches when added 60 min after venom. The addition of the combination of Varespladib and antivenom 60 min after venom failed to produce further improvements than Varespladib alone. This demonstrates that the window of time in which antivenom remains effective is relatively short compared to Varespladib and small-molecule inhibitors may be effective in abrogating some activities of Chinese *D. siamensis* venom.

Keywords: neurotoxicity; myotoxicity; Russell's viper; *Daboia siamensis*; Varespladib; antivenom

Key Contribution: This study provides further evidence that antivenoms are unlikely to reverse already established pre-synaptic neurotoxicity and myotoxicity caused by Chinese *D. siamensis* venom. Varespladib, a specific PLA_2 inhibitor, can both prevent and delay in vitro PLA_2-mediated pre-synaptic neurotoxicity, supporting the further use of specific toxin inhibitors against venom toxicity.

Citation: Lay, M.; Liang, Q.; Isbister, G.K.; Hodgson, W.C. In Vitro Efficacy of Antivenom and Varespladib in Neutralising Chinese Russell's Viper (*Daboia siamensis*) Venom Toxicity. *Toxins* **2023**, *15*, 62. https://doi.org/10.3390/toxins15010062

Received: 19 December 2022
Revised: 8 January 2023
Accepted: 9 January 2023
Published: 11 January 2023

1. Introduction

The Russell's viper (*Daboia siamensis*) has a vast but disjointed distribution across many regions of South and South-East Asia. It is found in parts of Pakistan, India, Bangladesh and Sri Lanka and, further east, in Taiwan, Thailand, Indonesia and Mainland China [1,2]. Across this distribution, *D. siamensis* is responsible for a large number of envenomings, which can result in marked local and systemic injuries in envenomed humans [3–7]. Although data regarding the incidence of bites in China are not available, clinical anecdotes suggest that they are relatively common. Combined with a lack of standardised and effective antivenom treatment, the overall fatality and disability rate is concerning [8]. Typically, envenoming by *Daboia* species is characterised by haemostatic disturbances, systemic bleeding, neurotoxicity, acute kidney injury, muscle damage, ecchymosis, swelling and pain [1,4,6–13].

Most viper venoms are dominated by three major protein families, including snake-venom metalloproteases (SVMPs), snake-venom serine proteases (SVSPs) and phospholipase A_2 (PLA$_2$) toxins [14]. Accordingly, these families represent more than 60% of all toxins found in viper venoms [14]. Of major interest are snake venom PLA$_2$ toxins, as they have a critical role in the early morbidity and mortality of snake-bite victims. PLA$_2$ toxins are a large family of enzymatic proteins that exert a wide variety of pharmacological and biochemical effects and have been isolated from both elapid and viperid species [15–20]. Those present in elapids are categorised within group I, whereas those found in Viperidae venoms are within group II [21,22]. Such toxins have been reported to be responsible for neurotoxicity (i.e., taipoxin from *Oxyuranus scutellatus*, Coastal taipan) [16,23], myotoxicity (i.e., crotoxin from *Crotalus durissus terrificus*, South American rattlesnake) [24,25], pain and oedema, cardiovascular disturbances, haemolysis and coagulation disorders, amongst other effects [18,26–30].

Proteomic studies of *Daboia* species have shown that the venoms contain large amounts of PLA$_2$ that display pre-synaptic neurotoxicity, myotoxicity or cytotoxicity in different models [5,8,12–14,18,31]. Previous biochemically characterised pre-synaptic PLA$_2$ toxins from *Daboia* species include Daboiatoxin (Myanmar), Viperitoxin F (Taiwan) and Russtoxin (Thailand) [31–33]. Our laboratory has previously characterised the major PLA$_2$ toxins, i.e., U1-viperitoxin-Dr1a and U1-viperitoxin-Dr1b, from Sri Lankan Russell's viper venom, which are postulated to be responsible for neurotoxicity and mild myotoxicity in envenomed patients [6,18]. While envenoming by Chinese *D. siamensis* has not been reported to produce clinically evident neurotoxicity, we have shown the venom displays pre-synaptic neurotoxicity and myotoxicity in the chick biventer cervicis nerve-muscle preparation [34]. These activities of the venoms from *Daboia* species have been attributed to the presence of PLA$_2$s, both clinically and in vitro [13,15,18,32,35–38].

Antivenoms are the main treatment for systemic envenoming in humans. However, challenges remain with the use of antivenoms, including poor efficacy against some local venom effects and the presence of immunoglobulins that do not bind all toxins. In addition, the need to administer antivenoms in healthcare facilities, the risk of adverse reactions and cold-storage requirements and supply chain issues, especially in rural/remote areas where snake bites are often widespread, remain [30,39,40]. Unless given early, antivenoms are generally considered ineffective against some effects of snake-bite envenoming, given that the effectiveness of antibodies is limited by their penetrability into peripheral sites [41]. In the case of venoms containing pre-synaptic neurotoxins, neuromuscular paralysis is generally irreversible, as these toxins result in damage to the nerve terminal after exposure to venom [23,42]. As a result, treatment with antivenoms or acetylcholinesterase inhibitors often has limited effectiveness [41]. This has been reported clinically [6,43] and replicated in studies using ex vivo neuromuscular junctions [16,36,42].

The high specificity of polyclonal antibodies means that antivenoms are normally highly efficacious against the venom for which they are raised, but generally display less efficacy against snake venoms of different genera or families [44]. There is currently no specific antivenom, commercially available, for envenoming by Chinese *D. siamensis*. Fortunately, a species-specific monovalent antivenom has been developed, which we have shown to be efficacious in preventing in vitro neurotoxic and myotoxic effects [34]. However, studies testing the efficacy of this antivenom in reversing venom-mediated effects and examining the window of time in which antivenom remains effective are required. Given that *D. siamensis* venom appears to contain pre-synaptic neurotoxins and myotoxins, both of which are considered clinically and experimentally difficult to reverse, the exploration of an alternative or a complementary treatment for *D. siamensis* envenoming is important.

As PLA$_2$ toxins are functionally critical and ubiquitous in many venoms, targeting this toxin class is an ideal target for treating many aspects of snake envenoming [14,19,40]. In this context, small-molecule inhibitors have been gaining attention as potential alternatives to immuno-based therapies due to their promising safety profile and general inhibitory broadness across snake species. Most notable is Varespladib (or LY315920), a PLA_2 inhibitor that shows pre-clinical efficacy against a wide variety of elapid and viper venoms [44,45]. Varespladib demonstrates inhibitory activity against a variety of snake venoms and toxins, including against the 'irreversible' pre-synaptic neurotoxins and myotoxins [23,45–47]. Lewin et al. [45] demonstrated that Varespladib and methyl-Varespladib had a wide breadth of efficacy against the venoms from 28 different snake species, all with variable PLA_2 activity. Additionally, the anti-pre-synaptic neurotoxic capability of Varespladib has been demonstrated against isolated toxins, such as taipoxin, in an in vitro neuromuscular preparation [23]. It has also been observed that Varespladib extends the window of time for the prevention or reversal of neuromuscular paralysis, muscle damage or venom-induced lethality in mice [23,48,49].

Given that PLA_2 toxins are present in *D. siamensis* venom, and there is currently not a specific antivenom available for clinical use, we examined the efficacy of Varespladib to inhibit PLA_2 toxicity of *D. siamensis* venom. We also compared the ability of Chinese *D. siamensis* monovalent antivenom to reverse the toxicity of Chinese *D. siamensis* venom alone and in combination with Varespladib.

2. Results

2.1. In Vitro Antivenom Reversal Studies

2.1.1. Effect of Chinese *D. siamensis* Monovalent Antivenom on Pre-Synaptic Neurotoxicity Induced by *D. siamensis* Venom

Chinese *D. siamensis* venom (10 µg/mL) in the absence of antivenom caused a reduction in indirect (nerve-mediated) twitches in the chick biventer cervicis nerve-muscle preparation, with a time to reach 50% twitch inhibition (i.e., t_{50} value) of 147 ± 12 min ($n = 11$–12, Figure 1a,b). Post-venom contractile responses to exogenous ACh, CCh or KCl were obtained to identify the site of neurotoxic activity, and were unaffected by the venom when compared to vehicle control, indicative of pre-synaptic neurotoxicity (Figure 1c).

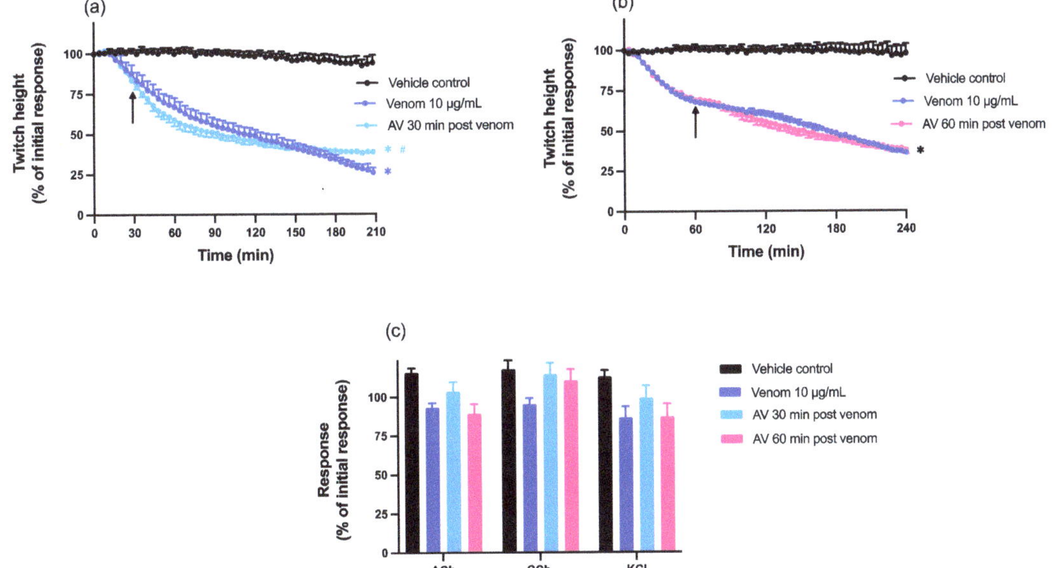

Figure 1. Effect of post-venom addition of antivenom (AV; 15 µL) on venom (10 µg/mL) induced pre-synaptic neurotoxicity in the chick biventer cervicis nerve-muscle preparation. Effect of AV added (**a**) 30 or (**b**) 60 min after venom on indirect twitch inhibition. Arrows indicate the times of addition

of AV; n = 5–6; * $p < 0.05$, significantly different from vehicle control at corresponding time point; # $p < 0.05$, significantly different from venom alone at corresponding time point, one-way ANOVA followed by Bonferroni's post hoc test. Effect of (**c**) vehicle, venom alone or AV added either 30 min or 60 min after venom on responses to exogenous agonists; ACh (1 mM), CCh (20 μM) and KCl (40 mM). n = 11–12, where n is the number of preparations from different animals; error bars indicate standard error of the mean.

Chinese *D. siamensis* antivenom (15 μL; 3x recommended concentration; [33]) added either 30 or 60 min after venom did not have a marked effect on the inhibitory action of the venom (Figure 1a,b). There was a slight reduction in twitch inhibition after 3 h when antivenom was added at the 30 min timepoint (Figure 1a). Antivenom added 30 or 60 min after venom had no significant effect on the agonist responses in the presence of venom (Figure 1c).

2.1.2. Effect of Chinese *D. siamensis* Monovalent Antivenom on Myotoxicity Induced by *D. siamensis* Venom

Chinese *D. siamensis* venom (10 μg/mL) caused a reduction in direct twitches in the chick biventer cervicis nerve-muscle preparation, with a time to reach 50% twitch inhibition (i.e., t_{50} value) of 115 ± 4 min (Figure 2a,c). Chinese *D. siamensis* venom (10 μg/mL) also caused an increase in resting baseline tension (Figure 2b,d), indicative of myotoxicity.

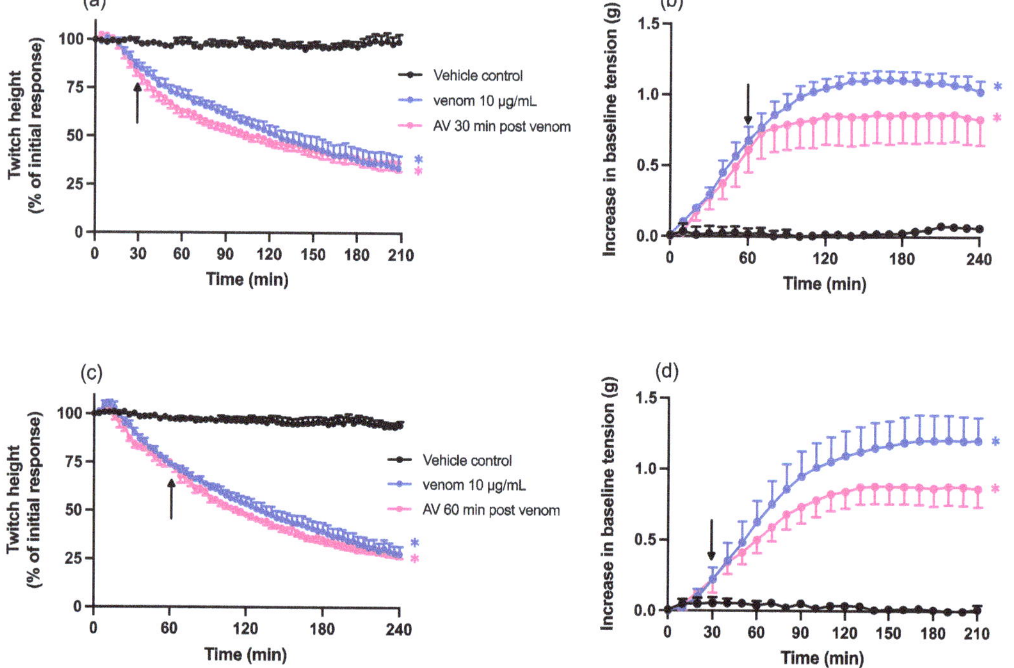

Figure 2. Effect of post-venom addition of antivenom (AV; 20 μL) on venom (10 μg/mL) induced myotoxicity in the chick biventer cervicis nerve-muscle preparation. Effect of AV added (**a**) 30 or (**c**) 60 min after venom on direct twitch inhibition. Arrows indicate the times of addition of AV; * $p < 0.05$, significantly different from vehicle control at corresponding time point. Effect of AV added at (**b**) 30 or (**d**) 60 min after venom on baseline tension, where * $p < 0.05$, significantly different from vehicle response, all one-way ANOVA followed by Bonferroni's post hoc test. n = 4–6, where n is the number of preparations from different animals; error bars indicate standard error of the mean.

Chinese *D. siamensis* antivenom (20 μL; 4x recommended concentration; [33]) added either 30 or 60 min after venom did not restore twitch height (Figure 2a,c) but did cause a slight reduction in the increase in baseline tension, at both time points, although this effect was not statistically significant compared to venom in the absence of antivenom (Figure 2b,d).

2.2. Phospholipase A₂ Assay

The sPLA$_2$ activity of Chinese *D. siamensis* venom was 865 ± 158 μmol/min/mg. The PLA$_2$ activity of venom was abolished by Varespladib (20 μM), i.e., undetectable when incubated with the inhibitor. Bee venom was used as a positive control and displayed an activity of 554 ± 6 μmol/min/mg.

2.3. In Vitro Protection Studies Using Varespladib
2.3.1. Effect of Varespladib on the Pre-Synaptic Neurotoxicity of *D. siamensis* Venom

Pre-incubation (15 min) of Chinese *D. siamensis* venom (10 μg/mL) with Varespladib (0.8 or 26 μM), before addition to the organ bath, prevented the effects of venom, i.e., a reduction in indirect twitches in the chick biventer cervicis nerve-muscle preparation (Figure 3a). Varespladib alone (0.8 or 26 μM) had no significant effect on twitch response or contractile responses to agonists (Figure 3a,b), although the combination of venom and Varespladib slightly potentiated contractile responses to ACh and CCh, but not KCl (Figure 3b).

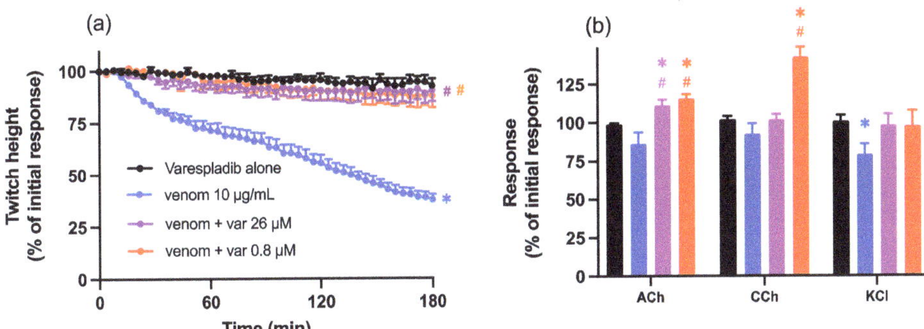

Figure 3. Effect of venom (10 μg/mL) in the presence or absence of Varespladib (0.8 μM or 26 μM) on (**a**) indirect (nerve-mediated) twitches and (**b**) contractile responses to exogenous agonists ACh (1 mM), CCh (20 μM) and KCl (40 mM) in the chick biventer cervicis nerve-muscle preparation. (**a**) Protective effect of Varespladib on venom-induced pre-synaptic neurotoxicity; * $p < 0.05$, significantly different to Varespladib alone; # $p < 0.05$, significantly different to venom in the absence of Varespladib, two-way ANOVA followed by Tukey's post hoc test. (**b**) Effect of Varespladib on responses to exogenous agonists; * $p < 0.05$, significantly different from pre-venom response to the same agonists, student's paired *t*-test; # $p < 0.05$, significantly different from venom alone, two-way ANOVA, followed by Tukey's post hoc test, $n = 5$–6, where n is the number of preparations from different animals; error bars indicate standard error of the mean.

2.3.2. Effect of Varespladib on the Myotoxicity of *D. siamensis* Venom

Chinese *D. siamensis* venom (10 μg/mL) caused a reduction in direct twitches in the chick biventer cervicis nerve-muscle preparation. Pre-incubation (15 min) of venom with Varespladib (0.8 or 26 μM) significantly inhibited the reduction in direct twitch height mediated by Chinese *D. siamensis* venom, with slightly lower neutralisation by the lower concentration of Varespladib (Figure 4a). Varespladib also abolished the increase in baseline tension when compared to venom alone, with no concentration-dependent difference between the effects of the two concentrations (Figure 4b).

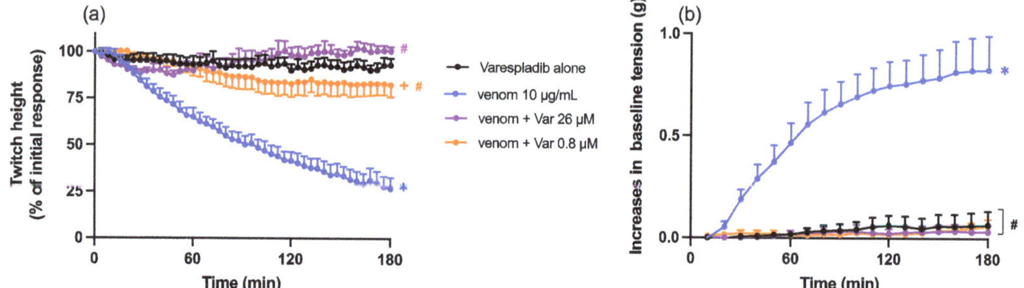

Figure 4. Effect of Chinese *D. siamensis* (10 µg/mL) venom in the presence or absence of Varespladib (0.8 or 26 µM) on (**a**) direct twitches and (**b**) baseline tension in the chick biventer cervicis nerve-muscle preparation. (**a**) Protective effect of Varespladib against venom-induced myotoxicity; * $p < 0.05$, significantly different to Varespladib alone; # $p < 0.05$, significantly different to venom in the absence of Varespladib; + $p < 0.05$, significantly different from venom + Var 26 µM; two-way ANOVA followed by Tukey's post hoc test. (**b**) Effect of Varespladib on venom-induced increases in baseline tension; * $p < 0.05$, significantly different from Varespladib alone; two-way ANOVA followed by Tukey's post hoc test; # $p < 0.05$, significantly different to venom in the absence of Varespladib $n = 5$–6, where n is the number of preparations from different animals; error bars indicate standard error of the mean.

2.4. In Vitro Reversal Studies by Varespladib

2.4.1. Effect of Post-Venom Addition of Varespladib on the Pre-Synaptic Neurotoxicity of *D. siamensis* Venom

To determine the ability of Varespladib to reverse the neurotoxic effects of Chinese *D. siamensis* venom, Varespladib (0.8 or 26 µM) was added 60 min after venom. Varespladib was able to partially restore the venom-mediated reduction in indirect twitches at both concentrations (Figure 5). This effect was transient, but Varespladib also prevented a further decline in twitch height.

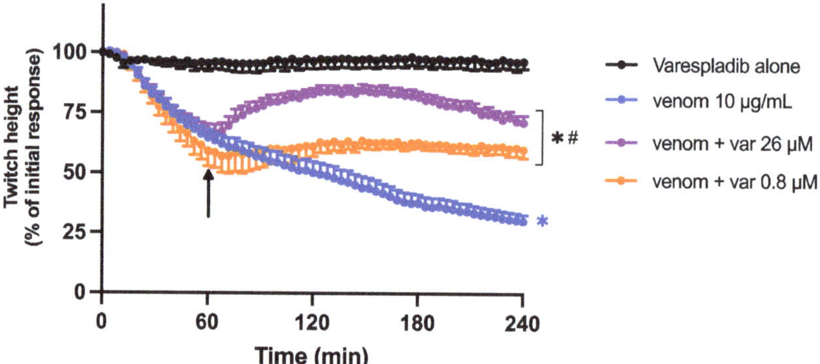

Figure 5. Effect of the addition of Varespladib (0.8 or 26 µM), 60 min post venom, on venom-induced pre-synaptic neurotoxicity (indirect twitches) in the chick biventer cervicis nerve-muscle preparation. * $p < 0.05$, significantly different to Varespladib alone; # $p < 0.05$, significantly different to venom in the absence of Varespladib; two-way ANOVA followed by Tukey's post hoc test. $n = 5$–6, where n is the number of preparations from different animals; error bars indicate standard error of the mean.

2.4.2. Effect of Post-Venom Addition of Varespladib on the Myotoxicity of *D. siamensis* Venom

Varespladib (26 µM), added 60 min after Chinese *D. siamensis* venom, prevented a further decline in direct twitches compared to venom alone (Figure 6a). However, a lower concentration of Varespladib (0.8 µM) did not cause a statistically significant delay in direct twitch inhibition. Both concentrations of Varespladib (i.e., 0.8 and 26 μM) reversed the venom-induced increase in baseline tension (Figure 6b).

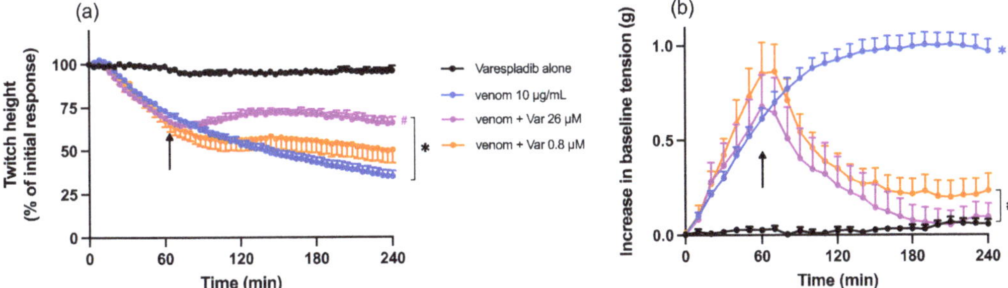

Figure 6. Effect of the addition of Varespladib (0.8 or 26 μM), 60 min post venom, on venom-induced myotoxicity. Arrows indicate the time of addition of Varespladib. (**a**) Effect of Varespladib on direct twitch reduction induced by venom; * $p < 0.05$, significantly different to Varespladib alone; # $p < 0.05$, significantly different to venom alone, two-way ANOVA followed by Tukey's post hoc test. (**b**) Effect of Varespladib on venom-induced increases of baseline tension; * $p < 0.05$, significantly different from Varespladib alone; # $p < 0.05$, significantly different to venom in the absence of Varespladib, two-way ANOVA followed by Tukey's post hoc. $n = 4$–6 where n is the number of preparations from different animals; error bars indicate standard error of the mean.

2.4.3. Effect of Post-Venom Addition of the Combination of Varespladib and Antivenom on the Pre-Synaptic Neurotoxicity of *D. siamensis* Venom

The combination of the lower concentration of Varespladib (0.8 μM) with Chinese *D. siamensis* antivenom (15 μL) prevented a further decline in twitch height when added 60 min post-venom (Figure 7). However, this effect was not significantly different from Varespladib (0.8 μM) alone, indicating no synergistic effect of the combination.

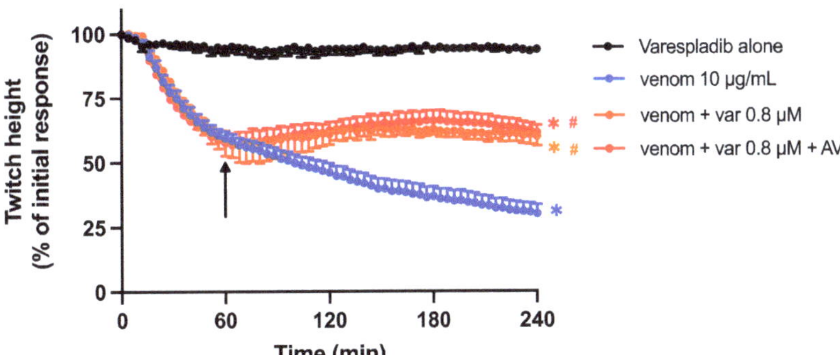

Figure 7. Effect of the addition of antivenom (AV; 15 μL) and Varespladib (0.8 μM), 60 min post venom, on the pre-synaptic neurotoxic effects of venom (10 μg/mL). Arrows indicate the time of addition of combined treatment. * $p < 0.05$, significantly different to Varespladib alone; # $p < 0.05$, significantly different to venom in the absence of treatment, one-way ANOVA followed by Bonferroni's post hoc test. $n = 4$–6 where n is the number of preparations from different animals; error bars indicate standard error of the mean.

2.4.4. Effect of Post-Venom Addition of the Combination of Varespladib and Antivenom on the Myotoxicity of *D. siamensis* Venom

The combination of the lower concentration of Varespladib (0.8 μM) with Chinese *D. siamensis* antivenom (20 μL) had a small but significant effect on venom-induced twitch inhibition (Figure 8a). However, this effect was not significantly different from Varespladib

(0.8 μM) alone. The combination of Varespladib (0.8 μM) and Chinese *D. siamensis* antivenom (20 μL) also reversed the increase in baseline tension induced by the venom, but this was not different from Varespladib (0.8 μM) alone, indicating no synergistic effect of the combination (Figure 8b).

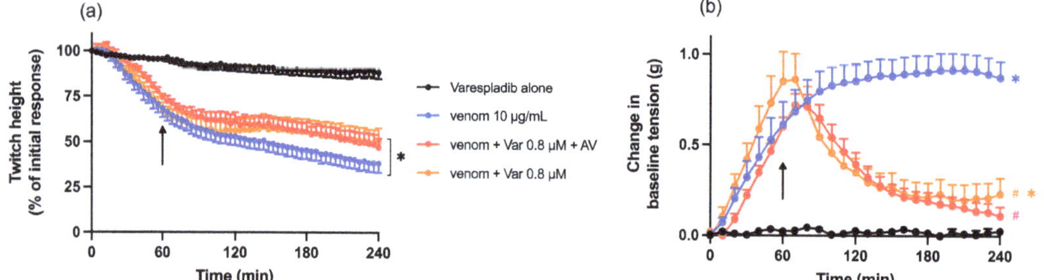

Figure 8. Effect of the addition of antivenom (AV; 20 μL) and Varespladib (0.8 μM), 60 min post venom, on the myotoxic effects of venom (10 μg/mL). Arrows indicate the time of addition of combined treatment. (**a**) Effect on direct twitches; * $p < 0.05$, significantly different to Varespladib alone; one-way ANOVA followed by Bonferroni's post hoc test. (**b**) Effect on baseline tension; * $p < 0.05$, significantly different from vehicle response; # $p < 0.05$, significantly different from venom alone; all one-way ANOVA followed by Bonferroni's post hoc test. $n = 4–6$ where n is the number of preparations from different animals; error bars indicate standard error of the mean.

3. Discussion

We previously reported that the in vitro pre-synaptic neurotoxic and myotoxic effects of Chinese *D. siamensis* venom are neutralised (i.e., prevented) by the pre-addition of specific Chinese *D. siamensis* monovalent antivenom, indicating the efficacy of this antivenom [34]. However, as these observations would not readily translate into clinical effectiveness, in the current study, we examined the ability of Chinese *D. siamensis* monovalent antivenom to prevent a further decline in and/or reverse the already established neurotoxicity and myotoxicity. We found that the antivenom was largely ineffective in preventing neuromuscular blockade when added 30 min or 60 min after venom. Additionally, the antivenom did not reverse the myotoxic effects when added at either time point. Therefore, the antivenom was unable to reverse the effects of the venom and prevent an ongoing decline in neuromuscular transmission, caused either by the action of neurotoxins (i.e., preventing neurotransmitter release) or myotoxins (i.e., damaging skeletal muscle). The limited ability of antivenoms to reverse pre-synaptic neurotoxicity and myotoxicity has been well established, both in clinical and experimental settings [6,18,23,42,43]. The lack of effectiveness of the antivenom, administered post-venom, indicates that the antibodies contained in the antivenom are unable to access the toxins, causing neuromuscular failure/damage, and/or that the effects of the 'irreversible' damage are likely to be occurring internally [6,19,20,36,41].

In agreement with our findings, an in vitro study on *Naja naja* (Indian Cobra) venom-induced muscle injury indicated a very short lag period before cytotoxins and myotoxins exerted irreversible injury, with the failure of Indian polyvalent antivenom to prevent further damage, even after 5 min [50]. The window of time in which antivenom remains effective, at least in in vitro experiments, such as these, may be very short. Indeed, we have shown that Australian polyvalent antivenom remains effective in neutralising *O. scutellatus* (Coastal taipan) venom or taipoxin-mediated pre-synaptic neurotoxicity only up to 10–15 min after addition of the venom/toxin [16,42]. This highlights the need for early antivenom administration for both neurotoxicity and myotoxicity, as has been described in case series where the severity of envenoming by various Australian elapids is largely reduced with antivenom administration within 2 to 6 h [51,52]. It has been suggested that local vascular alterations, i.e., haemorrhage and oedema, associated with local tissue and muscle damage, which are hallmarks of many viperid envenomings, including

D. siamensis, can affect the pharmacokinetics of antivenoms. Irrespective of molecular masses of either IgG or F(ab')^2 fragments, these types of venom-induced injuries favour the extravasation of antibodies and may contribute to the difficulty in neutralising or reversing locally acting myotoxins [53–55]. The irreversibility of the pathophysiological effects of some toxins limits the effectiveness of antivenoms, although this is primarily due to the nature of the venoms/toxins, rather than an issue with the efficacy of the antivenoms per se [53,56]. Apart from endocytosed toxins with an intracellular site of action, toxins that target membrane-bound receptors (e.g., neurotoxins), located in the extracellular matrix (e.g., metalloproteases) or in the bloodstream (e.g., pro-coagulants), have a higher probability of binding with circulating antibodies [53]. Moreover, the mismatch between low-molecular-weight toxins and antivenom pharmaco-kinetics and -dynamics means that antivenoms are ineffective at neutralizing both locally acting toxins (e.g., myotoxins and cytotoxins) and those with an intracellular site of action [53]. This can be explained by the fact that by the time antibodies and fragments reach the site of action, significant tissue damage has occurred. Antibodies cannot cross the blood/tissue barrier, as previously mentioned, when treating local and microvascular damage, while also having reduced antibody binding and neutralisation interactions [22,41].

Immunotherapy of snake envenoming presents a difficult challenge; hence, it has been proposed that the addition of potent enzyme inhibitors may circumvent the problem of poor antivenom efficacy, especially against poorly immunogenic toxins with low molecular weight and faster absorption rate [22]. In our study, we found that pre-incubation with Varespladib prevented both the neurotoxic and myotoxic effects of *D. siamensis* venom, indicating the key role PLA$_2$ toxins play in these effects. Our findings support previous reports showing that Varespladib prevents PLA$_2$-mediated neurotoxicity and myotoxicity across a diverse range of snake venoms [23–25,45,47,48,57–60]. The important role of PLA$_2$ toxins in *Daboia spp.* envenoming is also supported by our previous work, showing that removal of the major PLA$_2$ toxins, i.e., U1-viperitoxin-Dr1a and U1-viperitoxin-Dr1b, from Sri-Lankan *D. russelii* venom resulted in a loss of in vitro neurotoxicity and myotoxicity [18,37].

Interestingly, Varespladib was also able to significantly reverse or, at least, prevent a further decline in indirect (nerve-mediated) and direct (muscle) twitches induced by venom, whereas antivenom did not display this ability. However, the administration of a combination of antivenom and Varespladib after venom addition did not result in an additive or synergistic inhibitory effect against the neurotoxic or myotoxic effects. A similar finding of greater efficacy of Varespladib over antivenom has also been reported against *O. scutellatus* venom, an elapid venom with PLA$_2$-mediated neurotoxicity [23], and *C. d. terrificus* venom, from a viperid with PLA$_2$-mediated neurotoxicity and myotoxicity [24,25]. These observations support in vivo experiments in mice, where Varespladib was able to inhibit the myotoxic effects of four Chinese snake venoms [44] as well as attenuate the myonecrotic and haemorrhagic activity induced by *Deinagkistrodon acutus* (sharp-nosed pit viper) venom [46]. In contrast, the concomitant administration of antivenom and Varespladib appeared to provide additional survival benefits following *O. scutellatus* [49], as well as *Micrurus corallinus* (Coral snake) envenoming [60] in mice and rats, respectively.

The positive effects of Varespladib raise the question of whether the morphological changes to the neuromuscular junction caused by pre-synaptic neurotoxins are, indeed, irreversible, or at least partially reversible in the time frame of the current study. The reversal studies demonstrate the inability of antivenom, both alone and in combination with Varespladib, to reverse/further prevent *D. siamensis* venom-mediated neurotoxicity and myotoxicity, suggesting that the antibodies are either unable to reach the target sites as efficiently as small-molecule inhibitors, have a slower onset of action or are unable to displace or bind to the antigenic site of toxins already bound to target sites [42,50,56]. Considering that the effectiveness of the treatment combination was no different to the inhibitory actions of Varespladib alone, it also indicates that the efficacy of antivenoms is limited by the bite-to-antivenom time disparity, as well as toxin–target/substrate interaction, as previously described. Furthermore, the efficacy of antivenoms also depends on a different mechanism, such as steric hindrance, by trapping the toxin in the central compartment and enhancing the efficiency for toxin removal [50,53,56].

In this regard, the higher efficacy of Varespladib over antivenoms to inhibit the activity of these toxins may be due to (1) higher potency of Varespladib to inhibit PLA_2 activity at low concentrations; (2) the ability of Varespladib to bind with high affinity to neurotoxins and myotoxins, possibly displacing even those already bound extracellularly to their target sites or reducing affinity between toxin and substrate; or (3) the ability of Varespladib to enter the nerve terminal or cells, perhaps inhibiting intracellular snake venom PLA_2s [20,23,24]. It is worth noting that while the progression of direct twitch inhibition induced by venom was halted, but not reversed, by Varespladib, baseline tension was restored to pre-venom levels. This suggests that changes in baseline tension are due to alterations that are not related to mechanical damage to contraction mechanisms of myofibrils.

An important limitation of our study is the extrapolation of these data into clinical settings. Considering that the inhibitory action of Varespladib is rapid but has a relatively short half-life in vivo [48], multiple dosing amounts or continuous infusion of Varespladib to maintain concentrations high enough for toxin neutralisation may be required. For future in vitro studies, repeated dosing may accommodate for the time gap between venom addition and antivenom administration, as well as the further abrogation of venom damage, mirroring a clinical situation.

4. Conclusions

Our findings suggest that while small-molecular inhibitors are unlikely to replace the use of antivenoms, Varespladib could be used for short-term, immediate treatment to delay the onset of venom toxicity and increase the window of opportunity for antivenom administration before efficacy is diminished. Our work is in line with the initiative of the World Health Organisation (WHO) to accelerate snake-bite prevention and control protocols, as well as reduce snake-bite-related cases in South-East Asia by 50% by 2030 [61].

5. Materials and Methods

5.1. Animals

5–10-day-old male brown chicks (White Leghorn crossed with New Hampshire) were used in this study and obtained from Wagner's Poultry, Coldstream, Victoria. Animals were housed with food and water ad libitum at Monash Animal Services (Monash University, Clayton). Animal experiments were approved by the Monash University Ethics Committee (approval number 22575).

5.2. Chemicals and Drugs

The following chemicals and drugs were purchased from Sigma-Aldrich, St Louis, MO, USA: acetylcholine (ACh), carbamylcholine (CCh), d-tubocurarine (dTC), bovine serum albumin (BSA), Varespladib (CAS:172732-68-2) and dimethyl sulfoxide (DMSO). Potassium chloride (KCl) was purchased from Merck (Darmstadt, Germany). All chemicals were dissolved in MilliQ water, except Varespladib, which was dissolved in DMSO.

5.3. Venoms and Antivenoms

The Chinese *Daboia siamensis* venom used in this study was a pooled, freeze-dried sample from snakes collected from Yunnan province and obtained from Orientoxin Co., Ltd. (Laiyang, Shandong, China). Venom was dissolved in 0.05% (*w/v*) BSA to a final concentration of 2 mg/mL and frozen at $-20\,^{\circ}C$ until required. Chinese *D. siamensis* monovalent antivenom (Batch number 20200401; expiry date: 1 April 2022) was purchased from Shanghai Serum Biological Technology Co., Ltd. (Shanghai, China). According to manufacturer's instructions, 1 µL of Chinese *D. siamensis* antivenom neutralises 6 µg of *D. siamensis* venom. Antivenom was filtered prior to use with a 10 kDa centrifugal filter unit and centrifuged at 8000 rpm for 5 min. The supernatant was discarded and the retentate was stored at $-20\,^{\circ}C$

until required. Based on the manufacturer's neutralisation ratio the amount of antivenom was adjusted be equivalent to the concentration of unfiltered antivenom.

5.4. Isolated Chick Biventer Cervicis Nerve-Muscle Preparation

Chicks were euthanised by exsanguination following CO_2 inhalation. Two biventer cervicis nerve-muscle preparations, from each chick, were dissected and suspended on wire tissue holders under 1 g resting tension in 5 mL organ baths. Tissues were bubbled with 95% O_2 and 5% CO_2 in physiological salt solution of the following composition: 118.4 mM NaCl, 4.7 mM KCl, 1.2 mM KH_2PO_4, 2.5 mM $CaCl_2$, 25 mM $NaHCO_3$ and 11.1 mM glucose, and maintained at a temperature of 34 °C.

Indirect (nerve-mediated) twitches of tissues were evoked by stimulating the motor nerve (0.1 Hz, 0.2 ms) at supramaximal voltage (10–20 V) using an LE series stimulator (ADInstruments, Pty Ltd., Bella Vista, Australia). Twitches were measured and recorded on a PowerLab system (ADInstruments, Pty Ltd., Bella Vista, Australia) via a Grass FT03 force-displacement transducer. The abolishment of twitches following addition of dTC (10 µM) ensured selective stimulation of the motor nerve. The twitches were restored to baseline levels by repeated washing of the preparation with physiological salt solution over the course of 20 min. Following this period, electrical stimulation was ceased and the tissue was allowed to rest for approximately 5 min. Contractile responses to exogenous ACh (1 mM for 30 s), CCh (20 µM for 60 s) and KCl (40 mM, 30 s) were then obtained. Electrical stimulation was then recommenced for 20–30 min or until twitches were stable. For experiments examining myotoxicity, the electrode was placed around the muscle, which was directly stimulated (0.1 Hz, 2 ms) at supramaximal voltage (20–30 V). Residual nerve-mediated responses were abolished by the addition of dTC (10 µM), which remained in the organ bath throughout the experiment. All preparations were stabilised for at least 20–30 min before commencement of the experiment.

5.5. Protection and Reversal Protocols

The neutralising ability of Varespladib was tested by either (1) pre-incubation of venom with Varespladib for 15 min prior to addition of the mixture to the organ bath or (2) by the addition of Varespladib 60 min after venom. The concentration and ratio of Varespladib used in this study was adapted from the protocol described in a previous study, Maciel et al. [25]. To determine the ability of antivenom to reverse venom-induced neurotoxicity or myotoxicity, antivenom was added to the organ bath 30 or 60 min after venom addition. Antivenom was used at 3× the recommended concentration for neurotoxicity experiments and 4× the recommended concentration for myotoxicity experiments based on our previous study where these concentrations were shown to almost abolish neurotoxicity/myotoxicity when added prior to the venom [33].

5.6. Phospholipase A₂ Activity

PLA_2 activity of *D. siamensis* venom was determined using a PLA_2 assay kit (Cayman Chemical, Ann Arbour, MI, USA) according to the manufacturer's instructions. Venom stock solution (1 mg/mL) was serially diluted to a final concentration of 7.8 µg/mL, and a pre-mixed solution containing Varespladib (20 µM), assay buffer and indicator DTNB [5,5′-dithio-bis-(2-nitrobenzoic acid)], was added to wells of a 96-well plate. Substrate solution diheptanoyl thio-PC was added to each well, and the plate was read every 2 min at a wavelength of 414 nm for 22 min. Absorbance values were measured to calculate PLA_2 activity, expressed as micromoles of phosphatidylcholine hydrolysed per min per mg of enzyme (µmol/min/mg) of each venom dilution sample, and values indicated are the mean of triplicate wells. Bee venom PLA_2 was used as a positive control. Varespladib concentration was chosen based on the neutralisation capabilities demonstrated in previous studies [45].

5.7. Data Analysis and Statistics

For chick biventer experiments, twitch height was measured every four minutes after venom addition and expressed as a percentage of the pre-venom twitch response. For neurotoxicity experiments, post-venom contractile responses to exogenous agonists ACh, CCh and KCl were expressed as a percentage of the corresponding pre-venom contractile response. For myotoxicity experiments, changes in baseline tension, were measured every 10 min after venom addition. Either one-way or two-way analysis of variance (ANOVA) was performed for comparisons between different treatments. Comparisons of responses to exogenous agonists before and after venom treatment were performed using a student's paired t-test. All ANOVAs were followed by Bonferroni's multiple comparison or Tukey's post hoc test, respectively. Data are presented as the mean $\pm$ standard error of mean (SEM) where n is the number of tissue preparations. All data and statistical analyses were performed using Prism 9.4.1 (GraphPad Software, San Diego, CA, USA, 2022). $p < 0.05$ was considered statistically significant for all analyses.

Author Contributions: Conceptualization, M.L., Q.L., G.K.I. and W.C.H.; methodology, M.L., Q.L., G.K.I. and W.C.H.; formal analysis, M.L. and W.C.H.; investigation, M.L.; data curation and writing—original draft preparation, M.L.; writing—review and editing and supervision, Q.L., G.K.I. and W.C.H.; funding acquisition, Q.L., G.K.I. and W.C.H.; project administration, G.K.I. and W.C.H. All authors have read and agreed to the published version of the manuscript.

Funding: This study was supported by an Australian National Health and Medical Research Council (NHMRC) Senior Research Fellowship (ID: 1154503) awarded to G.K.I., a NHMRC Centres for Research Excellence Grant (ID:1110343) awarded to G.K.I. and W.C.H., a 2021 Guangzhou Science and Technology Program City and University (Hospital) joint funded Grant (ID: 202102010282) and a 2022 General project Grant of Natural Science Foundation of Guangdong basic and applied basic research foundation (ID:2022A1515012095) awarded to Q.L.

Institutional Review Board Statement: This study was conducted according to the NHMRC Australian Code for the Care and Use of Animals for Scientific Purposes and approved by the Monash University Animal Ethics Committee. Animal ethics number: 22575. Approval date: December 2019.

Informed Consent Statement: Not applicable.

Data Availability Statement: Not applicable.

Conflicts of Interest: The authors declare no conflict of interest. The funders had no role in the design of the study; in the collection, analyses, or interpretation of data; in the writing of the manuscript; or in the decision to publish the results.

References

1. Wüster, W. The genus *Daboia* (Serpentes: Viperidae): Russell's Viper. *Hamadryad* **1998**, *23*, 33–40.
2. Thorpe, R.S.; Pook, C.E.; Malhotra, A. Phylogeography of the Russell's viper (*Daboia russelii*) complex in relation to variation in the colour pattern and symptoms of envenoming. *Herpetol. J.* **2007**, *17*, 209–217.
3. Chaisakul, J.; Alsolaiss, J.; Charoenpitakchai, M.; Wiwatwarayos, K.; Sookprasert, N.; Harrison, R.A.; Chaiyabutr, N.; Chanhome, L.; Tan, C.H.; Casewell, N.R. Evaluation of the geographical utility of Eastern Russell's viper (*Daboia siamensis*) antivenom from Thailand and an assessment of its protective effects against venom-induced nephrotoxicity. *PLoS Negl. Trop. Dis.* **2019**, *13*, e0007338. [CrossRef]
4. Hung, D.Z.; Wu, M.L.; Deng, J.F.; Lin-Shiau, S.Y. Russell's viper snakebite in Taiwan: Differences from other Asian countries. *Toxicon* **2002**, *40*, 1291–1298. [CrossRef]
5. Risch, M.; Georgieva, D.; von Bergen, M.; Jehmlich, N.; Genov, N.; Arni, R.K.; Betzel, C. Snake venomics of the Siamese Russell's viper (*Daboia russelli siamensis*)—Relation to pharmacological activities. *J. Proteom.* **2009**, *72*, 256–269. [CrossRef]
6. Silva, A.; Maduwage, K.; Sedgwick, M.; Pilapitiya, S.; Weerawansa, P.; Dahanayaka, N.J.; Buckley, N.A.; Siribaddana, S.; Isbister, G.K. Neurotoxicity in Russell's viper (*Daboia russelii*) envenoming in Sri Lanka: A clinical and neurophysiological study. *Clin. Toxicol.* **2016**, *54*, 411–419. [CrossRef]
7. Warrell, D.A. Snake venoms in science and clinical medicine* 1. Russell's viper: Biology, venom and treatment of bites. *Trans. R. Soc. Trop. Med. Hyg.* **1989**, *83*, 732–740. [CrossRef]
8. Experts Group of Snake-bites Rescue and Treatment Consensus in China Expert Consensus on China Snake-bites rescue and Treatment. *Chin. J. Emerg. Med.* **2018**, *27*, 1315–1322. [CrossRef]

9. Belt, P.J.; Malhotra, A.; Thorpe, R.S.; Warrell, D.A.; Wüster, W. Russell's viper in Indonesia: Snakebite and systematics. *Symp. Zool. Soc. Lond.* **1997**, *70*, 219–234.

10. Phillips, R.E.; Theakston, R.D.G.; Warrell, D.A.; Galigedara, Y.; Abeysekera, D.T.D.J.; Dissanayaka, P.; Hutton, R.A.; Aloysius, D.J. Paralysis, Rhabdomyolysis and Haemolysis Caused by Bites of Russell's Viper (*Vipera russelli pulchella*) in Sri Lanka: Failure of Indian (Haffkine) Antivenom. *Q. J. Med.* **1988**, *68*, 691–716.

11. Pochanugool, C.; Wilde, H.; Bhanganada, K.; Chanhome, L.; Cox, M.J.; Chaiyabutr, N.; Sitprija, V. Venomous snakebite in Thailand II: Clinical Experience. *Mil. Med.* **1998**, *163*, 318–323. [CrossRef]

12. Tan, K.Y.; Tan, N.H.; Tan, C.H. Venom proteomics and antivenom neutralization for the Chinese eastern Russell's viper, *Daboia siamensis* from Guangxi and Taiwan. *Sci. Rep.* **2018**, *8*, 8545. [CrossRef]

13. Tan, N.H.; Fung, S.Y.; Tan, K.Y.; Yap, M.K.K.; Gnanathasan, C.A.; Tan, C.H. Functional venomics of the Sri Lankan Russell's viper (*Daboia russelii*) and its toxinological correlations. *J. Proteom.* **2015**, *128*, 403–423. [CrossRef]

14. Tasoulis, T.; Isbister, G.K. A Review and Database of Snake Venom Proteomes. *Toxins* **2017**, *9*, 290. [CrossRef]

15. Chaisukal, J.; Khow, O.; Wiwatwarayos, K.; Rusmili, M.R.A.; Prasert, W.; Othman, I.; Abidin, S.A.Z.; Charoenpitakchai, M.; Hodgson, W.C.; Chanhome, L.; et al. A Biochemical and Pharmacological Characterization of Phospholipase A$_2$ and Metalloproteinase Fractions from Eastern Russell's Viper (*Daboia siamensis*) Venom: Two Major Components Associated with Acute Kidney Injury. *Toxins* **2021**, *13*, 521. [CrossRef]

16. Herrera, M.; Collaço, R.d.C.d.O.; Villalta, M.; Segura, Á.; Vargas, M.; Wright, C.E.; Paiva, O.K.; Matainaho, T.; Jensen, S.D.; León, G.; et al. Neutralization of the neuromuscular inhibition of venom and taipoxin from the taipan (*Oxyuranus scutellatus*) by F(ab')$_2$ and whole IgG antivenoms. *Toxicol. Lett.* **2016**, *241*, 175–183. [CrossRef]

17. Angulo, Y.; Lomonte, B. Differential susceptibility of C2C12 myoblasts and myotubes to group II phospholipases A$_2$ myotoxins from crotalid snake venoms. *Cell Biochem. Funct.* **2005**, *23*, 307–313. [CrossRef]

18. Silva, A.; Kuruppu, S.; Othman, I.; Goode, R.J.; Hodgson, W.C.; Isbister, G.K. Neurotoxicity in Sri Lankan Russell's Viper (*Daboia russelii*) Envenoming is Primarily due to U1-viperitoxin-Dr1a, a Pre-Synaptic Neurotoxin. *Neurotox. Res.* **2017**, *31*, 11–19. [CrossRef]

19. Gutiérrez, J.M.; Lomonte, B. Phospholipase A2: Unveiling the secrets of a functionally versatile group of snake venom toxins. *Toxicon* **2013**, *62*, 27–39. [CrossRef]

20. Šribar, J.; Oberčkal, J.; Križaj, I. Understanding the molecular mechanisms underlying the presynaptic toxicity of secreted phospholipases A$_2$: An update. *Toxicon* **2014**, *89*, 9–16. [CrossRef]

21. Lomonte, B.; Gutiérrez, J.M. Phospholipases A$_2$ From *Viperidae* Snake Venoms: How do They Induce Skeletal Muscle Damage. *Acta Chim. Slov.* **2011**, *58*, 647–659. [PubMed]

22. Gutiérrez, J.M.; Ownby, C.L. Skeletal muscle degeneration induced by venom phospholipases A$_2$: Insights into the mechanisms of local and systemic myotoxicity. *Toxicon* **2003**, *42*, 915–931. [CrossRef] [PubMed]

23. Oliveira, I.C.F.; Gutiérrez, J.M.; Lewin, M.R.; Oshima-Franco, Y. Varespladib (LY315920) inhibits neuromuscular blockade induced by *Oxyuranus scutellatus* venom in nerve-muscle preparation. *Toxicon* **2020**, *187*, 101–104. [CrossRef] [PubMed]

24. de Souza, J.; Fontana Oliveria, I.C.F.; Yoshida, E.H.; Cantuaria, N.M.; Cogo, J.C.; Torres-Bonilla, K.A.; Hyslop, S.; Silva, N.J., Jr.; Floriano, R.S.; Gutiérrez, J.M.; et al. Effect of the phospholipase A$_2$ inhibitor Varespladib and its synergism with crotalic antivenom, on the neuromuscular blockade induced by *Crotalus durissus terrificus* venom (with and without crotamine) in mouse neuromuscular preparations. *Toxicon* **2022**, *214*, 54–61. [CrossRef]

25. Maciel, F.V.; Pinto, E.K.R.; Souza, N.M.V.; Gonçalves de Abreu, T.A.; Ortolani, P.L.; Fortes-Dias, C.L.; Calvalcante, W.L.G. Varespladib (LY315920) prevents neuromuscular blockage and myotoxicity induced by crotoxin on mouse neuromuscular preparations. *Toxicon* **2021**, *202*, 40–45. [CrossRef]

26. Bittenbinder, M.A.; Zdenek, C.N.; Op den Brouw, B.; Youngman, N.J.; Dobson, J.S.; Naude, A.; Vonk, F.J.; Fry, B.G. Coagulotoxic cobras: Clinical implications of strong anticoagulant actions of African spitting *Naja* venoms that are not neutralised by antivenom but are by LY315920 (Varespladib). *Toxins* **2018**, *10*, 516. [CrossRef] [PubMed]

27. Youngman, N.J.; Walker, A.; Naude, A.; Coster, K.; Sundman, E.; Fry, B.G. Varespladib (LY315920) neutralises phospholipase A$_2$ mediated prothrombinase-inhibition induced by *Bitis* snake venoms. *Comp. Biochem. Physiol. C Toxicol. Pharmacol.* **2020**, *236*, 108818. [CrossRef]

28. Kini, R.M. Excitement ahead: Structure, function and mechanism of snake venom phospholipase A$_2$ enzymes. *Toxicon* **2003**, *42*, 827–840. [CrossRef]

29. Kini, R.M.; Evans, H.J. A model to explain the pharmacological effects of snake venom phospholipases A$_2$. *Toxicon* **1989**, *27*, 613–635. [CrossRef]

30. Albulescu, L.O.; Xie, C.; Ainsworth, S.; Alsolaiss, J.; Crittenden, E.; Dawson, C.A.; Softley, R.; Bartlett, K.E.; Harrison, R.A.; Kool, J.; et al. A therapeutic combination of two small molecule toxin inhibitors provides broad preclinical efficacy against viper snakebite. *Nat. Commun.* **2020**, *11*, 6094. [CrossRef]

31. Sanz, L.; Quesada-Bernat, S.; Chen, P.Y.; Lee, C.D.; Chiang, J.R.; Calvete, J.J. Translational Venomics: Third-Generation Antivenomics of Anti-Siamese Russell's Viper, *Daboia siamensis*, Antivenom Manufactured in Taiwan CDC's Vaccine Center. *Trop. Med. Infect. Dis.* **2018**, *3*, 66. [CrossRef] [PubMed]

32. Maung-Maung-Thwin; Gopalakrishnakone, P.; Yuen, R.; Tan, C.H. A Major Lethal factor of the Venom of Burmese Russell's Viper (*Daboia russelli siamensis*): Isolation, N-Terminal sequencing and biological activities of daboiatoxin. *Toxicon* **1995**, *33*, 63–76. [CrossRef] [PubMed]

33. Tsai, I.H.; Lu, P.J.; Su, J.C. Two types of Russell's viper revealed by variation in Phospholipases A_2 from venom of the subspecies. *Toxicon* **1996**, *34*, 99–109. [CrossRef]

34. Lay, M.; Liang, Q.; Isbister, G.K.; Hodgson, W.C. In Vitro Toxicity of Chinese Russell's Viper (*Daboia siamensis*) Venom and Neutralisation by Antivenoms. *Toxins* **2022**, *14*, 505. [CrossRef] [PubMed]

35. Gopalan, G.; Maung-Maung-Thwin; Gopalakrishnakone, P.; Swaminathan, K. Structural and pharmacological comparison of daboiatoxin from *Daboia russelli siamensis* with viperotoxin F and vipoxin from other vipers. *Acta Crystallogr. Sect. D* **2007**, *63*, 722–729. [CrossRef]

36. Thakshila, P.; Hodgson, W.C.; Isbister, G.K.; Silva, A. In Vitro Neutralization of the Myotoxicity of Australian Mulga Snake (*Pseudechis australis*) and Sri Lankan Russell's Viper (*Daboia russelii*) Venoms by Australian and Indian Polyvalent Antivenoms. *Toxins* **2022**, *14*, 302. [CrossRef]

37. Silva, A.; Johnston, C.; Kuruppu, S.; Kneisz, D.; Maduwage, K.; Kleifeld, O.; Smith, A.I.; Siribaddana, S.; Buckley, N.A.; Hodgson, W.C.; et al. Clinical and Pharmacological Investigation of Myotoxicity in Sri Lankan Russell's Viper (*Daboia russelii*) Envenoming. *PLoS Negl. Trop. Dis.* **2016**, *10*, e0005172. [CrossRef]

38. Wang, Y.M.; Lu, P.J.; Ho, C.L.; Tsai, I.H. Characterization and molecular cloning of neurotoxic phospholipases A_2 from Taiwan viper (*Vipera russelli formosensis*). *Eur. J. Biochem.* **1992**, *209*, 635–641. [CrossRef]

39. Patikorn, C.; Ismail, A.K.; Abidin, S.A.Z.; Blanco, F.B.; Blessmann, J.; Choumlivong, K.; Comandante, J.D.; Doan, U.V.; Zainalabidin, M.I.; Khine, Y.Y.; et al. Situation of snakebite, antivenom market and access to antivenoms in ASEAN countries. *BMJ Glob. Health* **2022**, *7*, e007639. [CrossRef]

40. Bulfone, T.C.; Samuel, S.P.; Bickler, P.E.; Lewin, M.R. Developing Small Molecule Therapeutics for the Initial and Adjunctive Treatment of Snakebite. *J. Trop. Med.* **2018**, *2018*, 4320175. [CrossRef]

41. Ranawaka, U.K.; Lalloo, D.G.; de Silva, H.J. Neurotoxicity in Snakebite-The Limits of Our Knowledge. *PLoS Negl. Trop. Dis.* **2013**, *7*, e2302. [CrossRef] [PubMed]

42. Madhushani, U.; Isbister, G.K.; Tasoulis, T.; Hodgson, W.C.; Silva, A. In-Vitro Neutralization of the Neurotoxicity of Coastal Taipan Venom by Australian Polyvalent Antivenom: The Window of Opportunity. *Toxins* **2020**, *12*, 690. [CrossRef] [PubMed]

43. Silva, A.; Maduwage, K.; Sedgwick, M.; Pilapitiya, S.; Weerawansa, P.; Dahanayaka, N.J.; Buckley, N.A.; Johnston, C.; Siribaddana, S.; Isbister, G.K. Neuromuscular Effects of Common Krait (*Bungarus caeruleus*) Envenoming in Sri Lanka. *PLoS Negl. Trop. Dis.* **2016**, *10*, e0004368. [CrossRef] [PubMed]

44. Wang, Y.; Zhang, J.; Zhang, D.; Xiao, H.; Xiong, S.; Huang, C. Exploration of the Inhibitory Potential of Varespladib for Snakebite Envenomation. *Molecules* **2018**, *23*, 391. [CrossRef]

45. Lewin, M.; Samuel, S.; Merkel, J.; Bickler, P. Varespladib (LY315920) Appears to Be a Potent, Broad-Spectrum, Inhibitor of Snake Venom Phospholipase A_2 and a Possible Pre-Referral Treatment for Envenomation. *Toxins* **2016**, *8*, 248. [CrossRef]

46. Xiao, H.; Li, H.; Zhang, D.; Li, Y.; Sun, S.; Huang, C. Inactivation of Venom PLA_2 Alleviates Myonecrosis and Facilitates Muscle Regeneration in Envenomed Mice: A Time Course Observation. *Molecules* **2018**, *23*, 1911. [CrossRef]

47. Bryan-Quirós, W.; Fernández, J.; Gutiérrez, J.M.; Lewin, M.R.; Lomonte, B. Neutralizing properties of LY315920 toward snake venom group I and II myotoxic phospholipase A_2. *Toxicon* **2019**, *157*, 1–7. [CrossRef]

48. Gutiérrez, J.M.; Lewin, M.R.; Williams, D.J.; Lomonte, B. Varespladib (LY315920) and Methyl Varespladib (LY333013) Abrogate or Delay Lethality Induced by Presynaptically Acting Neurotoxic Snake Venoms. *Toxins* **2020**, *12*, 131. [CrossRef]

49. Lewin, M.R.; Gutiérrez, J.M.; Samuel, S.P.; Herrera, M.; Bryan-Quirós, W.; Lomonte, B.; Bickler, P.E.; Bulfone, T.C.; Williams, D.J. Delayed Oral LY333013 Rescues Mice from Highly Neurotoxic, Lethal Doses of Papuan Taipan (*Oxyuranus scutellatus*) Venom. *Toxins* **2018**, *10*, 380. [CrossRef]

50. Madhushani, U.; Thakshila, P.; Hodgson, W.C.; Isbister, G.K.; Silva, A. Effect of Indian Polyvalent Antivenom in the Prevention and Reversal of Local Myotoxicity Induced by Common Cobra (*Naja naja*) Venom from Sri Lanka In Vitro. *Toxins* **2021**, *13*, 308. [CrossRef]

51. Johnston, C.I.; Brown, S.G.A.; O'Leary, M.A.; Currie, B.J.; Greenberg, R.; Taylor, M.; Barnes, C.; White, J.; Isbister, G.K.; the ASP investigators. Mulga snake (*Pseudechis australis*) envenoming: A spectrum of myotoxicity, anticoagulant, coagulopathy, haemolysis and the role of early antivenom therapy—Australian Snakebite Project (ASP-19). *Clin. Toxicol.* **2013**, *51*, 417–424. [CrossRef] [PubMed]

52. Johnston, C.I.; Ryan, N.M.; O'Leary, M.A.; Brown, S.G.A.; Isbister, G.K. Australian taipan (*Oxyuranus* spp.) envenoming: Clinical effects and potential benefits of early antivenom therapy—Australian Snakebite Project (ASP—25). *Clin. Toxicol.* **2017**, *55*, 115–122. [CrossRef] [PubMed]

53. Gutiérrez, J.M.; Leon, G.; Lomonte, B. Pharmacokinetic-Pharmacodynamic Relationships of Immunoglobulin Therapy for Envenomation. *Clin. Pharmacokinet.* **2003**, *42*, 721–741. [CrossRef] [PubMed]

54. Isbister, G.K.; Maduwage, K.; Saiao, A.; Buckley, N.A.; Jayamanne, S.F.; Seyed, S.; Mohamed, F.; Chathuranga, U.; Mendes, A.; Abeysinghe, C.; et al. Population Pharmacokinetics of an Indian F(ab')$_2$ Snake Antivenom in Patients with Russell's Viper (*Daboia russelii*) Bites. *PLoS Negl. Trop. Dis.* **2015**, *9*, e0003873. [CrossRef] [PubMed]

55. Riviére, G.; Choumet, V.; Saliou, B.; Debray, M.; Bon, C. Absorption and Elimination of Viper Venom after Antivenom Administration. *J. Pharmacol. Exp. Ther* **1998**, *285*, 490–495. [PubMed]
56. Silva, A.; Isbister, G.K. Current research into snake antivenoms, their mechanisms of action and applications. *Biochem. Soc. Trans.* **2020**, *48*, 537–546. [CrossRef]
57. Salvador, G.H.M.; Borges, R.J.; Lomonte, B.; Lewin, M.R.; Fontes, M.R.M. The synthetic varespladib molecule is a multi-functional inhibitor for PLA$_2$ and PLA$_2$-like ophidic toxins. *Biochim Biophys Acta Gen. Subj.* **2021**, *1865*, 129913. [CrossRef]
58. Salvador, G.H.M.; Gomes, A.A.S.; Bryan-Quirós, W.; Fernández, J.; Lewin, M.R.; Gutiérrez, J.M.; Lomonte, B.; Fontes, M.R.M. Structural basis for phospholipase A$_2$-like toxin inhibition by the synthetic compound Varespladib (LY315920). *Sci. Rep.* **2019**, *9*, 17203. [CrossRef]
59. Tan, C.H.; Lingam, T.M.C.; Tan, K.Y. Varespladib (LY315920) rescued mice from fatal neurotoxicity caused by venoms of five major Asiatic kraits (*Bungarus* spp.) in an experimental envenoming and rescue model. *Acta Trop.* **2022**, *227*, 106289. [CrossRef]
60. Silva-Carvalho, R.; Gaspar, M.Z.; Quadros, L.H.B.; Lobo, L.G.G.; Rogério, L.M.; Santos, N.T.S.; Zerbinatti, M.C.; Santarém, C.L.; Silva, E.O.; Gerez, J.R.; et al. In Vivo Treatment with Varespladib, a Phospholipase A$_2$ Inhibitor, Prevents the Peripheral Neurotoxicity and Systemic Disorders Induced by Micrurus Corallinus (Coral Snake) in Rats. *Toxicol. Lett.* **2022**, *356*, 54–63. [CrossRef]
61. World Health Organization. *Regional Action Plan for Prevention and Control of Snakebite Envenoming in the South-East Asia 2022–2030*; Regional Office for South-East Asia, World Health Organization: New Delhi, India, 2022.

Article

A Long-Read Genome Assembly of a Native Mite in China *Pyemotes zhonghuajia* Yu, Zhang & He (Prostigmata: Pyemotidae) Reveals Gene Expansion in Toxin-Related Gene Families

Yan-Fei Song [1,2,†], Li-Chen Yu [3,†], Mao-Fa Yang [2,4], Shuai Ye [2], Bin Yan [2], Li-Tao Li [3], Chen Wu [5] and Jian-Feng Liu [1,2,*]

1   State Key Laboratory Breeding Base of Green Pesticide and Agricultural Bioengineering, Key Laboratory of Green Pesticide and Agricultural Bioengineering, Ministry of Education, Guizhou University, Guiyang 550025, China
2   Institute of Entomology, Guizhou University, Guizhou Provincial Key Laboratory for Agricultural Pest Management of the Mountainous Region, Scientific Observing and Experiment Station of Crop Pest Guiyang, Ministry of Agriculture, Guiyang 550025, China
3   Changli Institute of Pomology, Hebei Academy of Agriculture and Forestry Sciences, Changli 066600, China
4   College of Tobacco Science, Guizhou University, Guiyang 550025, China
5   The New Zealand Institute for Plant and Food Research Limited, Auckland 1142, New Zealand
*   Correspondence: jfliu3@gzu.edu.cn
†   These authors contributed equally to this work.

Citation: Song, Y.-F.; Yu, L.-C.; Yang, M.-F.; Ye, S.; Yan, B.; Li, L.-T.; Wu, C.; Liu, J.-F. A Long-Read Genome Assembly of a Native Mite in China *Pyemotes zhonghuajia* Yu, Zhang & He (Prostigmata: Pyemotidae) Reveals Gene Expansion in Toxin-Related Gene Families. *Toxins* 2022, 14, 571. https://doi.org/10.3390/toxins14080571

Received: 13 July 2022
Accepted: 17 August 2022
Published: 21 August 2022

Publisher's Note: MDPI stays neutral with regard to jurisdictional claims in published maps and institutional affiliations.

Abstract: *Pyemotes zhonghuajia* Yu, Zhang & He (Prostigmata: Pyemotidae), discovered in China, has been demonstrated as a high-efficient natural enemy in controlling many agricultural and forestry pests. This mite injects toxins into the host (eggs, larvae, pupae, and adults), resulting in its paralyzation and then gets nourishment for reproductive development. These toxins have been approved to be mammal-safe, which have the potential to be used as biocontrol pesticides. Toxin proteins have been identified from many insects, especially those from the orders Scorpions and Araneae, some of which are now widely used as efficient biocontrol pesticides. However, toxin proteins in mites are not yet understood. In this study, we assembled the genome of *P. zhonghuajia* using PacBio technology and then identified toxin-related genes that are likely to be responsible for the paralytic process of *P. zhonghuajia*. The genome assembly has a size of 71.943 Mb, including 20 contigs with a N50 length of 21.248 Mb and a BUSCO completeness ratio of 90.6% (n = 1367). These contigs were subsequently assigned to three chromosomes. There were 11,183 protein coding genes annotated, which were assessed with 91.2% BUSCO completeness (n = 1066). Neurotoxin and dermonecrotic toxin gene families were significantly expanded within the genus of *Pyemotes* and they also formed several gene clusters on the chromosomes. Most of the genes from these two families and all of the three agatoxin genes were shown with higher expression in the one-day-old mites compared to the seven-day-pregnant mites, supporting that the one-day-old mites cause paralyzation and even death of the host. The identification of these toxin proteins may provide insights into how to improve the parasitism efficiency of this mite, and the purification of these proteins may be used to develop new biological pesticides.

Keywords: *Pyemotes*; genome annotation; protein; toxin

Key Contribution: *Pyemotes zhonghuajia* is an important biocontrol agent against Isoptera, Homoptera, Hymenoptera, Lepidoptera, and Coleoptera pests. However, the genetic resources are largely lacking in the genus of *Pyemotes*; which has impeded our understanding of the parasitic and physiological mechanism of those venomous ectoparasitic mites at the molecular level. In this study, we have provided a genome assembly of *P. zhonghuajia* with a specific description of the toxin-related gene families. Our resources have provided opportunities for understanding the molecular mechanism underlying the use of toxins in this mite in order to improve its parasitism efficiency.

1. Introduction

Pyemotes zhonghuajia Yu, Zhang & He (Prostigmata: Pyemotidae) is a native viviparous mite initially collected from *Sinoxylon japonicum* Lesne and *Phloeosinus hopehi* Schedl by Lichen Yu in the 1990s [1]. It distributes naturally in Shanxi, Xinjiang, Ningxia, Hebei, Tianjin, and Beijing, China [2]. *Pyemotes zhonghuajia* is a dominant efficient ectoparasitic mite and is regarded as an important natural enemy in controlling many agricultural and forest pests [3]. This mite can inject toxins through puncturing the intersegmental cuticle using its mouthpart to paralyze a large number of pests including Isoptera, Homoptera, Hymenoptera, Lepidoptera, and Coleoptera [4–7]. After the host is completely paralyzed, the mite finds an optimal position to settle down and then obtains nourishment for reproductive development [7]. The gland of *P. zhonghuajia* is located at the junction of the head and neck, and those toxins are produced in secretory cells in the gland. The toxins produced by the mite are highly efficient in paralysis. A one-day-old mite *P. zhonghuajia* female can kill a third *Spodoptera litura* instar larva heavier than 680,000 times its own weight [7–9]. One single *P. zhonghuajia* female can lead to over 50% mortality rate of the first to third instar larvae of *Mythimna separata* Walker and *Spodoptera frugiperda* (Smith) [3,6]. The toxins are also found to be safe to mammals, which have the potential to be used as biological pesticides [6]. Nowadays, *Pyemotes zhonghuajia* is mass-produced and commercialized to control *M. separata*, *Aphis citricola* van der Goot, *S. frugiperda* Smith, *Sinoxylon japonicus* (Motschulsky), *Monochamus alternatus* Hope, and *Zeuzera leuconotum* Bulter [3,4,6,10–13].

Toxin proteins from the order Scorpiones and Araneae have been widely studied, some of which are now used for pest control. The scorpion insect-specific neurotoxins *AaIT* and *LqhIT2* target the insect's voltage-gated sodium channels that bind to the host's motor nerve branches to play critical roles in electrical signaling in stimulating skeletal muscles [14]. Identified from the highly toxic scorpion *Androctonus australis* [15], *AaIT* is a single-chain polypeptide containing 70 amino acids and four disulfide bridges [16]. As it is safe to mammals, *AaIT* can be used for pest control [17,18]. *AaIT*-expressed baculovirus has shown to reduce the survival time of the pests coupled with a significantly enhanced infection efficiency of this virus [19]. For example, *AaIT*-expressed *Bombyx mori* nucleopolyhedrovirus (BmNPV) can reduce feeding damage from silkworms to the host, and introducing *AaIT* gene into entomopathogen *Beauveria bassiana* can enhance its virulence to mosquitoes [14]. *LqhIT2* is another protein known with the potential of pest control. It was isolated from the scorpion *Leiurus quinquestriatus hebraeus*, which comprises 61 amino acids with four disulphide bridges [17]. The feeding capacity of the rice leafroller (*Cnaphalocrocis medinalis* Guenee) was shown to be decreased in the *LqhIT2*-inserted transgenic rice compared to the wild type [17]. In Araneae, more than 800 toxins have been isolated and described [20]. Dermonecrotic toxins, well-characterized from the Brown spider (*Genus Loxosceles*), are biochemical constituents in spider crude venom, and it can also induce necrotic and dermonecrotic lesions on rabbits and mice [21]. Agatoxins are neurotoxins, identified from *Agelenopsis aperta* and classified into three classes (α-, µ-, and ω- agatoxins), specifically targeting three classes of ion channels (voltage-activated calcium channels, transmitter-activated cation channels, and voltage-activated sodium channels), respectively [22]. The α-agatoxins have an acylpolyamine structure and can induce immediate but reversible paralysis. The µ-agatoxins do not have this structure and cause immediate but irreversible paralysis. The ω- agatoxins divide into four types (ω-Aga-1A, ω-Aga-IIA, Type III ω-Agatoxins, and Type IV ω-Agatoxins), and inhibit voltage-activated calcium channels in nerve terminals. Some enzymes also play important roles in arachnid biotoxins. Identified from spider venoms, enzymes mainly serve two important functions: (1) lysing polymers in the extracellular matrix and (2) binding to the compounds in the membrane [23,24].

There are only a few studies on toxin proteins in mites. In straw itch mite (*Pyemotes tritici*), a low molecular weight protein TxP was identified and showed inducing a rapid, muscle-contracting paralysis [25]. The toxicity of TxP -1 has shown to be comparable to *AaIT* or even stronger [26]. The TxP proteins were found to be translated by a range of cDNAs with variable length that are homologs to the insect-selective paralytic neurotoxin *tox34* [25]. It is notable that more than 18 recombinant baculoviruses engineered with *tox34* have been used for pest control [19,26]. In *P. zhonghuajia*, 12 *tox34* homologs were identified with sequence similarity ranging from 84.21% to 90.42% compared with those found in *P. tritici* [9]. However, other than these sequences, there is little genetic information about toxin-related proteins in the genus of *Pyemotes* that contains many venomous ectoparasitic mites.

Here, we sequenced the genome of *P. zhonghuajia* using Pacific Bioscience technology on the single-molecule real-time (SMRT) platform. The assembled genome was annotated with protein-coding genes, repeats, and non-coding RNAs (ncRNAs). We analyzed gene family evolution across Arachnida (1 Araneae, 1 Scorpiones, and 10 Acarina) with the main focus on the toxin-related genes.

2. Results and Discussion

2.1. Genome Assembly

A total of ~13 Gb and ~11 Gb Illumina short reads (150 bp) and PacBio long reads (average length of 9699.12 bp and N50 of 12.537 kb) were generated for genome assembly. The k-mer analysis based on short reads estimated genome size of 69.33 Mb comprising 7.09 Mb repetitive regions with genome heterozygosity of 0.04% (Supplementary Table S1). The *P. zhonghuajia* genome was assembled into 19 scaffolds containing 71.943 Mb with N50 of 21.248 Mb (Table 1), and 68.511 Mb was assigned to three pseudo-chromosomes. This assembled genome has a size comparable to the k-mer estimation and achieved a BUSCO complete gene ratio of 90.6% with duplicated and missing gene ratios of 1.3% and 7.9%, respectively. The mapping back rates from short and long reads as well as RNA-Seq data were 98.44%, 96.72%, and 92.84%, respectively. Compared with the other two mite genome assemblies containing thousands of scaffolds/contigs, our *P. zhonghuajia* genome assembly is much more continuous (Table 1). This genome assembly has a comparable size with the other two mite genomes.

Table 1. Genome Assembly and Annotation Statistics of *P. zhonghuajia*, *Tetranychus urticae*, and *Dermatophagoides pteronyssinus*.

Elements	Pyemotes zhonghuajia	Stratiolaelaps scimitus	Tetranychus urticae
Genome assembly			
Assembly size (Mb)	71.943	426.50	89.6
Number of scaffolds/contigs	19/20	158	-/2035
Longest scaffold/contig (Mb)	25.136/22.128	31.29	7/0.929
N50 scaffold/contig length (Mb)	22.128/21.248	7.66	10/120
GC (%)	25.02	45.85	-
Gaps (%)	0.00	0.00	-
BUSCO completeness (%)	90.6	93.1	-
Annotation			
Protein-coding genes	11,183	13,305	18,414
Mean protein length (aa)	480.97	500.59	-
Mean gene length (bp)	3243.42	7870.13	2652
Exons/introns per gene	3.59/2.45	6.24/-	3.82/-
Exon (%)	26.77	7.25	-
Mean exon length	475.53	372.35	178
Intron (%)	24.07	5.02	-
Mean intron length	626.12	1105.66	400
BUSCO completeness (%)	91.2	95.8	-

2.2. Genome Annotation

There was 11.4% of the assembly annotated as repetitive regions, which contained 77.050 repeats taking up ~8.2 Mb. The most abundant repeat class was LTR elements, taking up 3.62% of the assembly, followed by simple repeats (2.92%), unclassified repeats (1.98%), low complexity repeats (1.42%), and DNA elements (0.79%) (Supplementary Table S2). There were 11,183 gene models annotated with the average length of 3243.42 bp and 3.59 exons per gene. The average lengths of exon and intron were 475.53 bp and 626.12 bp, respectively. The BUSCO result showed 91.2% complete genes with duplicated and missing genes being 2.5% and 6.5%, showing that most of the annotated genes are likely to have complete lengths. It is noticed that the BUSCO complete gene ratio resulted from the annotated gene set is slightly higher than the genome assembly. It is likely because BUSCO only employs AUGUSTUS for gene prediction and has less power to predict complete genes correctly compared with our gene annotation method that was supported by RNA-Seq data. [27]. There were 218 ncRNAs annotated in the genome assembly, including eight miRNA, 48 rRNA, nine snRNA, and 112 tRNA. The annotated snRNAs included five spliceosomal RNAs (U1, U2, U4, U5 and U6), one minor splicesomal RNA (U6atac), and three C/D box snoRNAs (U3 and snoR38) (Supplementary Table S3). The tRNAs Supres and SelCys were absent in the annotation. This is the first non-coding RNA set annotated in a mite genome.

2.3. Species Phylogeny and Gene Family Evolution

There were 12 species including *P. zhonghuajia* selected for phylogenetic construction. In total, 180,720 genes were clustered into 17,407 gene families. In *P. zhonghuajia*, a total of 11,183 genes were analyzed, and there were 8388 genes assigned into 6224 gene families with species-specific gene families and genes being 178 and 761, respectively (Figure 1).

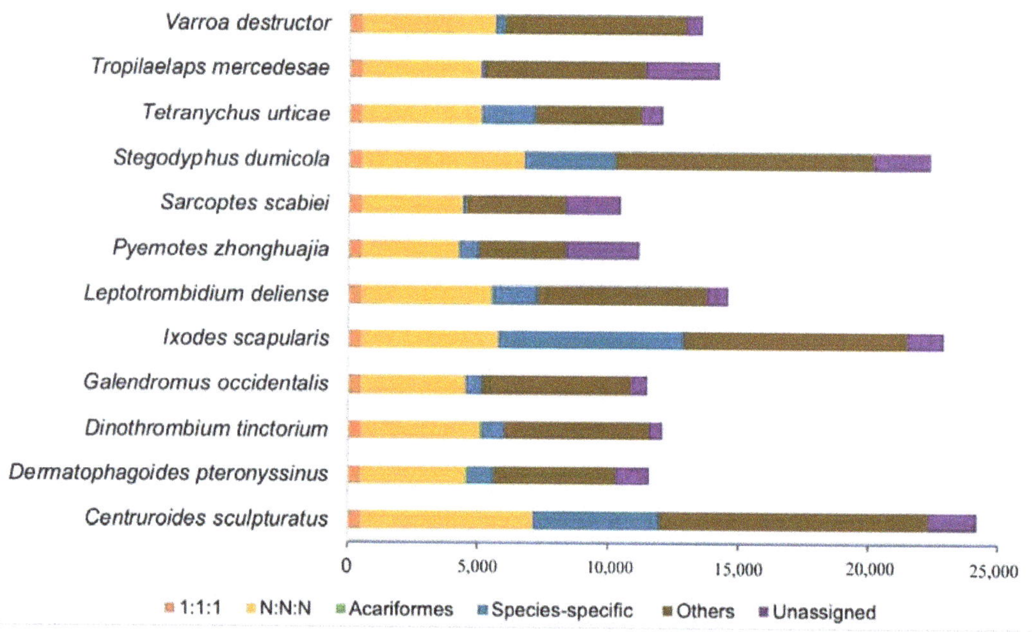

Figure 1. Histogram shows the number of genes assigned to different groups. The "1:1:1" and "N:N:N" groups represent single- and multi-copy genes found in all the species. The group "Acariformes" represents orthologs unique to Acariformes. The "Others" group indicates other orthologs which do not belong to any above-mentioned ortholog categories. The group "Unassigned" represent the orthologs which can't be assigned to any orthogroups.

There were 527 single-copy genes found in all the species, 473 of which contained 143,759 amino acid sites were used to construct a phylogenetic tree. All node supports were 100/100 (SH-aLRT support /ultrafast bootstrap support). The phylogenetic tree shows that *P. zhonghuajia* is sister to the two-spotted spider mite (*T. urticae*) and they are clustered with the other two mite species: chigger mite (*Leptotrombidium delicense*) and red velvet mite (*Dinothrombium tinctorium*) (Figure 2). The phylogenetic tree was consistent with the published classifications [1] and our calculation indicated *P. zhonghuajia* together with *T. urticae* emerged during Triassic (223.43~247.66 Mya).

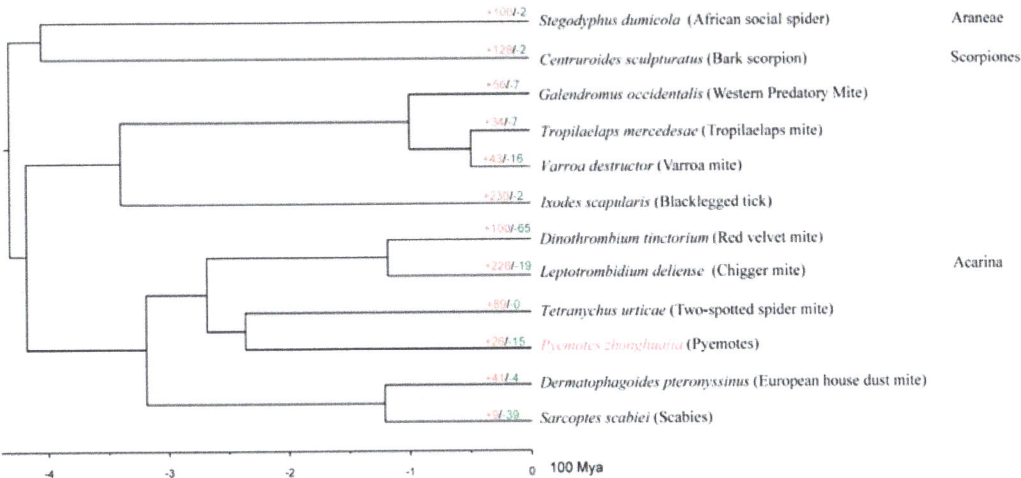

Figure 2. Dating tree with node values representing the number of expanded, contracted.

Compared with the closely related *T. urticae* that has over one thousand gene families identified as expanded families, *P. zhonghuajia* only has 621 (containing 1224 genes) gene families that were calculated as expanded families (Figures 2 and 3). These families are related to digestion, detoxification, and toxins, and those such as ABC transporter, Lipase, Trypsin, Dermonecrotic toxin, CD36 family, and neurotoxins (Supplementary Table S4). The families of dermonecrotic toxins and neurotoxins are mostly likely to participate in the parasitic mechanism [25,28–31]. There were 3927 genes lost in 3664 contracted families, including 15 gene families that were identified as significantly contracted families. These contracted families might be associated with viviparous reproduction in pyemotid mite; no free-living stages of larvae and nymph occur during the life cycle of *P. zhonghuajia* (Supplementary Table S5) [32].

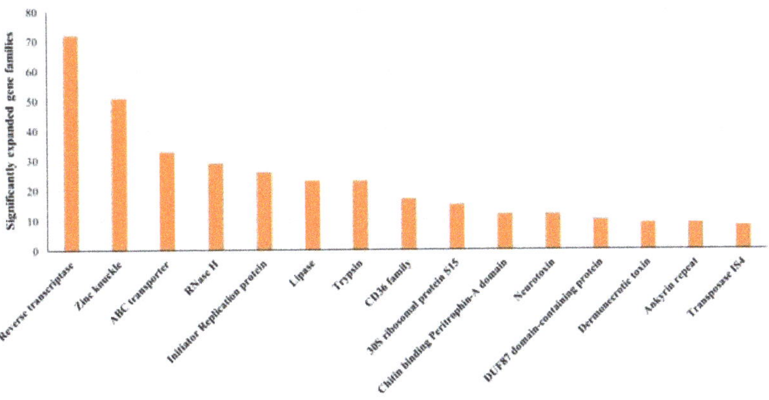

Figure 3. Top fifteen significantly expanded families with gene numbers of the families shown above the bars.

The results from the enrichment analysis of GO terms and KEGG pathways also showed the expanded gene families belong to the categories of digestion, such as lipid metabolic process and phosphatidic acid biosynthetic process (Supplementary Figure S1A), and PPAR signaling pathway and Biosynthesis of unsaturated fatty acids (Supplementary Figure S1B).

This suggests some of the gene family expansions are likely to be involved in the extensive feeding habits. The expanded chitin-binding families (Supplementary Figure S1A) may be related to intense enlargement of the parasitoid body during the reproductive period [33].

2.4. Neurotoxin, Dermonecrotic Toxin and Agatoxin Genes

It is interesting that we found significant expansion of neurotoxin and dermonecrotic toxin gene families in the *P. zhonghuajia* genome. Neurotoxins in spiders were found as the main component of the venom that targets the prey's ion channels leading to paralysis or death [34]. The *P. zhonghuajia* neurotoxin gene family was expanded within the Acari clade (Figure 4B). The 12 neurotoxin genes were present on all the three chromosomes with a six-gene locus being located near the end of chromosome 1 (Figure 4A). *PzNT3 and 4* were located nearby and closely related on the phylogenetic tree with nearly 85% sequence similarity (Supplementary Table S6), suggesting they were recently duplicated under a tandem duplication event. Similar observations were from *PzNT5 and 6*. Located on the two different chromosomes, a recent duplication event might have also occurred between *PzNT7 and 12* as they were closely related on the phylogenetic tree with high sequence similarity. *PzNT2* was present as a homologous sequence to *P. tritici* TxP2-1 (80% sequence similarity) and they were closely related to *P. tritici* TxP-1 (endcoded by *Tox34*) that was found playing a role in paralyzing and even killing insects [26,35]. Most neurotoxin genes (except *PzNT1 and 11*) have a higher number of reads aligned from RNA-Seq data obtained from the one-day-old mites compared with the seven-day-pregnant mites, indicating higher levels of gene expression (Supplementary Figure S2). *PzNT1* was located distantly from the six-gene locus on chromosome 1 and had shown to be distantly related to the rest of the neurotoxin genes on the phylogenetic tree. It is possible that there was a functional divergence of this gene compared with the rest of the family members.

Dermonecrotic toxins have been demonstrated to cause dermonecrotic lesions on the prey in spiders [21] and it is likely that the dermonecrotic toxins of *P. zhonghuajia* may also cause the dermonecrotic lesion of the host. In *P. zhonghuajia*, all nine dermonecrotic toxin genes were present roughly close to each other on the second half of chromosome 3 (Figure 4A). Our phylogenic tree showed these genes formed a single clade with the gene family expansion occurring after Pyemotes separated from Leptotrombidium and Dinothrombium (Figure 4C). *PzDNT1 and 2* were located next to each other and showed to be closely related on the phylogenetic tree with over 97% sequence similarity (Supplementary Table S8), suggesting a recent tandem gene duplication event. However, their expression divergence (higher expression of *PzDNT1* in the one-day-old mites compared with seven-day-pregnant mites and *PzDNT2* is opposite) suggests a functional difference. A recent duplication event was also shown between *PzDNT6 and 7*, both of which exhibited similar expression patterns in the two adult forms. Similar to *PzDNT2*, *PzDNT5* was also shown with higher expression level in the seven-day-pregnant mites compared with the one-day-old mites.

α-agatoxins was found to paralyze the prey in funnel web spiders [22]. There were three agatoxin genes identified in the *P. zhonghuajia* that formed a gene cluster on chromosome 2, which, with each other, has shown relatively distantly related on the phylogenetic tree (Figure 4D). *PzAg2 and 3* were separated from *PzAg1* at a very early stage during evolution. All three genes exhibited higher expression in the one-day-old mites compared with seven-day-pregnant mites, suggesting a role in paralyzing the hosts.

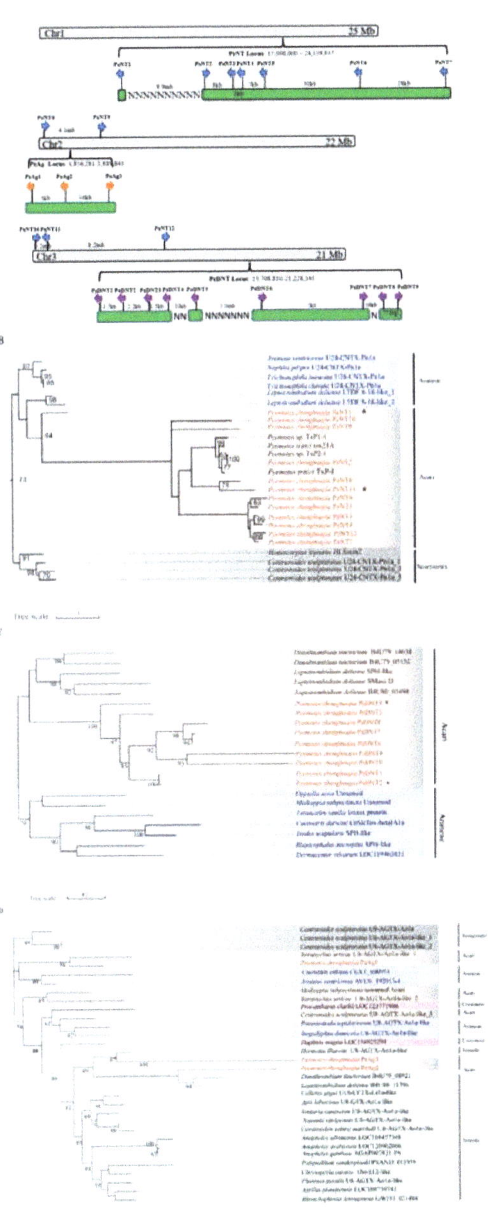

Figure 4. Distribution of toxin genes on the chromosomes and phylogenetic trees of the three toxin gene classes. (**A**) Distribution of toxin genes on the chromosomes. *Blue*: neurotoxin genes; *Orange*: agatoxin genes; *Purple*: dermonecrotic toxin genes. (**B–D**) Phylogenetic trees of neurotoxin genes, dermonecrotic toxin genes, and agatoxin genes. The "*" indicates the gene was shown with higher expression in the seven-day-pregnant mites compared with the one-day-old mites. The accession numbers of all the sequences used in the phylogenies are listed in Supplementary Table S12.

3. Conclusions

Pyemotidae is a significant family with several species being natural enemies, such as *P. tritici* and *P. zhonghuajia*. They can paralyze and even cause the death of the stored product insects and agriculture and forestry insects. Understanding the composition and function of toxin-related proteins of *P. zhonghuajia* is crucial to improve its predation efficiency and it can also provide the knowledge for the potential of transferring the mite toxin-related genes into crop genomes for pest control. In this study, we have generated a chromosome-level genome assembly of *P. zhonghuajia*, which is the first whole-genome assembly in Pyemotidae that provides an important genomic resource for the study of biocontrol potential and ecological importance. The two gene families encoding neurotoxins and dermonecrotic toxins were found with significant expansion within the genus of *Pyemotes*, which also formed several gene clusters on the chromosomes. It is possible that gene expansion provides a high dosage of toxin proteins that are released during parasitic process. Gene expansion might also result in a highly diverged population of toxins that enable the mite to have a wide range of hosts. Several recent gene duplication events that we observed from the two gene phylogenies indicate they may underlie key adaptive events in the evolution of *P. zhonghuajia*. All the toxin-related genes including the three agatoxin genes were shown with expression in the adults, and most of them exhibited higher level of expression in the one-day-old mites compared with seven-day-pregnant mites. This matches our observations on the one-day-old mite forms paralyzing and killing the hosts such as *S. frugiperda*, *M. separata*, and *S. litura* [3,6,7]. Future research will focus on confirming the presence of the proteins encoded by these genes through proteomic studies and functional characterization of the proteins through protein purification approaches and feeding experiments in *P. zhonghuajia*. Similar studies will also be conducted in other venomous ectoparasitic species from *Pyemotes*.

4. Materials and Methods

4.1. Sample Collection and Sequencing

Colonies of *P. zhonghuajia* were reared on mature larvae of *Sitotroga cerealella* (Oliver) (Lepidoptera: Gelechiidae) with wheat bran in a climate chamber at $25 \pm 1\,°C$ with $60 \pm 5\%$ relative humidity (RH) at Changli Institute of Pomology, Hebei Academy of Agriculture and Forestry Sciences. There were 2000 seven-day-pregnant mites used for Illumina whole-genome and PacBio sequencing, respectively. Genomic DNA was extracted using the QIAGEN DNeasy Blood & Tissue kit, which was then used to construct a 350 bp insert-size library using the Truseq DNA PCR-free kit for sequencing on the Illumina NovaSeq 6000 platform and a 15 kb insert-size library using the SMRTbell™ Template Prep Kit 2.0 for sequencing on the PacBio Sequel II platform. The whole-individual transcriptome was performed using RNA-Seq from 2000 one-day-old mites and 2000 seven-day-pregnant mites with three biological replicates for each group. Total RNA was extracted using the TRIzol™ Reagent kit and the RNA-Seq library was constructed using TruSeq RNA v2 kit. DNA/RNA extraction, library construction, and sequencing were performed at Berry Genomic (Beijing, China).

4.2. Genome Assembly

Illumina raw reads were cleaned using two tools under BBTools v38.67 [36] with the following steps: (1) removing duplicated reads; (2) trimming low-quality reads; (3) removing poly-A/G/C tails; (4) filtering reads less than 15 bp; and (5) correcting reads based on overlapping ends between pairs. The tool Clumpify was used for step (1) and steps (2) to (5) were performed using BBDuk with parameters "qtrim = rl trimq = 20 minlen = 15 ecco = t maxns = 5 trimpolya = 10 trimpolyg = 10 trimpolyc = 10". K-mer analysis based on Illumina short reads was performed using BBNorm (*k-mer*: 21) and the k-mer profile was visualized using the online version of Genomescope v2.0 [37] with parameters "-k 21 -p 2m 1000". A preliminary PacBio long-read assembly was performed using Flye v2.7.1 [38] with parameters "-i 2-m 3000". Purge_Dups v1.0.0 [39] was used to remove allelic contigs based on the read depth with a minimum alignment score of 70 after the long reads were mapped back to the assembly with Minimap2 (v2.17) [40]. Illumina short reads were used for two rounds of contig polishing performed by NextPolish (v1.1.0) [41] after mapping the reads back to the assembly using Minimap2. The contaminated contigs were assessed and removed using "blastn" from BLAST+ (v2.9.1) [42] with the sequence similarity search against *nt* and UniVec databases (both were downloaded in December 2020). The cleaned contigs were then uploaded to NCBI for an additional check of contamination. Assembly completeness was estimated using BUSCO (v3.1.0) [43] with the sequence similarity search against the arthropod single-copy gene set (arthropoda_odb9: n = 1367). To estimate the mapping rate from raw reads, both short and long genomic reads as well as RNA-Seq short reads were aligned back to the genome assembly using Minimap2.

4.3. Genome Annotation

Three essential genomic elements of *P. zhonghuajia* genome: repetitive elements, non-coding RNAs (ncRNAs), and protein-coding genes were annotated. To annotate repeats, RepeatModeler v2.0.1 [44] with LTR search process (-LTRStruct) was used to generate a de novo repeat library, which was then combined with Dfam 3.1 [45] and RepBase-20181026 [46] databases to form a custom repeat library. The repeat-masked genome assembly was produced using RepeatMasker v4.0.9 [47].

The MAKER v2.31.10 [48] pipeline was used to predict protein-coding genes by integrating ab initio, transcript- and homology-based evidence. The ab initio prediction was generated by a BRAKER v2.1.5 [49] pipeline to train Augustus v3.3.3 [50] and GeneMark-ES/ET/EP 4.48_3.60_lic [51] with the utilization of RNA-Seq data and protein sequences to increase prediction accuracy. The input alignments from mapping RNA-seq data to the genome assembly were produced using HISAT2 v2.2.0 [52] and the arthropod protein sequences were obtained from OrthoDB10 v1 database [53]. The genome-guided assembler StringTie v2.1.2 [54] was used to assemble transcripts as transcriptome evidence to integrate in MAKER. The protein sequences that were also utilized by MAKER for final prediction were downloaded from NCBI including the sequences from *Drosophila melanogaster* (Insecta), *Daphnia magna* (Crustacea), *Ixodes scapularis*, *Varroa destructor*, *Tetranychus urticae*, and *Dermatophagoides pteronyssinus* (Acari). Gene functional annotation was performed using Diamond v0.9.24 [55] from searching against the UniProtKB database with the sensitive mode "–more-sensitive -e 1e^{-5}". InterProScan 5.41–78.0 [56] was used to search against the databases including Pfam [57], SMART [58], Gene3D [59], Superfamily [60] and CDD [61]. The protein domains, gene ontology (GO), and gene pathways (KEGG, Reactome) were annotated using ggnog-mapper v2.0.1 [62] from searching against ggnog v5.0 [63].

NcRNAs including rRNA, snRNA, and miRNA were identified using infernal v1.1.3 [64] by searching sequence similarity against Rfam database. TRNAs were predicted using tRNAscan-SE v2.0.6 [65] and the high-confident sequences were maintained as the final tRNA set by the tRNAscan-SE script "EukHighConfidenceFilter".

4.4. Gene Ontology Analysis and Species Evolution

The protein sequences from 11 species across three orders (Araneae: Stegodyphus dumicola; Scorpiones: Centruroides sculpturatus; Arachnoidea: Dermatophagoides pteronyssinus, Dinothrombium tinctorium, Galendromus occidentalis, Ixodes scapularis, Leptotrombidium deliense, Sarcoptes scabiei, T. urticae, Tropilaelaps mercedesae, and Varroa destructor) were downloaded from NCBI in December 2020 together with the annotated protein set from *P. zhonghuajia* for gene family orthology inference using OrthoFinder v2.3.8 [66] after aligning sequences using Diamond. The resulting single-copy orthologs from OrthoFinder were used for phylogenetic analysis. The sequence alignment as input for phylogenetic construction was generated using the following steps: (1) aligning orthologous protein sequences using MAFFT v7.394 [67] with "L-INS-I"; (2) filtering ambiguous aligned regions using BMGE v1.12 [68] with the parameters "-m BLOSUM90 -h 0.4"; (3) Concatenating all the protein alignments generated above using FASconCAT-G v1.04 [69] as the input for phylogenetic construction. The species phylogeny was constructed using IQ-TREE v2.0-rc2 [70] with the parameters "-m MFP –mset LG –msub nuclear–rclusterf 10 -B 1000 –alrt 1000 –symtest-remove-bad –symtestpval 0.10". The estimated time of species divergence was calculated using MCMCTree from PAML v4.9j [71] with parameters "clock = 2, BDparas = 1 1 0.1, kappa_gamma = 6 2, alpha_gamma = 1 1, rgene_gamma = 2 20 1, sigma2_gamma = 1 10 1". There were four fossil evidences downloaded from the PBDB database (https://www.paleobiodb.org/navigator/, accessed on 14 August 2022) as calibrations used in this estimation above: Allopalaeophonus caledonicus (4.305–4.438) from the order Scorpiones as the root, Pseudoprotacarus scoticus (4.076–4.192) from Arachnida, Carbolohmannia maimaiphilus (3.114–3.232) from Acariformes, and Deinocroton draculi (0.935–1.455) from Mesostigmata.

4.5. Identification of Gene Family Expansion and Contraction

The gene family expansion and contraction in *P. zhonghuajia* genome compared with the 11 species used for phylogenetic construction were estimated using CAFÉ v4.2.1 [72] with the model of single birth–death parameter lambda and a significance level of 0.01 ($p = 0.01$). The identified significantly expanded gene families were then assigned with GO and KEGG categories using R package clusterProfiler v3.10.1 [73] with the default parameters ($p = 0.01$ and $q = 0.05$).

4.6. Phylogeny Construction and Gene Expression of Toxin-Related Gene Families

The *P. zhonghuajia* neurotoxin, dermonecrotic toxin, and agatoxin protein sequences were predicted from the genome assembly using BITACORA v1.2 [74] based on homology searches using the sequences from Chelicerata and Myriapoda downloaded from NCBI RefSeq (Supplementary Table S11) and confirmed by searching against protein database using the online blastp (https://blast.ncbi.nlm.nih.gov/Blast.cgi?PROGRAM=blastp&PAGE_TYPE=BlastSearch&LINK_LOC=blasthome, accessed on 14 August 2022). The HMM profiles generated from HMMER v3.2.1 [75] using "hmmbuild" were used in BITACORA. Multiple alignments were performed using Geneious prime V 2021.1.1 (created by Biomatters. Available from https://www.geneious.com, accessed on 5 November 2020) with the method of Clustal Omega. FastTree [76] was used to construct neurotoxin, dermonecrotic toxin, and agatoxin protein phylogenies based on the maximum likelihood method. We used RNA sequencing data to detect the expression of toxin-related genes in seven-day-pregnant mites and one-day- old mites. The estimation of gene expression levels was performed using Salmon [77] with the Gaussian axial fluctuation (GAF) model to generate normalized read counts. The gene expression heatmap was generated using the heatmap function (heatmap) from the R package NMF [78] (Supplementary Figure S2).

Supplementary Materials: The following are available online at https://www.mdpi.com/xxx/s1, Table S1: Genome survey based on k-mer distribution for *pyemotes zhonghuajia* (excel file), Table S2: Repeat annotation in the *Pyemotes zhonghuajia* genome (excel file), Table S3: Annotations of noncoding RNAs in the *Pyemotes zhonghuajia* genome (excel file), Table S4: Significantly expanded gene families (excel file), Table S5: Expended and contracted gene families in each species. Table S6: The identity of neurotoxins of *P. zhonghuajia* with *P. tritici* (excel file), Table S7: The identity of agatoxins of *P. zhonghuajia* with other species (excel file), Table S8: The identity of dermonecrotic of *P. zhonghuajia* with other species (excel file), Table S9: Information of the 24 toxin-related genes, Table S10:Transcript per million (TPM) values (excel file), Table S11: Gene IDs of the reference species, Table S12: The accession numbers of all the sequences used in the phylogenies (excel file). Figure S1: GO (A) and KEGG (B) function enrichment of significantly expanded gene families. Figure S2: Heatmap of the identified toxin-related genes (pdf file).

Author Contributions: Conceptualization, Y.-F.S. and J.-F.L.; methodology, Y.-F.S., M.-F.Y., J.-F.L. and S.Y.; software, Y.-F.S., C.W. and B.Y.; resources, Y.-F.S., S.Y., L.-T.L., M.-F.Y., L.-C.Y. and J.-F.L.; writing—original draft preparation, Y.-F.S.; writing—review and editing, J.-F.L., C.W. and Y.-F.S.; supervision, J.-F.L. and C.W.; project administration, J.-F.L.; funding acquisition, J.-F.L. All authors have read and agreed to the published version of the manuscript.

Funding: This research was funded by the National Natural Science Foundation of China (32060637 and 31770694), Guizhou Province Science and Technology Innovation Talent Team Project (Qian Ke He Pingtai Rencai-CXTD [2021]004), High-level Talent Innovation and Entrepreneurship Funding Project in Guizhou Province ([2021]01), the Growth Project of Youth Talent in Ordinary Universities in Guizhou Province ([2021]079), Natural Science Special Project in Guizhou University (Special post, [2020]-02) and Hebei Provincial Key R&D Program (20326517D).

Institutional Review Board Statement: Not applicable.

Informed Consent Statement: Not applicable.

Data Availability Statement: Genome assembly and raw sequencing data have been deposited at the NCBI under the accessions JACCHO000000000 and SRR12261228—SRR12261230, respectively. Genome annotations are available at the Figshare under the link: https://figshare.com/articles/dataset/A_Long-Read_Genome_Assembly_of_a_Native_Mite_in_China_Pyemotes_Zhonghuajia_Yu_Zhang_He_Prostigmata_Pyemotidae_Item/20518008 (accessed on 18 August 2022).

Conflicts of Interest: The authors declare no conflict of interest.

References

1. Yu, L.; Zhang, Z.Q.; He, L. Two New Species of *Pyemotes* Closely Related to *P. Tritici* (Acari: *Pyemotidae*). *Zootaxa* **2010**, *2723*, 1–40. [CrossRef]
2. He, L.M.; Jiao, R.; Xu, C.X.; Hao, B.F.; Han, J.C.; Yu, L.C. Application of mtDNA COI Gene Sequence in Indentification of *Pyemotes*. *J. Hebei Agric. Sci.* **2010**, *14*, 46–50.
3. Liu, J.F.; Tian, T.A.; Li, X.L.; Chen, Y.C.; Yu, X.F.; Tan, X.F.; Zhu, Y.; Yang, M.F. Is *Pyemotes Zhonghuajia* (Acari: *Pyemotidae*) a Suitable Biological Control Agent against the Fall Armyworm *Spodoptera Frugiperda* (Lepidoptera: *Noctuidae*)? *Syst. Appl. Acarol.* **2020**, *25*, 649–657. [CrossRef]
4. Li, L.; He, L.; Yu, L.; He, X.Z.; Xu, C.; Jiao, R.; Zhang, L.; Liu, J. Preliminary Study on the Potential of *Pyemotes Zhonghuajia* (Acari: *Pyemotidae*) in Biological Control of *Aphis Citricola* (Hemiptera: *Aphididae*). *Syst. Appl. Acarol.* **2019**, *24*, 1116–1120. [CrossRef]
5. De Sousa, A.H.; Mendonça, G.R.Q.; Lopes, L.M.; Faroni, L.R.D. Widespread Infestation of *Pyemotes Tritici* (Acari: *Pyemotidae*) in Colonies of Seven Species of Stored-Product Insects. *Genet. Mol. Res.* **2020**, *19*, 1–5. [CrossRef]
6. Tian, T.A.; Yu, L.; Sun, G.J.; Yu, X.F.; Li, L.; Wu, C.X.; Chen, Y.C.; Yang, M.F.; Liu, J.F. Biological Control Efficiency of an Ectoparasitic Mite *Pyemotes Zhonghuajia* on Oriental Armyworm *Mythimna Separata*. *Syst. Appl. Acarol.* **2020**, *25*, 1683–1692.
7. Chen, Y.C.; Tian, T.A.; Chen, Y.H.; Yu, L.C.; Hu, J.F.; Yu, X.F.; Liu, J.F.; Yang, M.F. The Biocontrol Agent *Pyemotes zhonghuajia* Has the Highest Lethal Weight Ratio Compared with Its Prey and the Most Dramatic Body Weight Change during Pregnancy. *Insects* **2021**, *12*, 490. [CrossRef]
8. Tomalski, M.D.; Bruce, W.A.; Travis, J.; Blum, M.S. Preliminary Characterization of Toxins from the Straw Itch Mite, *Pyemotes Tritici*, Which Induce Paralysis in the Larvae of a Moth. *Toxicon* **1988**, *26*, 127–132. [CrossRef]
9. Han, J.C.; He, L.M.; Jiao, R.; Hao, B.F.; Xu, Z.X.; Yu, L.C. Analysis of the toxin gene analogs cloned from *Pyemotes phloeosinus* sp. nov. *J. Hebei Agric. Sci.* **2008**, *12*, 72–74.
10. Lu, H.L.; Li, L.T.; Yu, L.C.; He, L.M.; Ouyang, G.C.; Liang, G.W.; Lu, Y.Y. Ectoparasitic mite, *Pyemotes zhonghuajia* (Prostigmata: *Pyemotidae*), for biological control of Asian citrus psyllid, *Diaphorina citri* (Hemiptera: *Liviidae*). *Syst. Appl. Acarol.* **2019**, *24*, 520–524. [CrossRef]
11. Guo, X.; Xu, Z.; Xiong, D.P. Study of utilizing *Pyemotes zhonghuajia* to control *Semanotus bifasciatus* beetles. *Chin. Bull. Entomol.* **2010**, *47*, 529–532.
12. Li, Y.C.; Huang, C.Y.; Xia, Z.R.; Zhou, J.S.; Li, J.S.; Sun, Y.Z.; Xu, Y.; Wu, S.Q.; Zhang, F.P. Preliminary Study on Biocontrol Potential of *Pyemotes Zhonghuajia* on *Monochamus Alternatus* Hope. *Genom. Appl. Biol.* **2019**, *38*, 2516–2521.
13. Zhang, Z.S.; Xiong, D.P.; Cheng, W. Control of Stem Borers by a Parasitoid, *Pyemotes trittci* Lagreze-Fossot & Montane. *Chin. J. Biol. Control.* **2008**, *01*, 1–6.
14. Deng, S.Q.; Chen, J.T.; Li, W.W.; Chen, M.; Peng, H.J. Application of the scorpion neurotoxin AaIT against insect pests. *Int. J. Mol. Sci.* **2019**, *20*, 3467. [CrossRef]
15. Harvey-Samuel, T.; Xu, X.; Lovett, E.; Dafa'alla, T.; Walker, A.; Norman, V.C.; Carter, R.; Teal, J.; Akilan, L.; Leftwich, P.T.; et al. Engineered expression of the invertebrate-specific scorpion toxin AaHIT reduces adult longevity and female fecundity in the diamondback moth *Plutella xylostella*. *Pest Manag. Sci.* **2021**, *77*, 3154–3164. [CrossRef]
16. Li, H.; Xia, Y. Improving the secretory expression of active recombinant AaIT in *Pichia pastoris* by changing the expression strain and plasmid. *World J. Microbiol. Biotechnol.* **2018**, *34*, 104. [CrossRef]
17. Tianpei, X.; Li, D.; Qiu, P.; Luo, J.; Zhu, Y.; Li, S. Scorpion peptide LqhIT2 activates phenylpropanoid pathways via jasmonate to increase rice resistance to rice leafrollers. *Plant Sci.* **2015**, *230*, 1–11. [CrossRef]
18. Zlotkin, E.; Eitan, M.; Bindokas, V.P.; Adams, M.E.; Moyer, M.; Burkhart, W.; Fowler, E. Functional duality and structural uniqueness of the depressant insect-selective neurotoxins. *Biochemistry* **1991**, *30*, 4814–4821. [CrossRef]

19. Kroemer, J.A.; Bonning, B.C.; Harrison, R.L. Expression, delivery and function of insecticidal proteins expressed by recombinant baculoviruses. *Viruses* **2015**, *7*, 422–455. [CrossRef]

20. Windley, M.J.; Herzig, V.; Dziemborowicz, S.A.; Hardy, M.C.; King, G.F.; Nicholson, G.M. Spider-venom peptides as bioinsecticides. *Toxins* **2012**, *4*, 191–227. [CrossRef]

21. Chaves-Moreira, D.; Senff-Ribeiro, A.; Wille, A.C.M.; Gremski, L.H.; Chaim, O.M.; Veiga, S.S. Highlights in the knowledge of brown spider toxins. *J. Venom. Anim. Toxins Incl. Trop. Dis.* **2017**, *23*, 6. [CrossRef] [PubMed]

22. Adams, M.E. Agatoxins: Ion channel specific toxins from the American funnel web spider, *Agelenopsis aperta*. *Toxicon* **2004**, *43*, 509–525. [CrossRef] [PubMed]

23. Nentwig, W.; Kuhn-Nentwig, L. Main components of spider venoms. In *Spider Ecophysiology*; Springer: Berlin/Heidelberg, Germany, 2013; pp. 191–202.

24. Kuhn-Nentwig, L.; Stöcklin, R.; Nentwig, W. Venom composition and strategies in spiders: Is everything possible? *Adv. Insect Phys.* **2011**, *40*, 1–86.

25. Tomalski, M.D.; Hutchinson, K.; Todd, J.; Miller, L.K. Identification and Characterization of Tox21A: A Mite CDNA Encoding a Paralytic Neurotoxin Related to TxP-I. *Toxicon* **1993**, *31*, 319–326. [CrossRef]

26. Burden, J.P.; Hails, R.S.; Windass, J.D.; Suner, M.M.; Cory, J.S. Infectivity, Speed of Kill, and Productivity of a Baculovirus Expressing the Itch Mite Toxin Txp-1 in Second and Fourth Instar Larvae of *Trichoplusia Ni*. *J. Invertebr. Pathol.* **2000**, *75*, 226–236. [CrossRef] [PubMed]

27. Yan, Y.; Zhang, N.; Liu, C.; Wu, X.; Liu, K.; Yin, Z.; Zhou, X.; Xie, L. A Highly Contiguous Genome Assembly of a Polyphagous Predatory Mite *Stratiolaelaps scimitus* (Womersley) (Acari: *Laelapidae*). *Genome Biol. Evol.* **2021**, *13*, evab011. [CrossRef] [PubMed]

28. Hollenstein, K.; Dawson, R.J.P.; Locher, K.P. Structure and mechanism of ABC transporter proteins. *Curr. Opin. Struct. Biol.* **2007**, *17*, 412–418. [CrossRef] [PubMed]

29. Jimenez-Acosta, F.; Planas, L.; Penneys, N. Demodex mites contain immunoreactive lipase. *Arch. Dermatol.* **1989**, *125*, 1436–1437. [CrossRef]

30. Ando, T.; Homma, R.; Ino, Y.; Ito, G.; Miyahara, A.; Yanagihara, T.; Kimura, H.; Ikeda, S.; Yamakawa, H.; Iwaki, M.; et al. Trypsin-like protease of mites: Purification and characterization of trypsin-like protease from mite faecal extract *Dermatophagoides farinae*. Relationship between trypsin-like protease and Der f III. *Clin. Exp. Allergy* **1993**, *23*, 777–784. [CrossRef]

31. Chaim, O.M.; Sade, Y.B.; Da Silveira, R.B.; Toma, L.; Kalapothakis, E.; Chávez-Olórtegui, C.; Mangili, O.C.; Gremski, W.; Von Dietrich, C.P.; Nader, H.B.; et al. Brown Spider Dermonecrotic Toxin Directly Induces Nephrotoxicity. *Toxicol. Appl. Pharmacol.* **2006**, *211*, 64–77. [CrossRef]

32. Krantz, G.W. Dolichocybe Keiferi, A New Genus and New Species of Pyemotid Mite, with a Description of a New Species of *Siteroptes* (Acarina: Pyemotidae). *Ann. Entomol. Soc. Am.* **1957**, *50*, 259–264. [CrossRef]

33. Shen, Z.; Jacobs-Lorena, M. Evolution of Chitin-Binding Proteins in Invertebrates. *J. Mol. Evol.* **1999**, *48*, 341–347. [CrossRef] [PubMed]

34. Peigneur, S.; Tytgat, J. Toxins in Drug Discovery and Pharmacology. *Toxins* **2018**, *10*, 126. [CrossRef] [PubMed]

35. Kuhn-Nentwig, L.; Langenegger, N.; Heller, M.; Koua, D.; Nentwig, W. The Dual Prey-Inactivation Strategy of Spiders-In-Depth Venomic Analysis of *Cupiennius salei*. *Toxins* **2019**, *11*, 167. [CrossRef]

36. Bushnell, B. BBtools. Available online: https://sourceforge.net/projects/bbmap/ (accessed on 4 April 2020).

37. Vurture, G.W.; Sedlazeck, F.J.; Nattestad, M.; Underwood, C.J.; Fang, H.; Gurtowski, J.; Schatz, M.C. GenomeScope: Fast reference-free genome profiling from short reads. *Bioinformatics* **2017**, *33*, 2202–2204. [CrossRef]

38. Kolmogorov, M.; Yuan, J.; Lin, Y.; Pevzner, P.A. Assembly of long errorprone reads using repeat graphs. *Nat. Biotechnol.* **2019**, *37*, 540–546. [CrossRef]

39. Guan, D.; McCarthy, S.A.; Wood, J.; Howe, K.; Wang, Y.; Durbin, R. Identifying and removing haplotypic duplication in primary genome assemblies. *Bioinformatics* **2020**, *36*, 2896–2898. [CrossRef]

40. Li, H. Minimap2: Pairwise alignment for nucleotide sequences. Minimap2: Pairwise alignment for nucleotide sequences. *Bioinformatics* **2018**, *34*, 3094–3100. [CrossRef]

41. Hu, J.; Fan, J.; Sun, Z.; Liu, S. NextPolish: A fast and efficient genome polishing tool for long read assembly. *Bioinformatics* **2020**, *36*, 2253–2255. [CrossRef]

42. Camacho, C.; Coulouris, G.; Avagyan, V.; Ma, N.; Papadopoulos, J.; Bealer, K.; Madden, T.L. BLAST +: Architecture and applications. *BMC Bioinform.* **2009**, *10*, 1. [CrossRef]

43. Waterhouse, R.M.; Seppey, M.; Simao, F.A.; Manni, M.; Ioannidis, P.; Klioutchnikov, G.; Kriventseva, E.V.; Zdobnov, E.M. BUSCO Applications from Quality Assessments to Gene Prediction and Phylogenomics. *Mol. Biol. Evol.* **2018**, *35*, 543–548. [CrossRef] [PubMed]

44. Flynn, J.M.; Hubley, R.; Goubert, C.; Rosen, J.; Clark, A.G.; Feschotte, C.; Smit, A.F. RepeatModeler2 for Automated Genomic Discovery of Transposable Element Families. *Proc. Natl. Acad. Sci. USA* **2020**, *117*, 9451–9457. [CrossRef] [PubMed]

45. Hubley, R.; Finn, R.D.; Clements, J.; Eddy, S.R.; Jones, T.A.; Bao, W.; Smit, A.F.A.; Wheeler, T.J. The Dfam Database of Repetitive DNA Families. *Nucleic Acids Res.* **2016**, *44*, D81–D89. [CrossRef]

46. Bao, W.D.; Kojima, K.K.; Kohany, O. Repbase Update, a database of repetitive elements in eukaryotic genomes. *Mob. DNA* **2015**, *6*, 11. [CrossRef]

47. Repeat Masker Open-4.0. 2013–2015. Available online: http://www.repeatmasker.org (accessed on 8 January 2020).

48. Holt, C.; Yandell, M. MAKER2: An annotation pipeline and genome- database management tool for second-generation genome projects. *BMC Bioinform.* **2011**, *12*, 491. [CrossRef]

49. Hoff, K.J.; Lange, S.; Lomsadze, A.; Borodovsky, M.; Stanke, M. BRAKER1: Unsupervised RNA-seq-based genome annotation with GeneMark-ET and AUGUSTUS. *Bioinformatics* **2016**, *32*, 767–769. [CrossRef] [PubMed]

50. Tanke, M.; Steinkamp, R.; Waack, S.; Morgenstern, B. AUGUSTUS: A web server for gene finding in eukaryotes. *Nucleic Acids Res.* **2004**, *32*, W309–W312.

51. Lomsadze, A.; Ter-Hovhannisyan, V.; Chernoff, Y.O.; Borodovsky, M. Gene identification in novel eukaryotic genomes by self-training algorithm. *Nucleic Acids Res.* **2005**, *33*, 6494–6506. [CrossRef] [PubMed]

52. Kim, D.; Paggi, J.M.; Park, C.; Bennett, C.; Salzberg, S.L. Graph-Based Genome Alignment and Genotyping with HISAT2 and HISAT-Genotype. *Nat. Biotechnol.* **2019**, *37*, 907–915. [CrossRef]

53. Kriventseva, E.V.; Kuznetsov, D.; Tegenfeldt, F.; Manni, M.; Dias, R.; Simão, F.A.; Zdobnov, E.M. OrthoDB V10: Sampling the Diversity of Animal, Plant, Fungal, Protist, Bacterial and Viral Genomes for Evolutionary and Functional Annotations of Orthologs. *Nucleic Acids Res.* **2019**, *47*, D807–D811. [CrossRef]

54. Kovaka, S.; Zimin, A.V.; Pertea, G.M.; Razaghi, R.; Salzberg, S.L.; Pertea, M. Transcriptome Assembly from Long-Read RNA-Seq Alignments with StringTie2. *Genome Biol.* **2019**, *20*, 278. [CrossRef] [PubMed]

55. Buchfink, B.; Xie, C.; Huson, D.H. Fast and Sensitive Protein Alignment Using DIAMOND. *Nat. Methods.* **2014**, *12*, 59–60. [CrossRef] [PubMed]

56. Finn, R.D.; Finn, R.D.; Attwood, T.K.; Babbitt, P.C.; Bateman, A.; Bork, P.; Bridge, A.J.; Chang, H.Y.; Dosztányi, Z.; El-Gebali, S.; et al. InterPro in 2017-beyond protein family and domain annotations. *Nucleic Acids Res.* **2017**, *45*, D190–D199. [CrossRef]

57. El-Gebali, S.; Mistry, J.; Bateman, A.; Eddy, S.R.; Luciani, A.; Potter, S.C.; Qureshi, M.; Richardson, L.J.; Salazar, G.A.; Smart, A.; et al. The Pfam Protein Families Database in 2019. *Nucleic Acids Res.* **2019**, *47*, D427–D432. [CrossRef] [PubMed]

58. Letunic, L.; Bork, P. 20 years of the SMART protein domain annotation resource. *Nucleic Acids Res.* **2018**, *46*, D493–D496. [CrossRef] [PubMed]

59. Lewis, T.E.; Sillitoe, I.; Dawson, N.; Lam, S.D.; Clarke, T.; Lee, D.; Orengo, C.; Lees, J. Gene3D: Extensive Prediction of Globular Domains in Proteins. *Nucleic Acids Res.* **2018**, *46*, D435–D439. [CrossRef] [PubMed]

60. Wilson, D.; Pethica, R.; Zhou, Y.; Talbot, C.; Vogel, C.; Madera, M.; Chothia, C.; Gough, J. SUPERFAMILY—Sophisticated Comparative Genomics, Data Mining, Visualization and Phylogeny. *Nucleic Acids Res.* **2009**, *37*, 380–386. [CrossRef]

61. Marchler-Bauer, A.; Bo, Y.; Han, L.; He, J.; Lanczycki, C.J.; Lu, S.; Chitsaz, F.; Derbyshire, M.K.; Geer, R.C.; Gonzales, N.R.; et al. CDD/SPARCLE: Functional Classification of Proteins via Subfamily Domain Architectures. *Nucleic Acids Res.* **2017**, *45*, D200–D203. [CrossRef]

62. Huerta-Cepas, J.; Forslund, K.; Coelho, L.P.; Szklarczyk, D.; Jensen, L.J.; von Mering, C.; Bork, P. Fast genome-wide functional annotation through orthology assignment by eggNOG-mapper. *Mol. Biol. Evol.* **2017**, *34*, 2115–2122. [CrossRef]

63. Huerta-Cepas, J.; Szklarczyk, D.; Heller, D.; Hernández-Plaza, A.; Forslund, S.K.; Cook, H.; Mende, D.R.; Letunic, I.; Rattei, T.; Jensen, L.J.; et al. EggNOG 5.0: A Hierarchical, Functionally and Phylogenetically Annotated Orthology Resource Based on 5090 Organisms and 2502 Viruses. *Nucleic Acids Res.* **2019**, *47*, D309–D314. [CrossRef]

64. Nawrocki, E.P.; Eddy, S.R. Infernal 1.1: 100-Fold Faster RNA Homology Searches. *Bioinformatics* **2013**, *29*, 2933–2935. [CrossRef] [PubMed]

65. Chan, P.P.; Lowe, T.M. tRNAscan-SE: Searching for tRNA genes in genomic sequences. In *Methods in Molecular Biology*; Humana Press: Totowa, NJ, USA, 2019; Volume 1962, pp. 1–14.

66. Emms, D.M.; Kelly, S. OrthoFinder: Phylogenetic Orthology Inference for Comparative Genomics. *Genome Biol.* **2018**, *20*, 238. [CrossRef] [PubMed]

67. Katoh, K.; Standley, D.M. MAFFT Multiple Sequence Alignment Software Version 7: Improvements in Performance and Usability. *Mol. Biol. Evol.* **2013**, *30*, 772–780. [CrossRef] [PubMed]

68. Criscuolo, A.; Gribaldo, S. BMGE (Block Mapping and Gathering with Entropy): A new software for selection of phylogenetic informative regions from multiple sequence alignments. *BMC Evol. Biol.* **2010**, *10*, 210. [CrossRef] [PubMed]

69. Kück, P.; Longo, G.C. FASconCAT-G: Extensive Functions for Multiple Sequence Alignment Preparations Concerning Phylogenetic Studies. *Front. Zool.* **2014**, *11*, 81. [CrossRef]

70. Minh, B.Q.; Schmidt, H.A.; Chernomor, O.; Schrempf, D.; Michael, D.; Haeseler, A.; Von Lanfear, R. IQ-TREE 2: New Models and Efficient Methods for Phylogenetic Inference in the Genomic Era. *Mol. Biol. Evol.* **2020**, *37*, 1530–1534. [CrossRef]

71. Yang, Z. PAML 4: Phylogenetic Analysis by Maximum Likelihood. *Mol. Biol. Evol.* **2007**, *24*, 1586–1591. [CrossRef]

72. Han, M.V.; Thomas, G.W.C.; Lugo-Martinez, J.; Hahn, M.W. Estimating Gene Gain and Loss Rates in the Presence of Error in Genome Assembly and Annotation Using CAFE 3. *Mol. Biol. Evol.* **2013**, *30*, 1987–1997. [CrossRef]

73. Yu, G.; Wang, L.G.; Han, Y.; He, Q.Y. ClusterProfiler: An R Package for Comparing Biological Themes among Gene Clusters. *OMICS A J. Integr. Biol.* **2012**, *16*, 284–287. [CrossRef]

74. Vizueta, J.; Sánchez-Gracia, A.; Rozas, J. BITACORA: A comprehensive tool for the identification and annotation of gene families in genome assemblies. *Mol. Ecol. Resour.* **2020**, *20*, 1445–1452. [CrossRef]

75. Potter, S.C.; Eddy, S.R.; Park, Y.; Lopez, R.; Finn, R.D.; Hmmer, T. HMMER Web Server: 2018 Update. *Nucleic Acids Res.* **2018**, *46*, 200–204. [CrossRef] [PubMed]

76. Price, M.N.; Dehal, P.S.; Arkin, A.P. FastTree 2–approximately maximum-likelihood trees for large alignments. *PLoS ONE* **2010**, *5*, e9490. [CrossRef] [PubMed]
77. Patro, R.; Duggal, G.; Love, M.I.; Irizarry, R.A.; Kingsford, C. Salmon provides fast and bias-aware quantification of transcript expression. *Nat. Methods* **2017**, *14*, 417–419. [CrossRef]
78. Gaujoux, R.; Seoighe, C. A Flexible R Package for Nonnegative Matrix Factorization. *BMC Bioinform.* **2010**, *11*, 367. [CrossRef]

MDPI
St. Alban-Anlage 66
4052 Basel
Switzerland
Tel. +41 61 683 77 34
Fax +41 61 302 89 18
www.mdpi.com

Toxins Editorial Office
E-mail: toxins@mdpi.com
www.mdpi.com/journal/toxins